Non-invasive Prenatal Screening (NIPS) in Clinical Practice

Riyaz Ahmad Rather • Subhas Chandra Saha
Editors

Non-invasive Prenatal Screening (NIPS) in Clinical Practice

Springer

Editors
Riyaz Ahmad Rather
Department of Biotechnology
Wachemo University
Hossana, Ethiopia

Subhas Chandra Saha
Department of Obstetrics and Gynecology
Post Graduate Institute of Medical
Education and Research
Chandigarh, India

ISBN 978-981-97-6401-3 ISBN 978-981-97-6402-0 (eBook)
https://doi.org/10.1007/978-981-97-6402-0

This Springer imprint is published by the registered company Springer Nature Singapore Pte Ltd.
The registered company address is: 152 Beach Road, #21-01/04 Gateway East, Singapore 189721, Singapore

In honor of the mentors who shape our understanding.

Preface

Since the discovery of cell-free fetal DNA, Non-Invasive Prenatal Screening (NIPS) has emerged as an essential tool in detecting various anomalies associated with the fetus. This pioneering breakthrough revealed that during pregnancy, small fragments of fetal DNA circulate freely in the maternal bloodstream, offering a noninvasive means of accessing genetic information about the developing fetus. In addition to accessing genetic information, NIPS can be utilized to screen anomalies associated with the fetus. Beyond identifying chromosomal pathological conditions, it also aids in detecting fetal gender, fetal RhD status, and other factors. Its promising efficiency has led certain countries to integrate this technology into routine clinical practice. As the landscape of prenatal screening rapidly evolves, this book aims to provide a comprehensive examination of NIPS within the context of clinical practice.

This volume offers a thorough overview of NIPS, encompassing advanced genomic approaches and operational strategies. It explores into the historical background and genesis of NIPS, its methodology, and clinical utility. The book is organized into three parts, navigating through the advent of NIPS, detection strategies, clinical implementation, and future prospects. Part I (Chaps. 1–4) introduces readers to the genesis of NIPS, while Part II (Chaps. 5–13) explores detection strategies and clinical implementation through in-depth chapters. Part III (Chaps. 14–16) addresses the challenges and future prospects of NIPS, providing insights into its evolving landscape.

Designed to fill a void in the literature, this book serves as a knowledge hub for specialists in prenatal medicine, reproductive specialists, genetic counselors, research scholars, and postgraduate medical students in Obstetrics and Gynecology. By accommodating chapters discussing the genesis of NIPS technology, approaches to clinical implementation, advanced genomics, challenges, and future avenues, this book aims to equip readers with a comprehensive understanding of NIPS and its implications for clinical practice.

Hossana, Ethiopia — Riyaz Ahmad Rather
Chandigarh, India — Subhas Chandra Saha

Acknowledgment

I extend my sincere gratitude to my co-editor, Professor SC Saha, for his unwavering contributions to this book. I am deeply thankful to all the contributing authors who, despite their prior commitments, dedicated their time and expertise to this project. A special acknowledgment goes to one of the contributors, Dr. Thomas Liehr, for his assistance in securing additional authors at the last moment when some had to withdraw from their commitments unexpectedly.

I am also grateful to Dr. Vijayalakshmi Srinivasan for her consistent support and invaluable critique of the manuscripts. Additionally, I extend my appreciation to the reviewers who provided valuable feedback during the chapter reviews.

I am indebted to my family and friends for their unwavering support throughout this endeavor. Lastly, I would like to express my heartfelt gratitude to my mentors, especially Professor Veena Dhawan, whose guidance and teachings have been instrumental in shaping my research journey.

Sincerely,

Riyaz Ahmad Rather

Contents

Part I

The Advent of NIPS

Evolution of Noninvasive Prenatal Screening: Past to Present 1

Riyaz Ahmad Rather and Subhas Chandra Saha

1.1 Introduction

The landscape of prenatal screening has undergone a transformative journey, evolving from conventional methods to the cutting-edge domain of non-invasive prenatal screening (NIPS). In the early annals of prenatal screening, the focus was predominantly on invasive techniques that carried inherent risks for both the mother and the developing fetus. These methods, such as amniocentesis and chorionic villus sampling (CVS), required the extraction of fetal tissue for genetic analysis [1, 2]. While they provided valuable insights into the genetic makeup of the fetus, the invasiveness of these procedures posed a considerable challenge, often leading to ethical dilemmas and concerns about potential harm to the fetus.

The genesis of non-invasive prenatal screening had a paradigm shift in the approach to fetal genetic testing. The key moment in this journey was marked by the discovery of cell-free fetal DNA (cffDNA) in maternal blood [3] (Fig. 1.1). The presence of fetal DNA circulating in the maternal bloodstream opened doors to a realm of non-invasive methodologies, paving the way for safer and more accessible prenatal screening. As we navigate through the historical milestones, it becomes evident that the initial steps in NIPS were cautious yet pioneering. Early methods focused on detecting chromosomal aneuploidies, providing a non-invasive alternative to traditional approaches [4]. The emergence of noninvasive techniques for gender determination and the detection of RhD status further expanded the scope of applications, offering expectant parents valuable information without the associated risks of invasive procedures [5–8].

R. A. Rather (✉)
Department of Biotechnology, College of Natural and Computational Science, Wachemo University, Hossana, Ethiopia

S. C. Saha
Department of Obstetrics and Gynecology, Post Graduate Institute of Medical Education and Research, Chandigarh, India

R. A. Rather, S. C. Saha (eds.), *Non-invasive Prenatal Screening (NIPS) in Clinical Practice*, https://doi.org/10.1007/978-981-97-6402-0_1

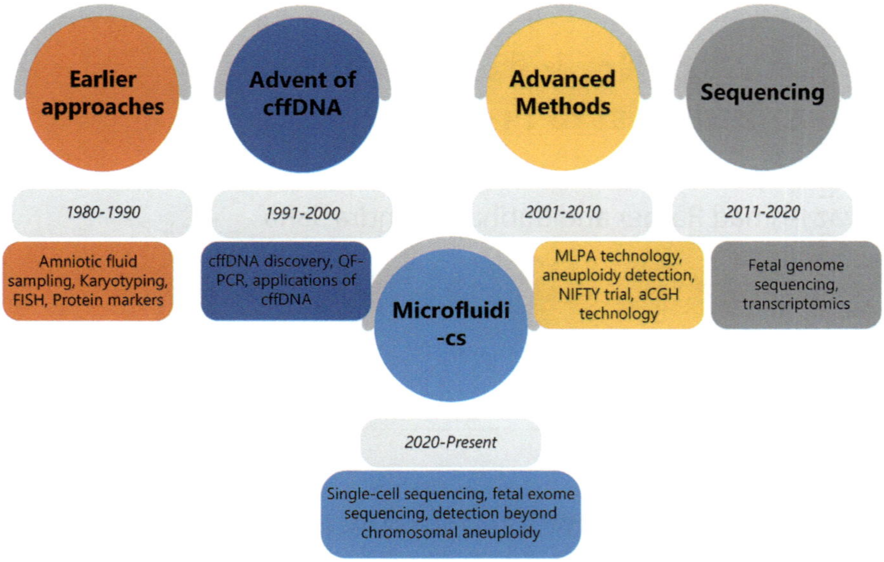

Fig. 1.1 Key milestones in prenatal screening. *FISH* fluorescence in-situ hybridization, *cffDNA* cell-free fetal DNA, *QF-PCR* quantitative fluorescence polymerase chain reaction, *MLPA* multiplex ligation-dependent probe amplification, *NIFTY* noninvasive fetal trisomy, *aCGH* array comparative genomic hybridization, *NGS* next-generation sequencing

The technological advancements have the positive role on the evolutionary trajectory of NIPS. The refinement of DNA sequencing technologies, coupled with the development of sophisticated bioinformatics tools, has empowered researchers and clinicians to delve deeper into the complexities of fetal genetic information [6]. This intersection of biology and technology has not only enhanced the accuracy of noninvasive screening but has also broadened the spectrum of detectable conditions, including monogenic disorders and X-chromosome-related anomalies [7]. As we embark on a journey through the evolution of NIPS, it is imperative to contextualize the historical milestones within the broader framework of ethical, social, and legal dimensions. The shift from invasive to noninvasive methods has not only reshaped the landscape of prenatal care but has also prompted reflections on issues of cost effectiveness, autonomy, informed consent, and the implications of extensive genetic information [9].

This chapter aims to discuss the layers of the past, present, and advancements in the realm of NIPS, setting the stage for a comprehensive exploration of the earlier methods of prenatal screening, current NIPS methods and applications, future directions, and ethical considerations that define this transformative field.

1.2 Early Methods of Prenatal Screening

During the initial phases of prenatal screening, the primary emphasis was on invasive methods that inherently risked both the well-being of the mother and the developing fetus. Procedures like amniocentesis and CVS involved the extraction of fetal

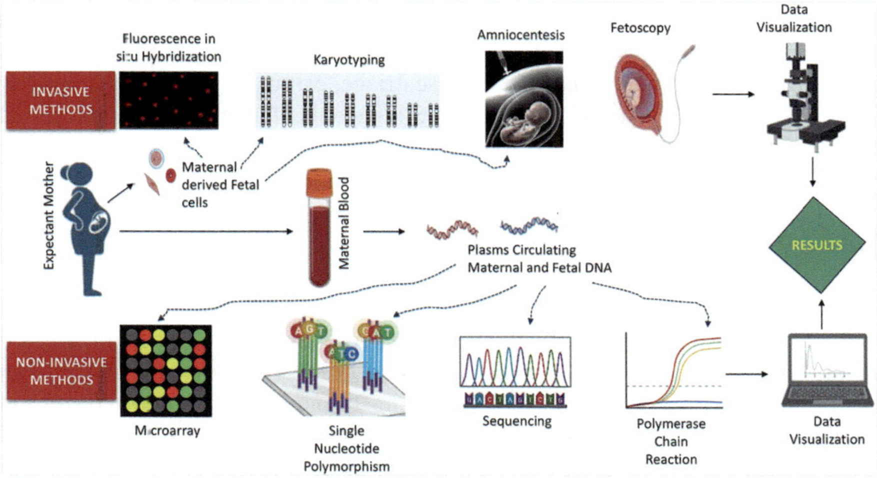

Fig. 1.2 Various invasive and noninvasive methods in prenatal screening. In a noninvasive prenatal screening test, a blood sample is collected from the mother, mother, wherein DNA is isolated from the plasma of the blood. After purification and enrichment, fetal DNA is employed through different genomic methods such as PCR, sequencing, microarray, and SNP. The data generated is visualized, interpreted, and conclusive results are then determined. On the other hand, invasive methods like amniocentesis, CVS, and karyotyping involve retrieving and processing a fetal sample, typically fetal cells, through these methods to obtain conclusive results

tissue for genetic analysis (Fig. 1.2). Although these techniques offered valuable genetic insights, their invasive nature presented significant challenges. In this section, we'll look at the early methods of invasive prenatal screening and how they evolved over time.

1.2.1 Amniocentesis

Amniocentesis, one of the oldest and most common invasive prenatal screening methods, was first introduced in the 1950s [10]. It entails inserting a needle through the abdominal wall and into the amniotic sac to collect a sample of amniotic fluid, which contains fetal cells that can be scrutinized for genetic abnormalities [11]. Amniocentesis is typically performed between 15th and 20th weeks of gestation, although it can be performed as early as 11th week. The procedure carries a risk of miscarriage, which is estimated to be between 0.06 and 0.1% [12]. The test can detect chromosomal abnormalities as well as other genetic disorders such as cystic fibrosis.

1.2.2 CVS

CVS is another invasive prenatal screening method that was first developed in the 1970s. In this procedure, a needle is inserted into the placenta through the abdominal wall, to collect a sample of chorionic villus tissue [13], which contains fetal cells suitable for genetic abnormality examination. CVS can be conducted earlier in pregnancy, usually between the 10th and 13th weeks of gestation. However, it comes with a slightly elevated risk of miscarriage, estimated to be approximately 0.5–1% [14]. Similar to amniocentesis, CVS is capable of identifying chromosomal abnormalities and other genetic disorders.

1.2.3 Fetoscopy

Fetoscopy is a prenatal endoscopic procedure designed to provide surgical access to the fetus, the amniotic cavity, the umbilical cord, and the fetal side of the placenta. This involves making a small incision (3–4 mm) in the abdomen, through which an endoscope is then inserted, passing through the abdominal wall and uterus into the amniotic cavity. Fetoscopy enables medical interventions like collecting tissue samples or laser occlusion of abnormal blood vessels [15]. This procedure was initially introduced in the 1960s and is employed for diagnosing various fetal abnormalities, including neural tube defects such as spina bifida [16]. It is typically performed between 18th and 22nd weeks of gestation and carries a small risk of complications such as bleeding and infection. While fetoscopy can provide valuable information about fetal development, it is not commonly used as a routine screening test due to its invasive nature. However, advances in technology and medical practice made these procedures safer and more accessible to a wider range of patients. For example, the use of ultrasound technology has made it possible to guide needles more accurately and visualize the fetal anatomy, reducing the risk of complications.

1.2.4 Fluorescence In Situ Hybridization (FISH)

Fluorescence in situ hybridization (FISH) is a molecular cytogenetic technique that is widely used in clinical and research settings to detect specific DNA sequences or chromosomal aberrations. FISH involves the hybridization of fluorescently labeled DNA probes to specific regions of chromosomes, allowing for the visualization of these regions under a fluorescence microscope [17]. FISH has several advantages over other cytogenetic techniques, including its high sensitivity, specificity, and speed. It can detect chromosomal abnormalities in interphase cells, allowing for rapid diagnosis of genetic disorders. Furthermore, FISH can be used to detect specific genetic mutations and copy number variations (CNVs), making it a valuable tool for both research and clinical settings [18]. Previously, FISH relied on invasive tissue sampling. Nowadays, circulating fetal cells from maternal plasma can be harvested and visualized by FISH, making its application a versatile two-way prospect. It is commonly used to detect certain chromosomal abnormalities such as Down syndrome, Turner syndrome, and Klinefelter syndrome [19]. This technique involves the use of specific FISH probes that target the chromosomal regions affected by these conditions. The probes are labeled with different fluorescent dyes to distinguish between the different chromosomes, and the hybridization of these probes to fetal DNA in maternal blood samples allows for the detection of abnormal chromosomal copy numbers.

1.2.5 Karyotyping

Karyotyping serves as a valuable tool in prenatal genetic screening, enabling the detection of chromosomal abnormalities in the developing fetus. It involves the analysis of chromosomes obtained from fetal cells acquired through invasive procedures such as CVS [20]. The chromosomes are stained and examined under a microscope to identify any structural abnormalities or aneuploidies, such as trisomy 21 [21]. It has been used for several decades as a standard method of prenatal genetic screening, particularly for detecting chromosomal abnormalities. Nonetheless, it does have certain limitations, notably the requirement for invasive procedures and a relatively lower resolution when compared to more modern molecular techniques [22].

Despite its limitations, karyotyping remains an important tool in prenatal genetic screening due to its ability to provide a comprehensive view of the fetal genome. It can also identify genetic abnormalities that are not detected by newer methods, such as balanced chromosomal rearrangements, which can cause birth defects or developmental delays in the fetus [20]. Although invasive testing methods have expanded the scope of data interpretation and visualization in prenatal screening, they have certain limitations, such as need of cytogenetic or clinical specialists, who can safely collect tissue samples without harming the fetus. There is an ongoing need for DNA-based molecular genetic techniques that can exploit fetal entities of maternal circulation as fair and viable molecules for prenatal screening. The past few years have produced many technologies beyond karyotyping that have greatly advanced the field of prenatal genetic screening.

1.3 Breaking Barriers in Prenatal Screening with Cell-Free Fetal DNA

1.3.1 cffDNA

cffDNA consists of small DNA fragments released from the placenta into the maternal bloodstream during pregnancy [3]. It is believed that the source of cffDNA is the placenta. The quantity of cffDNA in maternal plasma increases with pregnancy progression, detectable as early as the seventh week of gestation [23]. Its significance lies in NIPS, where it detects various fetal aneuploidies

[24]. However, it is important to note that NIPS is a screening test and not a diagnostic test, and further testing is typically recommended to confirm the diagnosis.

The breakthrough discovery of cffDNA was first reported in 1997 by Lo et al. and has since been widely studied and applied in various aspects of prenatal care [3]. The existence of cffDNA in maternal plasma has expanded the horizons of NIPS, allowing for the identification of chromosomal abnormalities, single-gene disorders, and even fetal sex determination [25]. Subsequent research has shown that cffDNA can be detected in maternal blood as early as the fifth week of gestation, with concentrations increasing throughout pregnancy [26]. Its identification in maternal circulation holds substantial implications for prenatal diagnosis by obviating the necessity for invasive procedures. These invasive procedures carry a slight yet noteworthy risk of fetal loss, and NIPS methods based on cffDNA bears the potential to substantially mitigate this risk.

Studies affirm the high accuracy of cffDNA based detection for chromosomal abnormalities, exceeding 99% sensitivity and specificity [27, 28]. However, NIPS testing based on cffDNA is not foolproof, allowing for false positives and negatives. In cases of high-risk indications, invasive testing is still recommended. Beyond chromosomal abnormalities, cffDNA excels in detecting single-gene disorders, with a near 100% accuracy rate [29]. Furthermore, determining fetal sex is now achieved as early as the ninth week through cffDNA, significant for sex-linked disorders or family planning preferences [30].

1.3.2 Advances in NIPS Technology

Significant advances in NIPS have enhanced accuracy and broadened the range of screenable conditions. Initially reliant on polymerase chain reaction (PCR), recent technological breakthroughs, such as next-generation sequencing (NGS), now enable the detection of a wider array of genetic conditions, including single-gene disorders, microdeletions, and duplications [25]. Among the major technological advancements, the utilization of targeted sequencing panels stands out, as it enables the precise detection of specific genetic conditions. For instance, these panels are invaluable in screening for common microdeletions, such as DiGeorge syndrome or 22q11.2 deletion syndrome [31]. These panels offer the flexibility to be customized for screening specific genetic conditions prevalent in certain populations, like sickle cell anemia in individuals of African descent. Complementing this, another significant technological stride in NIPS is the adoption of single nucleotide polymorphism (SNP) analysis, which extends the capability of detection to encompass copy number variations and other genetic conditions beyond the reach of NGS alone [32]. This is particularly advantageous in cases where a known genetic mutation exists within the family, as SNP analysis provides the means for targeted detection of the mutation in the fetus.

Over the years, various NIPS technologies have been developed and improved, with real-time PCR (rt-PCR) and NGS emerging as the most promising and widely used methods. rt-PCR is a widely used technique in NIPS due to its sensitivity, accuracy, and rapid turnaround time [33]. rt-PCR has demonstrated accurate detection of fetal chromosomal aneuploidies, fetal RhD status, gender, and single-gene disorders, boasting high sensitivity and specificity [34]. One of its primary advantages lies in the ability to analyze multiple targets simultaneously using a single reaction, leading to a substantial reduction in turnaround time and testing costs. Additionally, rt-PCR is a relatively straightforward and cost-effective technique, easily applicable in most clinical laboratories, rendering it a highly accessible and practical tool in NIPS. Digital PCR has been effectively utilized for detecting fetal chromosomal abnormalities with remarkable sensitivity and specificity [35].

NGS, on the other hand, is a more recent technology that has rapidly gained popularity due to its high-throughput capacity, accuracy, and ability to detect a wide range of genetic abnormalities [36]. NGS works by sequencing millions of DNA fragments in parallel, generating massive amounts of data that can be analyzed using sophisticated bioinformatics tools. This technique has been shown to accurately detect fetal trisomies, CNVs, and single-gene disorders, as well as fetal-specific SNPs [37]. One of the main advantages of NGS is its ability to detect a wide range of genetic abnormalities simultaneously, making it a highly efficient and cost-effective approach to testing. Moreover, NGS has the potential to detect new and rare genetic variants that may be missed by other techniques, providing a more comprehensive and accurate assessment of fetal health. Due to the promising accuracy, sensitivity, and specificity, and no risk of miscarriage, NIPS is currently recommended as a first-line screening test for fetal aneuploidies by various professional organizations, including the American College of Obstetricians and Gynecologists (ACOG) and the International Society for Prenatal Diagnosis [38].

1.4 The Global Spread of NIPS

In the United States of America (USA), NIPS has been available since 2011 and is now widely used as a primary screening tool for fetal genetic abnormalities [39]. The American Society of Clinical Pathologists has recommended that NIPS be offered to all pregnant women, regardless of age or risk factors [40]. Several NIPS-based tests, such as MaterniT21, Harmony, and Panorama, are commercially available in the USA and have been shown to have high sensitivity and specificity for detecting fetal genetic abnormalities [39].

In China, NIPS has become a widely adopted routine prenatal screening test, backed by the government's recommendation for universal provision to all pregnant women, including its integration into the national prenatal screening program [41]. The implementation of NIPS in China holds particular significance due to the long-standing one-child policy, enabling couples to detect fetal genetic abnormalities early and make informed decisions about their pregnancies.

In Europe, the adoption of NIPS varies across countries. The United Kingdom offers NIPS to women at high risk for fetal genetic abnormalities or those with abnormal results from previous screening tests [42]. In France, although not routinely available, NIPS is offered in certain private clinics and hospitals [43]. Germany provides NIPS to women at high risk for fetal genetic abnormalities, yet it is not widely accessible as a routine screening test [44].

In India, NIPS is still a relatively new technology, with slow adoption attributed to its high cost and stringent government laws such as the Pre-Natal Diagnostic Techniques Act of 1994 [45]. Despite this, several NIPS-based tests like PraenaTest and NIFTY are now available in India and gaining traction in urban healthcare settings.

Over the years, the increasing accuracy and non-invasive nature of NIPS have led to its widespread adoption in numerous countries however, it is essential to note that many African countries have yet to fully integrate it. In these regions, NIPS remains minimally present in healthcare settings, and its role in research is notably understated or virtually non-existent.

1.5 Current Clinical Applications of NIPS

NIPS has undergone significant evolution and has garnered widespread clinical utilization in prenatal screening. Initially promising for chromosomal aneuploidy detection, its applications have expanded beyond that, propelled by advancing genomic methods. Over time, its sensitivity and specificity have increased. The following section provides a brief overview of the current clinical applications of NIPS, noting that these applications are not exhaustive.

1.5.1 Trisomy Screening

NIPS serves as a valuable tool for screening common chromosomal abnormalities such as trisomy 21, 18, and 13, which represent the most prevalent chromosomal anomalies in live births. Its high sensitivity and specificity rates make it a reliable option for prenatal screening [46].

1.5.2 Sex Chromosome Abnormalities

NIPS can also identify sex chromosome abnormalities, including Turner syndrome (45,X), Klinefelter syndrome (47,XXY), and Triple X syndrome (47,XXX) [47]. While sex chromosome abnormalities exhibit lower detection rates than trisomy screening, they remain valuable for screening sex chromosome aneuploidies.

1.5.3 Microdeletions

The NIPS test can also be employed for screening rarer trisomies, copy number variants, and microdeletions [48], such as DiGeorge syndrome (22q11.2 deletion), Cri-du-chat syndrome (15p deletion), Prader-Willi syndrome (15q deletion), and Jacobsen syndrome (11q deletion) [49]. However, it is important to note that the sensitivity and positive predictive value for these conditions are notably lower, necessitating further testing for confirmation of a diagnosis.

1.5.4 RhD Genotyping

RhD genotyping is another noteworthy aspect of NIPS [50]. RhD, an antigen situated on the surface of red blood cells, plays a pivotal role in blood transfusion practice. In cases where there is incompatibility in RhD antigens between the mother and fetus, it can result in hemolytic disease of the newborn (HDN), a potentially life-threatening condition. In this scenario, NIPS assumes a crucial role by enabling the determination of the fetus's RhD status early in the pregnancy, thus facilitating the implementation of appropriate measures for managing HDN.

1.6 Ethical Considerations and Controversies Surrounding NIPS

As NIPS undergoes rapid evolution, it is crucial to acknowledge that along with this growth, there are ethical considerations and controversies that demand careful analysis and resolution. These issues include aspects such as informed consent, the accuracy and reliability of results, and potential implications for pregnancy termination [51, 52].

Informed consent is an important aspect of any medical procedure, and NIPS is no exception. Patients need to be fully informed about the benefits and limitations of NIPS, as well as the potential risks and consequences of obtaining certain results. In particular, patients need to understand that NIPS is a screening test and not a diagnostic test, and that confirmation of positive results may require further invasive testing [46]. The accuracy and reliability of NIPS results are also important ethical considerations. While NIPS has been shown to have high sensitivity and specificity for certain conditions, there is still a possibility of false positive or false negative results [53]. Patients need to be aware of these limitations and the potential consequences of acting on inaccurate results. One of the most controversial aspects of NIPS is the potential implications for pregnancy termination. NIPS can provide information about the likelihood of a fetus having certain genetic conditions, some of which may be severe or life threatening. This information can be used to inform decisions about pregnancy continuation or termination. However, there are ethical concerns about using this information for selective abortion or eugenics. Additionally, the accuracy of NIPS results can influence a patient's decision to terminate a pregnancy, which highlights the need for accurate and reliable testing.

There are also concerns about the potential for NIPS to reinforce societal stigmas and biases towards certain conditions, such as Down syndrome [54]. The availability and use of NIPS could potentially lead to increased selective abortion of fetuses with certain conditions, perpetuating negative attitudes towards these conditions and the individuals who have them.

1.7 Future Prospects

1.7.1 Integration of Advanced Genomic Methods

Currently, NIPS advantages sequencing technology to enhance its results, boosting both sensitivity and specificity. Whole-genome sequencing (WGS) stands out as one such technique with broad applications in detecting genetic changes in fetus [55]. WGS has the potential to provide a comprehensive and accurate screening for all fetal chromosomal abnormalities, including rare genetic conditions that were previously undetectable [56]. Several studies have demonstrated the feasibility and accuracy of WGS-based NIPS, but the high cost and computational complexity of the method have limited its widespread clinical adoption. WGS-based NIPS has several advantages over traditional targeted approaches. It can provide comprehensive screening for all chromosomal abnormalities, including those that are not detectable by current NIPS methods, such as mosaicism and chromosomal structural abnormalities. In addition, WGS-based method can also detect single-gene disorders that are caused by large structural variations or point mutations.

Recent studies affirm the feasibility and accuracy of WGS-based NIPS. For example, a large-scale clinical study showed that WGS-based NIPS had a high sensitivity and specificity for detecting fetal chromosomal abnormalities, including microdeletions and sex chromosome aneuploidies [57]. Another study by Lake et al. demonstrated the feasibility of WGS-based NIPS for detecting rare genetic disorders, such as lysosomal storage disorders and glycogen storage disorders [58].

1.7.2 Integration of NIPS into Routine Clinical Practice

As NIPS continues to advance, its integration into routine prenatal care is becoming increasingly crucial [59]. The integration has several benefits. First and foremost, NIPS is noninvasive and poses no risk to the fetus or the mother [60]. Unlike invasive prenatal tests, NIPS is a simple blood test that can be performed as early as tenth weeks of pregnancy. This makes NIPS a safer and more convenient option for high-risk pregnancies and women who may be hesitant to undergo invasive prenatal testing [61].

Another benefit of integrating NIPS into routine prenatal care is the increased accuracy and sensitivity of the test. It has a high detection rate for fetal aneuploidies and other chromosomal abnormalities, with some studies reporting detection rates of up to 99% [62]. This increased accuracy can provide expectant parents with greater peace of mind and help them make informed decisions about their pregnancy. With the development of new technologies, such as WGS, NIPS may become even more accurate and sensitive, and its potential applications may expand beyond the detection of aneuploidies and other chromosomal abnormalities [63]. However, as NIPS continues to evolve, it is important that healthcare providers and patients are informed about the benefits and limitations of the technology and that efforts are made to address the challenges associated with its integration into routine prenatal care.

1.8 Conclusion

NIPS is a rapidly evolving technology that has developed the field of prenatal screening. The discovery of cffDNA in maternal circulation in the 1990s opened up new possibilities for noninvasive prenatal testing. Over the years, the accuracy and reliability of NIPS have improved significantly, making it a valuable tool for the detection of fetal chromosomal abnormalities. Initially, NIPS was limited to the detection of trisomy 21, but now it can identify other chromosomal abnormalities such as trisomy 18 and 13, sex chromosome aneuploidies, and microdeletions. The introduction of newer technologies such as PCR, rt-PCR, and NGS has led to the development of more accurate and comprehensive NIPS assays. Several clinical studies have shown that NIPS has high sensitivity and specificity for the detection of fetal chromosomal abnormalities, making it an attractive option for prenatal screening. As a result, NIPS has been integrated into routine prenatal care in many countries around the world, especially for women at high chance of having a child with chromosomal abnormalities.

However, the integration of NIPS into routine prenatal care has raised several ethical and social concerns. The possibility of using NIPS for sex selection and the potential for false positives leading to unnecessary invasive testing are some of the ethical considerations associated with NIPS. Its high cost and the potential for increased anxiety among pregnant women also pose challenges to its integration into routine prenatal care.

In summary, NIPS has undergone substantial evolution in recent decades, now boasting remarkable accuracy and dependability in detecting fetal chromosomal abnormalities. While its integration into standard prenatal care has greatly benefitted high chance pregnancies, ethical dilemmas persist. Ongoing research and development in this area will undoubtedly shape the future of prenatal diagnosis and healthcare.

References

1. Rhoads GG, Jackson LG, Schlesselman SE, de la Cruz FF, Desnick RJ, Golbus MS, et al. The safety and efficacy of chorionic villus sampling for early prenatal diagnosis of cytogenetic abnormalities. N Engl J Med. 1989;320(10):609–17.
2. Hecher K, Snijders R, Nicolaides K. Screening for fetal chromosomal abnormalities by maternal serum biochemistry and ultrasound examination of fetal morphology. Curr Opin Obstet Gynecol. 1993;5(2):170–8.
3. Lo YM, Corbetta N, Chamberlain PF, Rai V, Sargent IL, Redman CW, Wainscoat JS. Presence of fetal DNA in maternal plasma and serum. Lancet. 1997;350(9076):485–7.
4. Rijnders RJ, Christiaens GC, Bossers B, van der Smagt JJ, van der Schoot CE, de Haas M. Clinical applications of cell-free fetal DNA from maternal plasma. Obstet Gynecol. 2004;103(1):157–64.
5. Lo YM. Non-invasive prenatal testing using massively parallel sequencing of maternal plasma DNA: from molecular karyotyping to fetal whole-genome sequencing. Reprod Biomed Online. 2013;27(6):593–8.
6. Chiu RW, Lo YM. Application of fetal DNA in maternal plasma for noninvasive prenatal diagnosis. Expert Rev Mol Diagn. 2002;2(1):32–40.
7. Taylor-Phillips S, Freeman K, Geppert J, Agbebiyi A, Uthman OA, Madan J, et al. Accuracy of non-invasive prenatal testing using cell-free DNA for detection of Down, Edwards and Patau syndromes: a systematic review and meta-analysis. BMJ Open. 2016;6(1):e010002.
8. Zhang H, Gao Y, Jiang F, Fu M, Yuan Y, Guo Y, Zhu Z, et al. Non-invasive prenatal testing for trisomies 21, 18 and 13: clinical experience from 146,958 pregnancies. Ultrasound Obstet Gynecol. 2015;45(5):530–8. Erratum in: Ultrasound Obstet Gynecol. 2015;46(1):130.
9. Breveglieri G, D'Aversa E, Finotti A, Borgatti M. Non-invasive prenatal testing using fetal DNA. Mol Diagn Ther. 2019;23(2):291–9.
10. Dacus JV, Wilroy RS, Summitt RL, Garbaciak JA, Abdella TN, Spinnato JA, et al. Genetic amniocentesis: a twelve years' experience. Am J Med Genet. 1985;20(3):443–52.
11. Nizard J. Amniocentesis: technique and education. Curr Opin Obstet Gynecol. 2010;22(2):152–4.
12. Schemmer G, Johnson A. Genetic amniocentesis and chorionic villus sampling. Obstet Gynecol Clin North Am. 1993;20(3):497–521.
13. Alfirevic Z, Navaratnam K, Mujezinovic F. Amniocentesis and chorionic villus sampling for prenatal diagnosis. Cochrane Database Syst Rev. 2017;9(9):CD003252.
14. Akolekar R, Beta J, Picciarelli G, Ogilvie C, D'Antonio F. Procedure-related risk of miscarriage following amniocentesis and chorionic villus sampling: a systematic review and meta-analysis. Ultrasound Obstet Gynecol. 2015;45(1):16–26.
15. Deprest JA, Gratacos E. Obstetrical endoscopy. Curr Opin Obstet Gynecol. 1999;11(2):195–203.
16. Yang EY, Adzick NS. Fetoscopy. Semin Laparosc Surg. 1998;5(1):31–9.
17. Bayani J, Squire JA. Fluorescence in situ hybridization (FISH). Curr Protoc Cell Biol. 2004;Chapter 22:Unit 22.4.
18. Gregg AR, Skotko BG, Benkendorf JL, Monaghan KG, Bajaj K, Best RG, Klugman S, et al. Noninvasive prenatal screening for fetal aneuploidy, 2016 update: a position statement of the American College of Medical Genetics and Genomics. Genet Med. 2016;18(10):1056–65.
19. Pergament E. The application of fluorescence in-situ hybridization to prenatal diagnosis. Curr Opin Obstet Gynecol. 2000;12(2):73–6.
20. Lichtenbelt KD, Knoers NV, Schuring-Blom GH. From karyotyping to array-CGH in prenatal diagnosis. Cytogenet Genome Res. 2011;135(3–4):241–50.
21. Bull MJ. Down syndrome. N Engl J Med. 2020;382(24):2344–52.
22. Gijsbers AC, Ruivenkamp CA. Molecular karyotyping: from microscope to SNP arrays. Horm Res Paediatr. 2011;76(3):208–13.

23. Fiorentino F, Bono S, Pizzuti F, Duca S, Polverari A, Faieta M, Baldi M, Diano L, Spinella F. The clinical utility of genome-wide non invasive prenatal screening. Prenat Diagn. 2017;37(6):593–601.
24. Brady P, Brison N, Van Den Bogaert K, de Ravel T, Peeters H, Van Esch H, et al. Clinical implementation of NIPT—technical and biological challenges. Clin Genet. 2016;89(5):523–30.
25. Suciu ID, Toader OD, Galeva S, Pop L. Non-invasive prenatal testing beyond trisomies. J Med Life. 2019;12(3):221–4.
26. Raj H, Yelne P. Cell-free fetal deoxyribonucleic acid (cffDNA) analysis as a remarkable method of non-invasive prenatal screening. Cureus. 2022;14(10):e29965.
27. Ashoor G, Syngelaki A, Wagner M, Birdir C, Nicolaides KH. Chromosome-selective sequencing of maternal plasma cell-free DNA for first-trimester detection of trisomy 21 and trisomy 18. Am J Obstet Gynecol. 2012;206(4):322.e1–5.
28. Committee opinion no. 640: cell-free DNA screening for fetal aneuploidy. Obstet Gynecol. 2015;126(3):e31–e37.
29. Allen S, Young E, Bowns B. Noninvasive prenatal diagnosis for single-gene disorders. Curr Opin Obstet Gynecol. 2017;29(2):73–9.
30. Lewis C, Hill M, Skirton H, Chitty LS. Non-invasive prenatal diagnosis for fetal sex determination: benefits and disadvantages from the service users' perspective. Eur J Hum Genet. 2012;20(11):1127–33.
31. Schmid M, Wang E, Bogard PE, Bevilacqua E, Hacker C, Wang S, et al. Prenatal screening for 22q11.2 deletion using a targeted microarray-based cell-free DNA test. Fetal Diagn Ther. 2018;44(4):299–304.
32. Zhou Z, Wang W, Wang LS, Zhang NR. Integrative DNA copy number detection and genotyping from sequencing and array-based platforms. Bioinformatics. 2018;34(14):2349–55.
33. Patsalis PC. A new method for non-invasive prenatal diagnosis of Down syndrome using MeDIP real time qPCR. Appl Transl Genom. 2016;1:3–8.
34. Zimmermann BG, Dudarewicz L. Real-time quantitative PCR for the detection of fetal aneuploidies. Methods Mol Biol. 2008;444:95–109.
35. D'Aversa E, Breveglieri G, Boutou E, Balassopoulou A, Voskaridou E, Pellegatti P, et al. Droplet digital PCR for non-invasive prenatal detection of fetal single-gene point mutations in maternal plasma. Int J Mol Sci. 2022;23(5):2819.
36. Thung DT, Beulen L, Hehir-Kwa J, Faas BH. Implementation of whole genome massively parallel sequencing for noninvasive prenatal testing in laboratories. Expert Rev Mol Diagn. 2015;15(1):111–24.
37. Bianchi DW, Platt LD, Goldberg JD, Abuhamad AZ, Sehnert AJ, Rava RP, MatErnal BLood IS Source to Accurately diagnose fetal aneuploidy (MELISSA) Study Group. Genome-wide fetal aneuploidy detection by maternal plasma DNA sequencing. Obstet Gynecol. 2012;119(5):890. Erratum in: Obstet Gynecol. 2012;120(4):957.
38. The American College of Obstetricians and Gynecologists Committee opinion no. 630. Screening for perinatal depression. Obstet Gynecol. 2015;125(5):1268–1271.
39. Agarwal A, Sayres LC, Cho MK, Cook-Deegan R, Chandrasekharan S. Commercial landscape of noninvasive prenatal testing in the United States. Prenat Diagn. 2013;33(6):521–31.
40. American Society for Clinical Pathology (ASCP). New ACOG guidelines recommend cfDNA-based NIPT for all pregnancies, not just risky ones. 2020. https://www.ascp.org/content/news-archive/news-detail/2020/09/24/new-acog-guidelines-recommend-cfdna-based-nipt-for-all-pregnancies-not-just-risky-ones#.
41. Allyse M, Minear MA, Berson E, Sridhar S, Rote M, et al. Non-invasive prenatal testing: a review of international implementation and challenges. Int J Womens Health. 2015;2015(7):113–26.
42. Gadsbøll K, Petersen OB, Gatinois V, Strange H, Jacobsson B, Wapner R, Vermeesch JR, NIPT-map Study Group, Vogel I. Current use of noninvasive prenatal testing in Europe, Australia and the USA: a graphical presentation. Acta Obstet Gynecol Scand. 2020;99(6):722–30.

43. Perrot A, Horn R. The ethical landscape(s) of non-invasive prenatal testing in England, France and Germany: findings from a comparative literature review. Eur J Hum Genet. 2022;30(6):676–81.
44. Bowman-Smart H, Wiesemann C, Horn R. Non-invasive prenatal testing in Germany: a unique ethical and policy landscape. Eur J Hum Genet. 2023;31(5):562–7.
45. Aggarwal S, Phadke SR. Medical genetics and genomic medicine in India: current status and opportunities ahead. Mol Genet Genom Med. 2015;3:160–7.
46. Benn P, Borrell A, Chiu RW, Cuckle H, Dugoff L, Faas B, et al. Position statement from the Chromosome Abnormality Screening Committee on behalf of the Board of the International Society for Prenatal Diagnosis. Prenat Diagn. 2015;35(8):725–34.
47. Zhang B, Lu BY, Yu B, Zheng FX, Zhou Q, Chen YP, et al. Noninvasive prenatal screening for fetal common sex chromosome aneuploidies from maternal blood. J Int Med Res. 2017;45(2):621–30.
48. Vora NL, O'Brien BM. Noninvasive prenatal testing for microdeletion syndromes and expanded trisomies: proceed with caution. Obstet Gynecol. 2014;123(5):1097–9.
49. Kucharik M, Gnip A, Hyblova M, Budis J, Strieskova L, Harsanyova M, et al. Non-invasive prenatal testing (NIPT) by low coverage genomic sequencing: detection limits of screened chromosomal microdeletions. PLoS One. 2020;15(8):e0238245.
50. Ontario Health (Quality). Noninvasive fetal RhD blood group genotyping: a Health technology assessment. Ont Health Technol Assess Ser. 2020;20(15):1–160.
51. Johnston M, Warton C, Pertile MD, Taylor-Sands M, Delatycki MB, Hui L, et al. Ethical issues associated with prenatal screening using non-invasive prenatal testing for sex chromosome aneuploidy. Prenat Diagn. 2023;43(2):226–34.
52. Newson AJ. Ethical aspects arising from non-invasive fetal diagnosis. Semin Fetal Neonatal Med. 2008;13(2):103–8.
53. Lutgendorf MA, Stoll KA, Knutzen DM, Foglia LM. Noninvasive prenatal testing: limitations and unanswered questions. Genet Med. 2014;16(4):281–5.
54. Deans Z, Newson A. Ethical considerations and practicalities of prenatal testing for Down syndrome: the opinions of parents of affected children. Am J Med Genet A. 2015;167A(7):1530–8.
55. Meier N, Bruder E, Lapaire O, Hoesli I, Kang A, Hench J, et al. Exome sequencing of fetal anomaly syndromes: novel phenotype-genotype discoveries. Eur J Hum Genet. 2019;27(5):730–7.
56. Zhou J, Yang Z, Sun J, Liu L, Zhou X, Liu F, et al. Whole genome sequencing in the evaluation of fetal structural anomalies: a parallel test with chromosomal microarray plus whole exome sequencing. Genes (Basel). 2021;12(3):376.
57. Fu F, Li R, Li Y, Nie ZQ, Lei T, Wang D, et al. Whole exome sequencing as a diagnostic adjunct to clinical testing in fetuses with structural abnormalities. Ultrasound Obstet Gynecol. 2018;51(4):493–502.
58. Lake BD, Young EP, Winchester BG. Prenatal diagnosis of lysosomal storage diseases. Brain Pathol. 1998;8(1):133–49.
59. Rose NC, Barrie ES, Malinowski J, Jenkins GP, McClain MR, LaGrave D, et al. Systematic evidence-based review: the application of noninvasive prenatal screening using cell-free DNA in general-risk pregnancies. Genet Med. 2022;24(7):1379–91. Erratum in: Genet Med 2022 Sep;(9):1992.
60. Rather RA, Saha SC. Reappraisal of evolving methods in non-invasive prenatal screening: discovery, biology and clinical utility. Heliyon. 2023;9(3):e13923.
61. Bedei I, Wolter A, Weber A, Signore F, Axt-Fliedner R. Chances and challenges of new genetic screening technologies (NIPT) in prenatal medicine from a clinical perspective: a narrative review. Genes (Basel). 2021;12(4):501.
62. Lannoo L, van Straaten K, Breckpot J, Brison N, De Catte L, Dimitriadou E, et al. Rare autosomal trisomies detected by non-invasive prenatal testing: an overview of current knowledge. Eur J Hum Genet. 2022;30(12):1323–30.
63. Emms A, Castleman J, Allen S, Williams D, Kinning E, Kilby M. Next generation sequencing after invasive prenatal testing in fetuses with congenital malformations: prenatal or neonatal investigation. Genes (Basel). 2022;13(9):1517.

Fetal Origin Circulating Cell-Free Nucleic Acids in Maternal Circulation and Their Clinical Importance

2

Riyaz Ahmad Rather

2.1 Introduction

Cell-free nucleic acids (cfNAs) are extracellular DNA or RNA molecules present in body fluids. These entities exhibit tremendous potential in the diagnosis and prognosis of various diseases. Among cfNAs, fetal-origin circulating nucleic acids, hereafter called as cell-free fetal nucleic acids (cffNAs), are released from fetal-associated tissues into the maternal bloodstream [1–3]. These molecules show significant promise in prenatal screening (Table 2.1). Detection of these molecules in maternal blood is possible as early as seventh weeks of gestation [4]. The concentration of circulating cffNAs in maternal blood increases with the progression of pregnancy [5]. These molecules play a crucial role in diagnosing various fetal conditions, including chromosomal abnormalities, genetic disorders, and monogenic disorders, as well as determining fetal sex [6]. Circulating cfNAs in maternal plasma consist of both maternal and fetal components. The origin of these nucleic acids has been a subject of significant confusion and debate. However, there is now a growing consensus that they likely originate from the placenta. Maternal cfNAs primarily originate from the mother's peripheral blood cells, apoptotic cells, and other physiological processes. Conversely, besides the placenta, cffNAs are also derived from the apoptotic cells of the developing fetus [7]. cffNAs represent a minority fraction within the total cfNAs pool but carry vital genetic information specific to the fetus. These cfNAs are believed to be released from fetal cells into the maternal bloodstream through different mechanisms, including placental microchimerism [8]. During pregnancy the placenta, a temporary organ, formed around the fetus, comprising cells from both the mother and the fetus. This results in the presence of fetal cells in the mother's bloodstream, facilitated by the placenta [9]. Furthermore, fetal cells

R. A. Rather (✉)
Department of Biotechnology, College of Natural and Computational Science, Wachemo University, Hossana, Ethiopia

R. A. Rather, S. C. Saha (eds.), *Non-invasive Prenatal Screening (NIPS) in Clinical Practice*, https://doi.org/10.1007/978-981-97-6402-0_2

Table 2.1 Some cfNAs and their clinical applications in prenatal screening

Type of cell-free fetal nucleic acid	Function	Prenatal screening application
Cell-free fetal DNA	Genetic information from the fetus present in maternal circulation	Noninvasive prenatal screening for chromosomal abnormalities, detection of single gene disorders, determination of fetal sex
Cell-free fetal RNA	Regulates gene expression	Detection of gene expression patterns, identification of potential genetic disorders
MicroRNAs	Regulates gene expression and plays a role in various biological processes	Potential biomarkers for prenatal screening, prediction of pregnancy-related complications
Long noncoding RNAs	Regulates gene expression, chromatin organization, and cellular processes	Investigation of epigenetic modifications, identification of potential biomarkers for prenatal diagnosis, understanding of gene regulation in fetal development
Circulating placental RNAs	Reflects placental function and health	Assessment of placental health, prediction of pregnancy-related complications, monitoring of placental disorders
Extracellular vesicle RNAs	Carry genetic material between cells and play a role in intercellular communication	Identification of biomarkers for pregnancy-related disorders, assessment of fetal-maternal communication
Y-chromosomal DNA	Indicates fetal sex	Determination of fetal sex, detection of sex-linked genetic disorders
Mitochondrial DNA	Maternally inherited DNA involved in energy production	Detection of mitochondrial disorders, assessment of mitochondrial function in the fetus
Circulating trophoblast DNA	Reflects placental and fetal trophoblast shedding	Evaluation of placental health and dysfunction, prediction of pregnancy complications

undergo natural cell death throughout pregnancy, releasing their DNA and RNA into the maternal bloodstream [10].

Moreover, fetal cells can also shed from the placenta and enter the maternal bloodstream [11]. These processes contribute to the presence of cfNAs in maternal blood, enabling their potential use in prenatal screening. It is reported that there is a rapid clearance of fetal DNA from maternal blood shortly after delivery, in stark contrast to the behavior of fetal cells that can survive for several weeks postpartum and a small fraction of which can even persist for years or decades [12].

Noninvasive prenatal screening (NIPS) based on cfNA analysis has primarily been utilized for detecting fetal chromosomal aneuploidies. However, due to advanced molecular techniques the scope of detection has expanded to include other fetal-related complications such as preeclampsia, gestational diabetes, intrauterine growth restriction, polyhydramnios, and single-gene disorders, among others [13–16]. Moreover, cfNA analysis has demonstrated promising results in the identification of various fetal genetic disorders. This opens up avenues for early diagnosis and subsequent management of these disorders, ultimately leading to improved outcomes for affected individuals and their families. This chapter broadly explore the clinical significance of circulating cfNAs of fetal origin in prenatal screening. By examining the detection techniques, clinical applications, and future prospects of cfNA analysis, we aim to gain valuable insights into its potential in enhancing prenatal care and genetic screening.

2.2 Cell-Free Nucleic Acids

2.2.1 Cell-Free Fetal DNA

Cell-free DNA (cfDNA) molecules are liberated into the bloodstream upon cell death. With the same phenomenon during pregnancy, cell-free fetal DNA (cffDNA) is released into short fragments into the maternal bloodstream through the natural turnover of placenta cells [17] (Fig. 2.1). These short DNA fragments circulates freely and offers a unique opportunity to assess the fetal genome without the need for invasive procedures. CffDNA is predominantly composed of short DNA fragments, ranging from 100 to 300 base pairs in length. Analysis studies focused on size determination have revealed that the predominant size of cffDNA is approximately 166 bp [18]. Interestingly, this size coincides with the length of DNA associated with nucleosomes. This observation suggests that the released cffDNA originates from the breakdown of chromatin into nucleosome units.

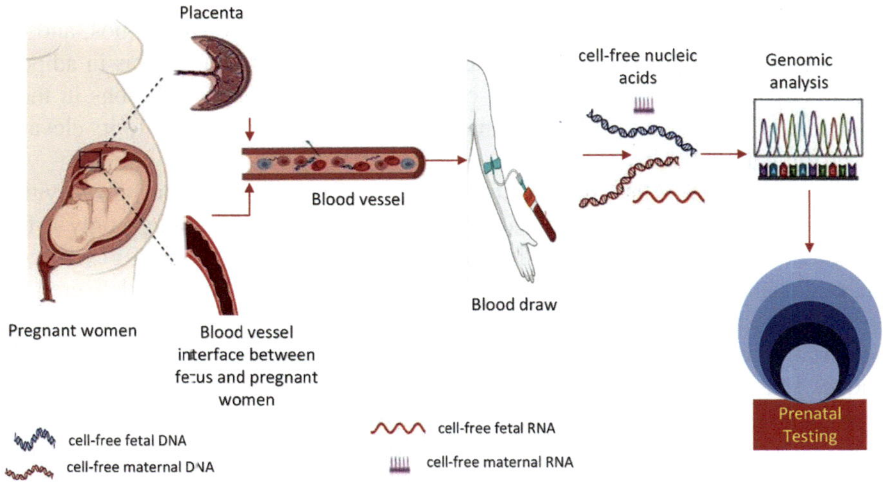

Fig. 2.1 The origin of cell-free nucleic acids (cfNAs) and their utilization in noninvasive prenatal screening. cfNAs, originating from the placenta, traverse into the maternal circulation of pregnant women. Additionally, fragments of cell-free DNA from cell necrosis or apoptosis are present in the woman's own circulation. Both maternal and fetal cfNAs are readily accessible in maternal circulation. A few milliliters of blood are drawn from the mother and processed for genomic analysis as part of prenatal screening or testing

Circulating fetal DNA mirrors the entire fetal genome, allowing for comprehensive genetic analysis. It carries paternally inherited alleles, enabling the identification of fetal genetic variations and mutations. According to the view of Chiu et al. (2021), the predominant consensus is that the placenta plays a crucial role in contributing "fetal" DNA to the maternal circulation [19]. Initial suspicions arose when fetal DNA was observed in maternal circulation, even in cases where embryo development had not occurred [20]. Furthermore, the presence of circulating fetal DNA has been detected in early pregnancy, prior to the development of fetal organs [21, 22]. The identification of placenta-specific methylation signatures in cfDNA molecules found in maternal plasma provided even more compelling evidence to support this notion [23]. These findings collectively reinforce the understanding that the placenta serves as a significant source of cffDNA in the maternal bloodstream.

cffDNA can be detected as early as the first trimester of pregnancy [6, 24]. During the early phases of pregnancy, there exist numerous genome-equivalents of maternal plasma, ranging from tens to hundreds [6]. As gestation progresses, the absolute quantity of fetal DNA increases. However, it is important to note that fetal DNA exists as a minor component within the background of maternal cfDNA. During the first and second trimesters, the fetal DNA volume, also known as the fetal fraction, accounts for approximately 10–20% of the maternal plasma cfDNA [25]. While the fetal fraction does increase as pregnancy advances, the rise is less significant compared to the overall increase in fetal DNA per unit volume of maternal plasma. This suggests that the amount of maternally derived DNA also increases with gestation [26].

The quantity of cffDNA can vary in specific pregnancy-related conditions. Elevated levels have been observed in pregnancies associated with preeclampsia, preterm labor, fetomaternal hemorrhage, cystic fibrosis, oligohydramnios, and trisomy 21 [19, 27–32]. In cases of maternal obesity, increased apoptosis in adipose tissue has been reported, potentially contributing to lower fetal fractions in these individuals [33]. Overall, various pathological conditions contribute to its elevated levels in maternal plasma.

So, what happens to cffDNA after delivery? Does it persist or vanish within hours? This question has been the subject of investigation, although only a few researchers have addressed this dilemma. For example, Lo et al. (1999) used polymerase chain reaction (PCR) to quantify cffDNA in women who had delivered via cesarean section [12]. The study concluded that cffDNA disappeared from the maternal circulation within 2 h after delivery. Subsequently, more advanced techniques like sequencing were employed to assess its lifespan after delivery, and it was found to persist in maternal plasma for up to one day [34]. This contradicts the previously held belief that cffDNA is rapidly cleared from maternal plasma after birth. It should be noted, however, that cffDNA does not survive subsequent pregnancies [35]. It has been suggested that clearance may occur at least in part through the renal route [34], although further research is needed to establish and reproduce these facts.

2.2.2 Detection Strategies

Detecting and analyzing cffDNA from maternal plasma requires advanced molecular techniques. The most commonly used method is quantitative PCR (qPCR) [36], which allows for the quantification of specific fetal DNA sequences. Another approach is digital PCR (dPCR), which enables the precise quantification of target DNA molecules [37]. Next-generation sequencing (NGS) is a powerful tool for analyzing cffDNA. It allows for the identification of fetal-specific sequences and the detection of chromosomal abnormalities, such as aneuploidies, microdeletions, and microduplications [38]. NGS-based approaches, including whole-genome sequencing (WGS) and targeted sequencing, provide all-inclusive and accurate assessment of fetal genetic information [39]. Additionally, single-nucleotide polymorphism (SNP) analysis can be employed to distinguish fetal DNA from maternal DNA [40]. By utilizing SNP markers, which exhibit differential patterns between the mother and fetus, fetal-specific gene sequences can be isolated and analyzed.

2.2.3 Difference Between Fetal and Maternal Cell-Free DNA

Within the maternal plasma, cfDNA consists of a mixture of fetal DNA and maternal DNA [41]. Understanding the differences between these two components is crucial for accurate fetal genetic analysis and reliable prenatal screening (Table 2.2). Fetal cfDNA, also known as cffDNA, originates or derives from placental cells and apoptotic debris [12, 33]. It enters the maternal circulation through placental shedding. Maternal cfDNA, on the other hand, originates from the maternal tissues, including peripheral blood cells and other physiological processes [42]. It represents the genetic milieu of the mother and serves as a reference for analyzing cffDNA. It carries genetic information from both parents, including paternally inherited alleles. It also provides insights into fetal-specific genetic variations and mutations, making it a valuable source for prenatal genetic testing. In contrast, maternal cfDNA reflects the genetic composition of the mother and includes her genetic variants.

Table 2.2 Differences between fetal and maternal cfDNA

Characteristic	Fetal cfDNA	Maternal cfDNA
Allelic representation	Paternally inherited alleles	Represents the genetic milieu of the mother
Allelic content	Wild-type alleles (low frequency of fetal mutations)	Mixture of wild-type and mutant alleles (heterozygous)
Size of DNA fragments	Short fragments (100–300 base pairs, predominant at ~166 bp)	Varied sizes, not as specific
Origin	Placental cells, apoptotic debris	Maternal tissues (peripheral blood cells, etc.)
Differentiation method	Methylation patterns, size fragment analysis	Methylation patterns, size fragment analysis, SNP markers

An important distinction between fetal and maternal cfDNA lies in their allelic representation. Maternal cfDNA consists of a mixture of both wild-type and mutant alleles, representing the heterozygous nature of the mother's genome [43]. In contrast, cffDNA carries only the wild-type alleles inherited from the father, as fetal mutations would be present at a low frequency [5]. Differentiating cffDNA from maternal cfDNA poses analytical challenges due to their similar composition and origin [44, 45]. Nonetheless, several methodologies have emerged to address these obstacles. One strategy involves investigating methylation patterns of cfDNA, which display distinct patterns in the mother and fetus. Through the identification and examination of methylation patterns, it becomes possible to differentiate cffDNA from maternal cfDNA [46, 47]. Another approach entails size fragment analysis, where a comparison of the mothers and circulation fetal DNA fragments are compared using a paired-end massively parallel shotgun sequencing approach. Sequencing enables the identification of fetal-specific genetic variations within the mixed pool of cfDNA [48].

2.2.4 Clinical Applications of cffDNA

Currently cffDNA analysis has numerous clinical applications in prenatal diagnostics. The most well-known application is NIPS of fetal aneuploidies, such as Down syndrome, Edwards syndrome, and Patau syndrome. NIPS based on cffDNA offers high sensitivity and specificity [49], significantly reducing the need for invasive procedures. Several countries have implemented comprehensive national programs for screening fetal anomalies [50]. These programs consist of two key phases: screening and diagnosis. During the screening phase, a combination of biochemical marker blood tests, ultrasound scans, and maternal age assessment is utilized to identify pregnancies at high risk. Following this, in the diagnosis phase, invasive procedures are conducted for pregnancies that exceed a certain risk threshold. While ultrasound scanning remains vital for detecting various fetal abnormalities, there is a possibility of replacing the current multistage screening approach for trisomies with a single noninvasive screening test using cffDNA in the early stages of pregnancy. This would offer a more streamlined and efficient process for prenatal screening. cffDNA analysis also enables the detection of fetal sex, particularly in cases where sex-linked disorders are a concern [51]. By examining specific sex chromosome sequences, such as the SRY gene for male fetuses, cffDNA analysis provides accurate determination of fetal sex. Furthermore, cffDNA analysis holds promise for the diagnosis of monogenic disorders and genetic mutations [39]. Targeted sequencing approaches allow for the identification of specific disease-causing variants, offering early detection and enabling appropriate genetic counseling and intervention. cffDNA analysis can also be utilized for monitoring the health and well-being of the fetus during pregnancy. It provides valuable information about placental function, fetal growth, and potential complications, such as pre-eclampsia or intrauterine growth restriction [1, 27, 31]. By monitoring specific biomarkers in cffDNA, healthcare providers can assess fetal well-being and make informed decisions regarding management and intervention. The clinical applications of cffDNA analysis, ranging from noninvasive aneuploidy screening to the diagnosis of monogenic disorders, have significantly improved prenatal care. Continued advancements in technology and research will further enhance its potential, ultimately leading to improved prenatal diagnostics and personalized medicine.

2.2.5 Cell-Free Fetal RNA

Cell-free fetal RNA (cffRNA) is a remarkable component of maternal circulation that holds immense potential for prenatal screening. It originates from the developing fetus [52] and carries valuable genetic information. cffRNA is derived from various sources, including placental trophoblasts, fetal organs, and fetal hematopoietic cells [53]. These cells release RNA molecules into the maternal bloodstream, making cffRNA a valuable resource for accessing fetal genetic information. The composition includes messenger RNA (mRNA), microRNA (miRNA), long noncoding RNA (lncRNA), and other small RNA species. One unique characteristic of cffRNA is its relative abundance compared to maternal RNA [54]. As fetal cells release RNA into the maternal circulation, the concentration of cffRNA increases throughout pregnancy, providing an ample source for analysis. Its detection and analysis in maternal blood offer a noninvasive approach to assess the fetal transcriptome and identify potential genetic abnormalities.

mRNA analysis allows for the assessment of gene expression patterns during fetal development [55]. In the context of prenatal screening, mRNA profiling has been employed to identify genetic disorders and predict phenotypic outcomes. For instance, amniotic fluid cffRNA transcriptome may reflect the fetal and placental development and therefore aid in the monitoring of normal and abnormal development [56].

miRNAs are small RNA molecules involved in posttranscriptional gene regulation and serve as potential biomarkers in prenatal screening. The dysregulation of specific miRNAs has been associated with genetic disorders and adverse pregnancy outcomes. In a study by Miura et al., miRNA expression profiles were analyzed in amniotic fluid samples, revealing differential expression of miRNAs in pregnancies affected by neural tube defects [57]. This highlights the diagnostic potential of miRNA in identifying fetal abnormalities.

lncRNAs, which play crucial roles in gene regulation, have recently gained attention in prenatal screening. Aberrant expression of specific lncRNAs has been linked to developmental disorders and pregnancy complications. For example, a study by Fu et al. found that the coexpression of lncRNA-miRNA-mRNA is affected by preeclampsia in pregnancies, suggesting its potential as a diagnostic marker for this condition [58]. Another study by Monteiro et al. identified dysregulated lncRNAs in placental tissues associated with fetal growth restriction, providing insights into the underlying molecular mechanisms [59].

2.3 Detection and Analysis of cfNAs in Maternal Circulation

The detection and analysis of cfNAs require efficient methods of sample collection, processing, and quality control to ensure accurate and reliable results. To obtain maternal blood samples, various blood collection techniques are employed. The most prevalent method is venipuncture, wherein a needle is introduced into a vein to draw blood into tubes with suitable anticoagulants [60]. Another method involves fingerstick, where the finger is pricked with a lancet to obtain a small blood sample, subsequently transferred to collection tubes. Specialized blood collection tubes, equipped with reagents for stabilizing and preserving cfNAs during blood collection, offer another avenue [61]. Following blood collection, the isolation and preservation become pivotal for maintaining sample integrity and ensuring accurate analysis. Standard methods of isolation and purification include plasma separation through centrifugation, extraction using techniques like column-based purification or magnetic bead-based methods, and employing commercial kits. To maintain cfNA integrity during storage, samples can be stored at low temperatures (−80 °C).

2.3.1 Quality Control Measures

To guarantee the reliability and precision, a series of quality control measures are systematically implemented. These include the assessment of sample integrity, where the quality of the sample is examined for factors such as hemolysis, contamination, or degradation, utilizing techniques like visual inspection, spectrophotometry, or specialized kits. Additionally, control measures involve the use of internal controls, such as spiked-in synthetic DNA or RNA [62] to monitor the efficiency of extraction and identify potential biases in downstream analysis. Quantification and size profiling are carried out through methods qPCR, dPCR, and NGS to precisely measure and analyze the size distribution of cfNA molecules. Recently, methods such as multi-frequency impedance cytometry and machine learning have been developed to measure the concentration of DNA in a sample [63]. In addition to the above-mentioned quality control measures, robust data validation procedures, stringent analysis, artifact filtration, and thorough statistical analysis are also important. These measures collectively ensure the reliable interpretation of cfNA analysis.

2.4 Cell-Free Nucleic Acid Detection Techniques

The detection and analysis of cffNA are essential in NIPS. Techniques such as PCR, NGS, microarray, and emerging technologies like artificial intelligence serve powerful tools for precise detection. These methods enable robust detection of nucleic acids, hence the identification of fetal aneuploidies, genetic disorders, and various fetal abnormalities. PCR, a widely utilized technique, is valued for its sensitivity and specificity [37, 38, 64]. It involves amplifying specific DNA sequences through cycles of denaturation, annealing, and extension. Techniques like qPCR and quantitative methylation-specific PCR have been used for cffNA analysis, providing real-time monitoring and detecting methylated DNA sequences as biomarkers for fetal aneuploidies [65]. Likewise, NGS allows high-throughput analysis by parallel sequencing millions of cffNA fragments, enabling comprehensive insights into the fetal genome through approaches like WGS and targeted sequencing [66]. WGS offers a global view of detecting genetic variations and aneuploidies, while targeted sequencing focuses on specific regions of interest. dPCR is a precise and sensitive technique, enabling absolute quantification of cffNA by partitioning the template into multiple reactions [67]. dPCR has eliminated the need for standard curves and reference samples, demonstrating success in detecting fetal aneuploidies and single-gene disorders. Similarly, microarray-based technique offers a high-throughput approach [68–70]. These arrays consist of immobilized probes that hybridize with the complementary cffNA sequences. By detecting the signal generated from the hybridization, the presence or absence of specific cffNA sequences can be determined. Microarray analysis allows the simultaneous detection of multiple cffNA targets and has been applied in the detection of fetal aneuploidies and genetic disorders.

2.4.1 Emerging Methods

Advancements in technology continue to drive the development of NIPS. Emerging technologies such as single-molecule sequencing, nanopore sequencing, droplet digital PCR (ddPCR), microfluidics, fetal RNA profiling, and artificial intelligence-enabled methods have demonstrated increased efficiency and productivity (Table 2.3). Single-molecule sequencing technologies provide long-read sequencing capabilities, enabling the detection of structural variations and complex genomic rearrangements in DNA [71]. Nanopore sequencing allows real-time sequencing by passing the cffNA through nanopores, enabling direct electrical detection [72]. ddPCR, similar to dPCR, partitions the cffNAs into numerous droplets for absolute quantification [73]. Fetal RNA profiling involves the analysis of fetal RNA transcripts in maternal circulation, offering potential for early detection of fetal disorders [74]. The combined artificial intelligence and whole-genome epigenomic analysis of circulating cfDNA in maternal blood holds the potential for detecting fetal congenital heart defects [75]. These techniques with ongoing advancements continually enhance sensitivity, accuracy and efficiency in prenatal screening.

2.5 Computational Insights and Data Analysis

Molecular methods generate extensive raw data related to cfNAs that necessitates processing to derive conclusive results. Given that this data is predominantly employed for disease screening, it is crucial for the analysis to be executed with precision; consequently, computational methods are frequently employed for this purpose. The interpretation of data and its analysis concerning the detection of

Table 2.3 Detection technologies for cfNAs in prenatal screening

Detection technology	Features	Advantages	Limitations
Digital PCR	Absolute quantification	High sensitivity and accuracy	Limited multiplexing capabilities
Microarray analysis	Simultaneous analysis of multiple targets	High-throughput screening capability	Limited dynamic range for low abundance targets
Emerging technologies	Novel and innovative approaches	Potential for improved sensitivity, specificity, and multiplexing capabilities	Limited validation and standardization
Next-generation sequencing	High-throughput sequencing	Comprehensive analysis of genetic variants and transcriptomes	High cost, complex data analysis, and longer turnaround time
Massively parallel sequencing	Massively parallel sequencing	Accurate detection of rare variants and structural variations	Requires specialized equipment and bioinformatics expertise
Other PCR variants	Various PCR variants (e.g., qPCR, RT-PCR)	Established and widely used methods for quantification and detection	Limited multiplexing capabilities, sensitivity dependent on primer design, and amplification efficiency

cffNAs in maternal plasma relies heavily on bioinformatics tools and algorithms. These tools play an important role in identifying fetal-specific genetic variations [76]. Effectively extracting meaningful information from the vast pool of sequencing data demands the application of sophisticated bioinformatics tools and algorithms. These tools are indispensable for identifying fetal-specific genetic variations, aneuploidies, and other pertinent genetic information. Developed to address various aspects of data analysis, including read alignment, variant calling, and copy number variation analysis, these tools contribute to a wide-ranging understanding of the genomic landscape [77]. Commonly used tools such as Bowtie and Burrows Wheeler Analysis ensure efficient and accurate alignment, facilitating the identification of fetal-specific sequences distinct from the maternal genome [78, 79]. Variant calling, essential for pinpointing genetic variations between fetal and maternal genomes, employs tools like GATK [80, 81]. Widely adopted in cffNA analysis, these tools identify single nucleotide variants (SNVs), insertions, deletions, and structural variations. Filtering strategies are then applied to discriminate genuine fetal variants from sequencing artifacts and maternal DNA contamination ensuring the precision and reliability of the results. Likewise, copy number variations play a crucial role in prenatal screening, and their detection requires specific algorithms. Tools like CNVnator and Control-FREEC are commonly used for copy number variation analysis [82, 83]. These algorithms detect changes in the copy number of DNA segments and provide insights into chromosomal aberrations associated with fetal abnormalities.

While bioinformatics tools and algorithms have aided in data analysis, they also pose certain challenges and limitations. For instance, the presence of a low fetal fraction in maternal plasma poses challenges in accurately detecting and interpreting cffNA [84]. Low fetal fraction results in reduced sensitivity and increase the likelihood of false-negative results. Secondly, cffNA data analysis is complicated by the inherent biological variability between pregnancies. Genetic variations and expression levels may vary significantly among individuals [85], making it challenging to establish robust reference databases and accurate interpretation criteria. Additionally, technical artifacts such as sequencing errors and biases can introduce false-positive or false-negative results [82]. Moreover, maternal DNA contaminates the cffNA sample, making it challenging to distinguish fetal-specific genetic variations [19].

2.6 Clinical Applications of Cell-Free Nucleic Acids in Fetal Screening

Single-gene disorders arising from mutations in a single gene and inherited in various patterns, such as autosomal dominant, autosomal recessive, or X-linked inheritance can be noninvasively detected through the use of cfNAs. Pioneering studies have demonstrated the successful detection of single gene disorders like cystic fibrosis, β-thalassemia, and Duchenne muscular dystrophy using fetal nucleic acids [29, 50, 71–73, 86–89]. Mendelian disorders, following Mendelian inheritance patterns and caused by mutations in specific genes, can also be identified through cfNA analysis. Diseases such as sickle cell anemia, phenylketonuria, and Tay-Sachs disease have been successfully detected using this approach [86, 90–94]. Autosomal dominant and recessive disorders requiring the identification of disease-causing mutations can be detected using targeted approaches like PCR or NGS. Few studies reported the successful detection of disorders including neurofibromatosis type 1, Huntington's disease, and cystic fibrosis using cfNA analysis [29, 95–99]. X-linked disorders affecting males can also be identified through cfNA analysis, with studies detecting mutations associated with disorders like hemophilia A, Duchenne muscular dystrophy, and fragile X syndrome [89, 100–102]. Moreover, NIPS using cffDNA has shown high sensitivity and specificity in screening for common chromosomal abnormalities, including trisomy 21, trisomy 18, and trisomy 13, making it a standard screening tool [103, 104]. Additionally, cfNA analysis provides an accurate and noninvasive method for determining fetal sex, aiding in the early detection of sex-linked disorders and offering valuable insights for genetic counseling [105].

Although the application of cffDNA in NIPS is extensive, the utilization of cffRNA remains minimal. However, applications of cffRNA in NIPS have been elucidated. For instance, the application of cell-free fetal mRNA in SNP for detecting trisomies using the allelic ratio of specific heterozygous SNP has been confirmed [106]. Similarly, there are placenta-specific miRNAs that are expressed differently in certain parts of the placenta and secreted from the trophoblast layer in varying concentrations during pregnancy [107]. These placenta-specific miRNAs are not expressed in other tissues and can be classified as placenta-specific, placenta-associated, and placenta-derived miRNAs [108]. cffRNAs have proven valuable in identifying pregnancy-related complications like preeclampsia, gestational diabetes, and intrauterine growth restriction. The expression levels of circular RNAs have been implicated in gestational diabetes [109]. Although the role of lncRNAs in NIPS is marginal, evidence suggests that there are a few, that are placenta-associated, and some have been identified as prognostic markers for congenital heart disease [110].

2.7 Emerging Trends in Detection Strategies

Advancements in technology have introduced new techniques, aimed at enhancing efficacy and efficiency and exemplified by ddPCR, a highly sensitive method enabling absolute quantification of cffNAs through sample partitioning into thousands of droplets for precise nucleic acid quantification [67, 73, 92]. Similarly, single-molecule real-time sequencing, with long-read capabilities, facilitates the detection of large genomic rearrangements and structural variations in cffNA, thereby offering high-resolution insights into the fetal genome and aiding in identifying rare genetic disorders [62, 71]. Likewise, massively parallel signature sequencing a high-throughput method provides comprehensive profiling of cffNA, including detailed information on gene expression, DNA methylation patterns, and copy number variations, crucial for detecting fetal genetic abnormalities [35]. Mass spectrometry-based approaches encompass mass spectrometric genotyping and analysis of DNA methylation and present a unique opportunities for cfNA detection and analysis, characterized by high-throughput and multiplex capabilities that allow simultaneous detection of multiple genetic variants and epigenetic modifications at protein level [111].

The utilization of microfluidic technology in NIPS is rapidly emerging. Immunoaffinity and size-based microfluidic methods are employed. These techniques have the capability to enhance the isolation and enrichment of fetal-originated cells. For example, immunoaffinity microfluidic methods can effectively enrich and isolate circulating fetal extravillous trophoblasts and fetal nucleated red blood cells [112]. Conversely, size-based microfluidic systems are utilized for isolating fetal extravillous trophoblasts.

2.8 Ethical Considerations and Future Research Areas

Prenatal screening is surrounded by diverse ethical considerations and challenges (Table 2.4). Informed consent is a critical prerequisite for conducting cffNA analysis, necessitating that pregnant women receive comprehensive information about the testing's purpose, benefits, limitations, and potential risks to make informed decisions autonomously. Privacy and confidentiality are paramount, requiring robust data protection measures, including secure storage, restricted access, and anonymization techniques, to safeguard the genetic information derived from cffNA.

In terms of future exploration, identifying novel biomarkers within cffNA, such as microRNAs, long noncoding RNAs, and other noncoding elements, holds

Table 2.4 Ethical considerations and challenges in cfNA analysis

Ethical considerations and challenges	Description	Potential solutions or recommendations
Informed consent	Ensuring that participants fully understand the purpose, risks, benefits, and potential implications of the research	Implement a robust informed consent process, providing clear and comprehensive information about the study, allowing participants to make informed decisions. Encourage ongoing communication and support for participants throughout the study
Privacy and data confidentiality	Protecting the privacy of individuals whose cfNA data is being collected and stored	Implement strict data security measures, including encryption and secure storage. Adhere to applicable data protection laws and regulations. Obtain appropriate consent for data sharing and use, ensuring anonymity and de-identification whenever possible
Access to genetic information	Addressing concerns regarding the access, use, and potential misuse of genetic information obtained through cfNA analysis	Develop clear policies and guidelines on who can access the genetic information, ensuring it is used solely for authorized purposes. Educate healthcare providers, researchers, and participants about the importance of responsible data handling and the potential implications of genetic information
Societal impact and discrimination	Considering the potential societal impact and implications of cfNA analysis results	Foster public awareness and education about cfNA analysis, genetic disorders, and associated ethical considerations. Advocate for nondiscriminatory policies and legislation to prevent genetic discrimination and stigmatization
Equity and accessibility	Ensuring equitable access to NIPS technologies, particularly in underserved populations	Develop strategies to reduce cost barriers, improve accessibility to NIPS in diverse populations, and address healthcare disparities. Advocate for policies that promote equitable access to NIPS technologies and prenatal care

promise for enhancing diagnostic and prognostic capabilities. Additionally, integrating cffNA analysis with other omics technologies, like proteomics and metabolomics, can provide an inclusive understanding of fetal health and disease, unraveling complex interactions and pathways associated with genetic disorders. Standardization efforts and the implementation of quality control measures are essential to ensure the reproducibility and accuracy of cffNA analysis across diverse laboratories and platforms, facilitating its widespread integration into clinical practice.

2.9 Conclusion

The presence of fetal derived cfNAs in maternal blood provides a noninvasive and accessible source of genetic information about the developing fetus. This has changed the field of prenatal screening by offering new opportunities for early detection and diagnosis of fetal genetic disorders. The clinical importance of cfNAs lies in its ability to provide valuable insights into fetal health and enable informed reproductive decision-making. The detection and analysis of cfNAs in maternal circulation offer a noninvasive approach with significant clinical importance in fetal diagnosis. The advancements in detection technologies, expansion of clinical applications, and ongoing research in cfNA analysis hold great promise for improving prenatal care and enhancing our ability to detect and manage fetal genetic disorders. By harnessing the potential of cfNAs, we can empower expectant mothers with valuable information about their fetus's health, leading to better outcomes and informed decision-making in prenatal care.

References

1. Taglauer ES, Wilkins-Haug L, Bianchi DW. Review: cell-free fetal DNA in the maternal circulation as an indication of placental health and disease. Placenta. 2014;35 Suppl(Suppl):S64–8.
2. Szilágyi M, Pös O, Márton É, Buglyó G, Soltész B, Keserű J, et al. Circulating cell-free nucleic acids: main characteristics and clinical application. Int J Mol Sci. 2020;21(18):6827.
3. Lo YM, Corbetta N, Chamberlain PF, Rai V, Sargent IL, Redman CW, Wainscoat JS. Presence of fetal DNA in maternal plasma and serum. Lancet. 1997;350(9076):485–7.
4. Hui L, Bianchi DW. Fetal fraction and noninvasive prenatal testing: what clinicians need to know. Prenat Diagn. 2020;40(2):155–63.
5. Bischoff FZ, Lewis DE, Simpson JL. Cell-free fetal DNA in maternal blood: kinetics, source and structure. Hum Reprod Update. 2005;11(1):59–67.
6. Lo YM, Tein MS, Lau TK, Haines CJ, Leung TN, Poon PM, et al. Quantitative analysis of fetal DNA in maternal plasma and serum: implications for noninvasive prenatal diagnosis. Am J Hum Genet. 1998;62(4):768–75.
7. Sharp AN, Heazell AE, Crocker IP, Mor G. Placental apoptosis in health and disease. Am J Reprod Immunol. 2010;64(3):159–69.
8. Rosner M, Kolbe T, Hengstschläger M. Fetomaternal microchimerism and genetic diagnosis: on the origins of fetal cells and cell-free fetal DNA in the pregnant woman. Mutat Res Rev Mutat Res. 2021;788:108399.
9. Hochstenbach R, Nikkels PG, Elferink MG, Oudijk MA, van Oppen C, van Zon P, et al. Cell-free fetal DNA in the maternal circulation originates from the cytotrophoblast: proof from an unique case. Clin Case Rep. 2015;3(6):489–91.
10. Grabuschnig S, Bronkhorst AJ, Holdenrieder S, Rosales Rodriguez I, Schliep KP, Schwendenwein D, et al. Putative origins of cell-free DNA in humans: a review of active and passive nucleic acid release mechanisms. Int J Mol Sci. 2020;21(21):8062.
11. Bianchi DW. Circulating fetal DNA: its origin and diagnostic potential—a review. Placenta. 2004;25 Suppl A:S93–S101.
12. Lo YM, Zhang J, Leung TN, Lau TK, Chang AM, Hjelm NM. Rapid clearance of fetal DNA from maternal plasma. Am J Hum Genet. 1999;64(1):218–24.
13. Jimbo M, Sekizawa A, Sugito Y, Matsuoka R, Ichizuka K, Saito H, et al. Placenta increta: postpartum monitoring of plasma cell-free fetal DNA. Clin Chem. 2003;49(9):1540–1.
14. Hahn S, Holzgreve W. Prenatal diagnosis using fetal cells and cell-free fetal DNA in maternal blood: what is currently feasible? Clin Obstet Gynecol. 2002;45:649–56; discussion 730–2.
15. Sugito Y, Sekizawa A, Farina A, Yukimoto Y, Saito H, Iwasaki M, et al. Relationship between severity of hyperemesis gravidarum and fetal DNA concentration in maternal plasma. Clin Chem. 2003;49(10):1667–9.
16. Wataganara T, LeShane ES, Farina A, Messerlian GM, Lee T, Canick JA, et al. Maternal serum cell-free fetal DNA levels are increased in cases of trisomy 13 but not trisomy 18. Hum Genet. 2003;112(2):204–8.

17. Van Opstal D, Veen S, Joosten M, Diderich KEM, Govaerts LCP, Polak J, et al. Placental studies elucidate discrepancies between NIPT showing a structural chromosome aberration and a differently abnormal fetal karyotype. Prenat Diagn. 2019;39(11):1016–25.
18. Lo YM, Chan KC, Sun H, Chen EZ, Jiang P, Lun FM, et al. Maternal plasma DNA sequencing reveals the genome-wide genetic and mutational profile of the fetus. Sci Transl Med. 2010;2(61):61ra91.
19. Chiu RWK, Lo YMD. Cell-free fetal DNA coming in all sizes and shapes. Prenat Diagn. 2021;41(10):1193–201.
20. Alberry M, Maddocks D, Jones M, Abdel Hadi M, Abdel-Fattah S, Avent N, et al. Free fetal DNA in maternal plasma in anembryonic pregnancies: confirmation that the origin is the trophoblast. Prenat Diagn. 2007;27:415–8.
21. Guibert J, Benachi A, Grebille AG, Ernault P, Zorn JR, Costa JM, et al. Kinetics of SRY gene appearance in maternal serum: detection by real time PCR in early pregnancy after assisted reproductive technique. Hum Reprod. 2003;18:1733–6.
22. Karakas B, Qubbaj W, Al-Hassan S, Coskun S, et al. Noninvasive digital detection of fetal DNA in plasma of 4-week-pregnant women following in vitro fertilization and embryo transfer. PLoS One. 2015;10:e0126501.
23. Chim SSC, Tong YK, Chiu RWK, Lau TK, Leung TN, Chan LY, et al. Detection of the placental epigenetic signature of the maspin gene in maternal plasma. Proc Natl Acad Sci U S A. 2005;102:14753–8.
24. Akolekar R, Farkas DH, Vanagtmael AL, Bombard AT, Nicolaides KH. Fetal sex determination using circulating cell-free fetal DNA (ccffDNA) at 11 to 13 weeks of gestation. Prenat Diagn. 2010;30:918–23.
25. Palomaki GE, Kloza EM, Lambert-Messerlian GM, Haddow JE, Neveux LM, Ehrich M, et al. DNA sequencing of maternal plasma to detect Down syndrome: an international clinical validation study. Genet Med. 2011;13:913–20.
26. Shi X, Zhang Z, Cram DS, Liu C. Feasibility of noninvasive prenatal testing for common fetal aneuploidies in an early gestational window. Clin Chim Acta. 2015;439:24–8.
27. Karapetian AO, Baev OR, Sadekova AA, Krasnyi AM, Sukhikh GT. Cell-free foetal DNA as a useful marker for preeclampsia prediction. Reprod Sci. 2021;28:1563. https://doi.org/10.1007/s43032-021-00466-w.
28. Lun FM, Tsui NB, Chan KC, Leung TY, Lau TK, Charoenkwan P, Chow KC, et al. Noninvasive prenatal diagnosis of monogenic diseases by digital size selection and relative mutation dosage on DNA in maternal plasma. Proc Natl Acad Sci U S A. 2008;105:19920–5.
29. Chandler NJ, Ahlfors H, Drury S, Mellis R, Hill M, McKay FJ, et al. Noninvasive prenatal diagnosis for cystic fibrosis: implementation, uptake, outcome, and implications. Clin Chem. 2020;66:207–16.
30. Yaron Y, Pauta M, Badenas C, Soler A, Borobio V, Illanes C, Paz-Y-Miño F, et al. Maternal plasma genome-wide cell-free DNA can detect fetal aneuploidy in early and recurrent pregnancy loss and can be used to direct further workup. Hum Reprod. 2020;35:1222–9.
31. Munchel S, Rohrback S, Randise-Hinchliff C, et al. Circulating transcripts in maternal blood reflect a molecular signature of early-onset preeclampsia. Sci Transl Med. 2020;12(550):eaaz0131.
32. Chandrasekharan S, Minear MA, Hung A, Allyse M, et al. Noninvasive prenatal testing goes global. Sci Transl Med. 2014;6(231):231fs15.
33. Haghiac M, Vora NL, Basu S, Johnson KL, Presley L, Bianchi DW, et al. Increased death of adipose cells, a path to release cell-free DNA into systemic circulation of obese women. Obesity (Silver Spring). 2012;20:2213–9.
34. Yu SC, Lee SW, Jiang P, Leung TY, Chan KC, Chiu RW, et al. High-resolution profiling of fetal DNA clearance from maternal plasma by massively parallel sequencing. Clin Chem. 2013;59:1228–37.
35. Benachi A, Steffann J, Gautier E, Ernault P, Olivi M, Dumez Y, et al. Fetal DNA in maternal serum: does it persist after pregnancy? Hum Genet. 2003;113:76–9.

36. Wei C, Saller DN, Sutherland JW. Detection and quantification by homogeneous PCR of cell-free fetal DNA in maternal plasma. Clin Chem. 2001;47(2):336–8.
37. Charoenkwan P, Traisrisilp K, Sirichotiyakul S, Phusua A, Sanguansermsri T, Tongsong T. Noninvasive prenatal diagnosis of beta-thalassemia disease by using digital PCR analysis of cell-free fetal DNA in maternal plasma. Fetal Diagn Ther. 2023;49(11–12):468–78.
38. Curnow KJ, Sanderson RK, Beruti S. Noninvasive detection of fetal aneuploidy using next generation sequencing. Methods Mol Biol. 2019;1885:325–45.
39. Zhang J, Li J, Saucier JB, Feng Y, Jiang Y, Sinson J, et al. Non-invasive prenatal sequencing for multiple Mendelian monogenic disorders using circulating cell-free fetal DNA. Nat Med. 2019;25(3):439–47.
40. Kantor V, Jelsema R, Xu W, DiNonno W, Young K, Demko Z, et al. Non-invasive prenatal screening for fetal triploidy using single nucleotide polymorphism-based testing: differential diagnosis and clinical management in cases showing an extra haplotype. Prenat Diagn. 2022;42(8):994–9.
41. Shi J, Zhang R, Li J, Zhang R. Size profile of cell-free DNA: a beacon guiding the practice and innovation of clinical testing. Theranostics. 2020;10(11):4737.
42. Pan M, Chen P, Lu J, Liu Z, Jia E, Ge Q. The fragmentation patterns of maternal plasma cell-free DNA and its applications in non-invasive prenatal testing. Prenat Diagn. 2020;40(8):911–7.
43. Lo YM. Fetal DNA in maternal plasma: biology and diagnostic applications. Clin Chem. 2000;46(12):1903–6.
44. Smith JF, Blumenfeld Y. Cell-free fetal DNA in maternal plasma: progress and potential. NeoReviews. 2008;9(8):e332–7.
45. Hill M, Barrett AN, White H, Chitty LS. Uses of cell free fetal DNA in maternal circulation. Best Pract Res Clin Obstet Gynaecol. 2012;26(5):639–54.
46. Tsui DW, Chiu RW, Lo YD. Epigenetic approaches for the detection of fetal DNA in maternal plasma. Chimerism. 2010;1(1):30–5.
47. Rather RA, Saha SC, Dhawan V. The most favourable procedure for the isolation of cell-free DNA from the plasma of iso-immunized RHD-negative pregnant women. J Circ Biomark. 2015;4:12.
48. Cirigliano V, Ordoñez E, Rueda L, Syngelaki A, Nicolaides KH. Performance of the neoBona test: a new paired-end massively parallel shotgun sequencing approach for cell-free DNA-based aneuploidy screening. Ultrasound Obstet Gynecol. 2017;49:460–4.
49. Wright CF, Burton H. The use of cell-free fetal nucleic acids in maternal blood for non-invasive prenatal diagnosis. Hum Reprod Update. 2009;15(1):139–51.
50. Shaw J, Scotchman E, Chandler N, Chitty L. Non-invasive prenatal testing for aneuploidy, copy number variants and single gene disorders. Reproduction. 2020;160(5):A1.
51. Hill M, Lewis C, Jenkins L, Allen S, Elles RG, et al. Implementing noninvasive prenatal fetal sex determination using cell-free fetal DNA in the United Kingdom. Expert Opin Biol Ther. 2012;12 Suppl 1:S119–26.
52. Rather RA, Saha SC. Reappraisal of evolving methods in non-invasive prenatal screening: discovery, biology and clinical utility. Heliyon. 2023;9(3):e13923.
53. Moufarrej MN, Bianchi DW, Shaw GM, Stevenson DK, Quake SR. Noninvasive prenatal testing using circulating DNA and RNA: advances, challenges, and possibilities. Annu Rev Biomed Data Sci. 2023;6:397–418.
54. Li N, Gu Y, Tang J, Li Y, Chen D, Xu Z. Circulating non-coding RNAs and exosomes: liquid biopsies for monitoring preeclampsia. Methods Mol Biol. 2023;2695:263–77.
55. Hui L, Wick HC, Edlow AG, Cowan JM, Bianchi DW. Global gene expression analysis of term amniotic fluid cell-free fetal RNA. Obstet Gynecol. 2013;121(6):1248–54.
56. Park HJ, Cho HY, Cha DH. The amniotic fluid cell-free transcriptome provides novel information about fetal development and placental cellular dynamics. Int J Mol Sci. 2021;22(5):2612.
57. Miura K, Miura S, Yamasaki K, Higashijima A, Kinoshita A, Yoshiura K, et al. Identification of pregnancy-associated microRNAs in maternal plasma. Clin Chem. 2010;56(10):1767–71.
58. Fu X, Wu M, Chen Y, Li L, Yin H, Liang K, et al. The expression profile of plasmatic exosomal lncRNAs in early-onset preeclampsia by sequencing. Am J Transl Res. 2022;14(6):3806–23.

59. Monteiro LJ, Peñailillo R, Sánchez M, Acuña-Gallardo S, Mönckeberg M, Ong J, et al. The role of long non-coding RNAs in trophoblast regulation in preeclampsia and intrauterine growth restriction. Genes (Basel). 2021;12(7):970.
60. Thébault P, Cailhier JF, Lapointe R. Blood sample processing and banking for functional and molecular analyses. Methods Mol Biol. 2023;2614:37–46.
61. Heatlie J, Chang V, Fitzgerald S, Nursalim Y, Parker K, Lawrence B, Print CG, et al. Specialized cell-free DNA blood collection tubes can be repurposed for extracellular vesicle isolation: a pilot study. Biopreserv Biobank. 2020;18(5):462–70.
62. Mandal S, Li Z, Chatterjee T, Khanna K, Montoya K, Dai L, et al. Direct kinetic fingerprinting for high-accuracy single-molecule counting of diverse disease biomarkers. Acc Chem Res. 2021;54(2):388–402.
63. Kokabi M, Sui J, Gandotra N, Pournadali Khamseh A, Scharfe C, Javanmard M. Nucleic acid quantification by multi-frequency impedance cytometry and machine learning. Biosensors (Basel). 2023;13(3):316.
64. Torres B, Stoffel TJ, Oro D, Rossi EM, de Miranda GB, Houenou LJ, Trott A. Early noninvasive fetal RHD genotyping and sex determination by conventional and multiplex PCR performing a rapid and low-cost cell-free fetal DNA extraction method. Eur J Obstet Gynecol Reprod Biol. 2012;165(2):370.
65. Sigalotti L, Covre A, Colizzi F, Fratta E. Quantitative methylation-specific PCR: a simple method for studying epigenetic modifications of cell-free DNA. Methods Mol Biol. 2019;1909:137–62.
66. Xu XP, Gan HY, Li FX, Tian Q, Zhang J, Liang RL, et al. A method to quantify cell-free fetal DNA fraction in maternal plasma using next generation sequencing: its application in non-invasive prenatal chromosomal aneuploidy detection. PLoS One. 2016;11(1):e0146997.
67. Shaw J, Scotchman E, Paternoster B, Ramos M, Nesbitt S, Sheppard S, et al. Non-invasive fetal genotyping for maternal alleles with droplet digital PCR: a comparative study of analytical approaches. Prenat Diagn. 2023;43(4):477–88.
68. Fan HC, Gu W, Wang J, Blumenfeld YJ, El-Sayed YY, Quake SR. Non-invasive prenatal measurement of the fetal genome. Nature. 2012;487(7407):320–4.
69. Chu T, Burke B, Bunce K, Surti U, Allen Hogge W, Peters DG. A microarray-based approach for the identification of epigenetic biomarkers for the noninvasive diagnosis of fetal disease. Prenat Diagn. 2009;29(11):1020–30.
70. Damour G, Baumer K, Legardeur H, Hall D. Early noninvasive prenatal paternity testing by targeted fetal DNA analysis. Sci Rep. 2023;13(1):12139.
71. Fatima N, Petri A, Gyllensten U, Feuk L, Ameur A. Evaluation of single-molecule sequencing technologies for structural variant detection in two Swedish human genomes. Genes (Basel). 2020;11(12):1444.
72. Jain M, Koren S, Miga KH, Quick J, Rand AC, Sasani TA, et al. Nanopore sequencing and assembly of a human genome with ultra-long reads. Nat Biotechnol. 2018;36(4):338–45.
73. D'Aversa E, Breveglieri G, Boutou E, Balassopoulou A, Voskaridou E, Pellegatti P, et al. Droplet digital PCR for non-invasive prenatal detection of fetal single-gene point mutations in maternal plasma. Int J Mol Sci. 2022;23(5):2819.
74. Rasmussen M, Reddy M, Nolan R, Camunas-Soler J, Khodursky A, Scheller NM, et al. RNA profiles reveal signatures of future health and disease in pregnancy. Nature. 2022;601(7893):422–7.
75. Bahado-Singh R, Friedman P, Talbot C, Aydas B, Southekal S, Mishra NK, et al. Cell-free DNA in maternal blood and artificial intelligence: accurate prenatal detection of fetal congenital heart defects. Am J Obstet Gynecol. 2023;228(1):76.e1–76.e10.
76. Jin S, Wang T, Wenying C, Wu Y, Huang S, Zeng P. Maternal and fetal origins of offspring blood pressure: statistical analysis using genetic correlation and genetic risk score-based Mendelian randomization. Int J Epidemiol. 2023;52(5):1360–76.
77. Boehm FJ, Zhou X. Statistical methods for Mendelian randomization in genome-wide association studies: a review. Comput Struct Biotechnol J. 2022;20:2338–51.

78. Langmead B, Trapnell C, Pop M, Salzberg SL. Ultrafast and memory-efficient alignment of short DNA sequences to the human genome. Genome Biol. 2009;10(3):R25.
79. Li H, Durbin R. Fast and accurate short read alignment with Burrows-Wheeler transform. Bioinformatics. 2009;25(14):1754–60.
80. McKenna A, Hanna M, Banks E, Sivachenko A, Cibulskis K, Kernytsky A, et al. The genome analysis toolkit: a MapReduce framework for analyzing next-generation DNA sequencing data. Genome Res. 2010;20(9):1297–303.
81. Li H, Handsaker B, Wysoker A, Fennell T, Ruan J, Homer N, et al. The sequence alignment/map format and SAMtools. Bioinformatics. 2009;25(16):2078–9.
82. Abyzov A, Urban AE, Snyder M, Gerstein M. CNVnator: an approach to discover, genotype, and characterize typical and atypical CNVs from family and population genome sequencing. Genome Res. 2011;21(6):974–84.
83. Boeva V, Popova T, Bleakley K, Chiche P, Cappo J, Schleiermacher G, et al. Control-FREEC: a tool for assessing copy number and allelic content using next-generation sequencing data. Bioinformatics. 2012;28(3):423–5.
84. Becking EC, Schuit E, van Baar de Knegt SME, Sistermans EA, Henneman L, Bekker MN, et al. Association between low fetal fraction in cell-free DNA screening and fetal chromosomal aberrations: a systematic review and meta-analysis. Prenat Diagn. 2023;43(7):838–53.
85. Spielman RS, Bastone LA, Burdick JT, Morley M, Ewens WJ, Cheung VG. Common genetic variants account for differences in gene expression among ethnic groups. Nat Genet. 2007;39(2):226–31.
86. Wynn J, Hoskovec J, Carter RD, Ross MJ, Perni SC. Performance of single-gene noninvasive prenatal testing for autosomal recessive conditions in a general population setting. Prenat Diagn. 2023;43(10):1344–54.
87. Liu D, Nong C, Lai F, Tang Y, Wang T. Research progress of cell-free fetal DNA in non-invasive prenatal diagnosis of thalassemia. Hemoglobin. 2023;47(2):80–4.
88. Suwannakhon N, Hemvuthiphan J, Pangeson T, Mahingsa K, Pingyod A, Bumrungpakdee W, et al. Non-invasive prenatal screening & diagnosis of β-thalassaemia in an affected foetus. Indian J Med Res. 2023;157(5):447–52.
89. Zaninović L, Bašković M, Ježek D, Katušić Bojanac A. Accuracy of non-invasive prenatal testing for Duchenne muscular dystrophy in families at risk: a systematic review. Diagnostics (Basel). 2023;13(2):183.
90. Alford B, Landry BP, Hou S, Bower X, Bueno AM, Chen D, et al. Validation of a non-invasive prenatal test for fetal RhD, C, c, E, K and Fya antigens. Sci Rep. 2023;13(1):12786.
91. Hanson B, Scotchman E, Chitty LS, Chandler NJ. Non-invasive prenatal diagnosis (NIPD): how analysis of cell-free DNA in maternal plasma has changed prenatal diagnosis for monogenic disorders. Clin Sci (Lond). 2022;136(22):1615–29.
92. Yan Y, Wang F, Zhang C, Jin X, Zhang Q, Feng X, et al. Evaluation of droplet digital PCR for non-invasive prenatal diagnosis of phenylketonuria. Anal Bioanal Chem. 2019;411(27):7115–26.
93. Peng D, Ganye Z, Gege S, Yanjie X, Ning L, Xiangdong K. Clinical application of non-invasive prenatal diagnosis of phenylketonuria based on haplotypes via paired-end molecular tags and weighting algorithm. BMC Med Genet. 2021;14(1):294.
94. Gonsalves Z, Klugman S. Noninvasive prenatal screening for single-gene disorders. Clin Obstet Gynecol. 2023;66(3):629–35.
95. Alyafee Y, Al Tuwaijri A, Umair M, Alharbi M, Haddad S, Ballow M, et al. Non-invasive prenatal testing for autosomal recessive disorders: a new promising approach. Front Genet. 2022;13:1047474.
96. Wu W, Zhou X, Jiang Z, Zhang D, Yu F, Zhang L, et al. Noninvasive fetal genotyping of single nucleotide variants and linkage analysis for prenatal diagnosis of monogenic disorders. Hum Genomics. 2022;16(1):28.
97. Kallionpää RA, Ahramo K, Aaltonen M, Pennanen P, Peltonen J, Peltonen S. Circulating free DNA in the plasma of individuals with neurofibromatosis type 1. Am J Med Genet A. 2021;185(4):1098–104.

98. Liautard-Haag C, Durif G, VanGoethem C, Baux D, Louis A, Cayrefourcq L, et al. Noninvasive prenatal diagnosis of genetic diseases induced by triplet repeat expansion by linked read haplotyping and Bayesian approach. Sci Rep. 2022;12(1):11423.
99. Piña-Aguilar RE, Simpson SA, Alshatti A, Clarke A, Craufurd D, Dorkins H, et al. 27 years of prenatal diagnosis for Huntington disease in the United Kingdom. Genet Med. 2019;21(7):1639–43.
100. Noda Y, Kato T, Kato A, Nishizawa H, Miyazaki J, Ito M, et al. Potentially effective method for fetal gender determination by noninvasive prenatal testing for X-linked disease. Congenit Anom (Kyoto). 2019;59(3):88–92.
101. Dardik R, Janczar S, Lalezari S, Avishai E, Levy-Mendelovich S, Barg AA, et al. Four decades of carrier detection and prenatal diagnosis in hemophilia a: historical overview, state of the art and future directions. Int J Mol Sci. 2023;24(14):11846.
102. Alvarez-Mora MI, Agusti I, Wijngaard R, Martinez-Barrios E, Barcos T, Borras A, et al. Evaluation of FMR4, FMR5 and FMR6 expression levels as non-invasive biomarkers for the diagnosis of fragile X-associated primary ovarian insufficiency (FXPOI). J Clin Med. 2022;11(8):2186.
103. Dungan JS, Klugman S, Darilek S, Malinowski J, Akkari YMN, Monaghan KG, et al. Noninvasive prenatal screening (NIPS) for fetal chromosome abnormalities in a general-risk population: an evidence-based clinical guideline of the American College of Medical Genetics and Genomics (ACMG). Genet Med. 2023;25(2):100336. Erratum in: Genet Med. 2023;25(8).
104. Liang D, Cram DS, Tan H, Linpeng S, Liu Y, Sun H, et al. Clinical utility of noninvasive prenatal screening for expanded chromosome disease syndromes. Genet Med. 2019;21(9):1998–2006.
105. Hill M, Finning K, Martin P, Hogg J, Meaney C, Norbury G, et al. Non-invasive prenatal determination of fetal sex: translating research into clinical practice. Clin Genet. 2011;80(1):68–75.
106. Zhou F, Liu S, Wang H. Noninvasively diagnosing for fetal trisomy 21 by examining heterozygous single nucleotide polymorphisms in the placental specific genes on chromosome 21. Eur J Obstet Gynecol Reprod Biol. 2019;233:19–25.
107. Xu P, Ma Y, Wu H, Wang YL. Placenta-derived microRNAs in the pathophysiology of human pregnancy. Front Cell Dev Biol. 2021;9:646326.
108. Morales-Prieto DM, Favaro RR, Markert UR. Placental miRNAs in feto-maternal communication mediated by extracellular vesicles. Placenta. 2020;102:27–33.
109. Wang H, She G, Zhou W, Liu K, Miao J, Yu B. Expression profile of circular RNAs in placentas of women with gestational diabetes mellitus. Endocr J. 2019;66(5):431–41.
110. Gu M, Zheng A, Tu W, Zhao J, Li L, Li M, et al. Circulating LncRNAs as novel, non-invasive biomarkers for prenatal detection of fetal congenital heart defects. Cell Physiol Biochem. 2016;38(4):1459–71.
111. Sharma VK, Vouros P, Glick J. Mass spectrometric based analysis, characterization and applications of circulating cell free DNA isolated from human body fluids. Int J Mass Spectrom. 2011;304(2–3):172–83.
112. Jou HJ, Lo PH, Ling PY. Recent advances of microfluidic platform for cell based non-invasive prenatal diagnosis. Int J Mol Sci. 2023;24(2):991.

Cell-Free Fetal DNA: Genesis and Clinical Utility

3

Mumtaz Anwar, Vijayalakshmi Srinivasan, Ajay Prakash, Zeenat Farooq, Bekele Chakiso, Erchafo Mohamed, Eyob Tilahun, and Riyaz Ahmad Rather

3.1 Introduction

Cell-free fetal DNA (cffDNA) comprises small fragments of fetal DNA generated from the fetus and freely circulating in the maternal bloodstream during pregnancy [1]. It is released into the maternal circulation through physiological processes. One primary source is the apoptosis of placental trophoblasts, fetal-derived cells crucial in placental development [2]. As these cells undergo programmed cell death, fragments of fetal DNA are released, entering the maternal bloodstream. Additionally, necrosis of fetal cells and shedding of placental DNA contribute to cffDNA presence in maternal plasma [3]. Cell-free DNA in maternal plasma is a mix of fetal and maternal DNA, with the fetal fraction a small proportion compared to maternal DNA. The fetal fraction varies throughout pregnancy, being higher in early gestation and gradually decreasing as pregnancy progresses [4]. cffDNA analysis allows the identification and characterization of specific fetal genetic sequences, providing

M. Anwar · B. Chakiso
Department of Pharmacology and Regenerative Medicine, College of Medicine, University of Illinois, Chicago, IL, USA

V. Srinivasan
IBDP—Biology, M.Ct.M Chidambaram Chettyar International School, Chennai, India

A. Prakash
Department of Medicine, Nigist Eleni Mohammed Memorial Hospital, Wachemo University, Hosanna, Ethiopia

Z. Farooq · E. Mohamed · E. Tilahun
Department of Biology, College of Natural and Computational Science, Wachemo University, Hosanna, Ethiopia

R. A. Rather (✉)
Department of Biotechnology, College of Natural and Computational Science, Wachemo University, Hosanna, Ethiopia

R. A. Rather, S. C. Saha (eds.), *Non-invasive Prenatal Screening (NIPS) in Clinical Practice*, https://doi.org/10.1007/978-981-97-6402-0_3

valuable insights into the genetic composition and health status of the developing fetus.

The concept of cffDNA was introduced in 1997 when researchers demonstrated fetal DNA's presence in maternal circulation [5]. This breakthrough sparked extensive research, paving the way for noninvasive prenatal testing (NIPT). Before cffDNA discovery, prenatal genetic testing mainly relied on invasive techniques, posing slight risks and limiting applicable pregnancies [6]. Its discovery has brought significant clinical relevance to prenatal care. The application of cffDNA in prenatal testing for various fetal anomalies is vast, and a detailed exploration of its numerous applications will be provided in the subsequent sections of this chapter.

3.2 Genesis of cffDNA

3.2.1 Discovery of cffDNA in Maternal Circulation

The exploration into the realm of cffDNA began in the late 1990s, marked by groundbreaking studies that paved the way for unraveling its existence and significance (Table 3.1). These pioneering investigations established the initial groundwork for comprehending the presence and potential applications of cffDNA in the field of prenatal screening [5, 7]. Researchers observed the presence of fetal DNA in maternal plasma, challenging the long-held belief that the maternal bloodstream was devoid of fetal genetic material. Seminal research by Lo et al. in 1997 provided the first evidence of cffDNA in maternal circulation, sparking widespread interest and subsequent investigations [5]. These early studies paved the way for exploring the potential clinical applications of cffDNA. To establish the presence of fetal DNA in maternal plasma, researchers employed various methodologies and techniques, and validation studies were conducted to confirm its origin and characteristics [8]. Researchers developed innovative approaches to quantify and characterize cffDNA, distinguishing it from maternal DNA. The use of polymerase chain reaction (PCR) and other molecular techniques played a crucial role in detecting and analyzing fetal-specific genetic sequences within the cffDNA [9]. These breakthroughs provided compelling evidence of the existence of fetal DNA in maternal plasma and set the stage for further advancements.

3.2.2 Origin and Release of cffDNA into Maternal Plasma

The origin and release of cffDNA into maternal plasma are intriguing processes that involve various mechanisms. One of the primary sources of cffDNA is the apoptosis, or programmed cell death, of placental trophoblasts [2, 3, 10–12]. These trophoblasts are fetal-derived cells that play a crucial role in the development of the placenta. As these cells naturally undergo apoptosis, small fragments of fetal DNA are released and enter the maternal bloodstream. In addition to apoptosis, necrosis of the cells originating from the fetus also contributes to its presence in maternal

Table 3.1 A comprehensive chronology of the evolution of cell-free fetal DNA discovery, applications, advantages, and challenges from last three decades

Year	Milestone	Description	Advantages	Disadvantages
1990–1992	Discovery of fetal DNA in maternal plasma	Presence of fetal DNA in maternal plasma was established	1. Noninvasive prenatal testing potential 2. Reduction in reliance on invasive procedures	1. Limited understanding of cffDNA characteristics 2. Technological limitations for isolation and analysis
1993–1995	Identification of fetal DNA origin	Researchers identify placental trophoblasts as the origin of cffDNA in maternal circulation	1. Clarification of cffDNA origin 2. Potential for targeted analysis	1. Limited knowledge on variations in trophoblast DNA release 2. Challenges in distinguishing maternal and fetal DNA
1996–1998	Demonstration of fetal-specific markers	Studies demonstrate the existence of fetal-specific markers in cffDNA	1. Improved accuracy in detecting fetal DNA 2. Enhanced specificity for fetal genetic characteristics	1. Challenges in identifying universal fetal markers 2. Need for refining marker selection techniques
1999–2001	Introduction of cffDNA in prenatal screening	Initial use of cffDNA in prenatal screening for chromosomal abnormalities	1. Noninvasive nature reduces procedural risks 2. Potential for early detection of genetic conditions	1. Limited scope for conditions beyond chromosomal abnormalities 2. Developmental stage limitations in cffDNA technology
2002–2004	Advancements in cffDNA isolation techniques	Introduction of advanced genomic methods for efficient cffDNA isolation	1. Improved precision and efficiency in cffDNA isolation 2. Increased accessibility in laboratory settings	1. Variability in isolation efficiency among different methods 2. Cost considerations in adopting advanced techniques
2005–2007	Expanded clinical applications	Exploration of cffDNA applications beyond chromosomal abnormalities	1. Increased utility in diverse prenatal assessments 2. Expanded information for comprehensive patient care	1. Ethical considerations regarding the extent of information revealed 2. Need for standardization in expanded applications
2008–2010	Large-scale clinical trials	Initiation of large-scale clinical trials assessing the accuracy and reliability of cffDNA-based testing	1. Establishment of statistical reliability through extensive trials 2. Increased confidence in cffDNA technology	1. Ethical considerations in large-scale trials involving diverse populations 2. Resource-intensive nature of large-scale trials

(continued)

Table 3.1 (continued)

Year	Milestone	Description	Advantages	Disadvantages
2011–2013	Commercialization and accessibility	Commercial availability of cffDNA testing through various kits and laboratories	1. Enhanced accessibility for healthcare providers and patients 2. Standardization of testing protocols	1. Commercial competition and variations in kit performance 2. Cost implications for patients without insurance coverage
2014–2016	Refinement of testing algorithms	Continuous refinement of testing algorithms for improved accuracy and reliability	1. Enhanced precision in detecting a broader range of genetic conditions 2. Improved risk assessment	1. Need for ongoing updates and education on evolving algorithms 2. Interpretation challenges in complex cases
2017–2019	Integration with multi-omic approaches	Exploration of combining cffDNA testing with other omics technologies for comprehensive fetal health assessment	1. Potential for a more holistic understanding of fetal health 2. Improved predictive power	1. Complexity in integrating and interpreting multi-omic data 2. Technological and resource challenges in combined approaches
2020–2023	Ongoing research for enhanced sensitivity	Current research focused on enhancing the sensitivity and specificity of cffDNA testing	1. Potential for increased detection rates and accuracy 2. Ongoing improvement in testing reliability	1. Need for ongoing validation studies 2. Integration challenges with rapidly evolving technologies

cffDNA cell-free fetal DNA

plasma [3]. Fetal cells can undergo necrosis due to various factors, such as stress or injury. When fetal cells undergo necrosis, they release their DNA into the maternal circulation, adding to the pool of cffDNA.

Another mechanism that leads to the release of cffDNA is the shedding of placental DNA. The placenta is a dynamic organ that constantly remodels itself throughout pregnancy. As the placenta undergoes changes, it sheds small fragments of DNA, including fetal DNA, into the maternal bloodstream [13]. The release of cffDNA into maternal blood occurs continuously throughout pregnancy. However, the amount of cffDNA present in maternal circulation varies at different stages of gestation [4]. During the initial stages of pregnancy, the proportion of cffDNA is typically more prominent, gradually diminishing as the pregnancy advances. Various factors, including the age of gestation, well-being of the placenta, and maternal weight, influence the quantities of cffDNA present in the maternal plasma.

3.2.3 Mechanisms of cffDNA Clearance and Degradation

cffDNA is not permanent in maternal plasma and undergoes clearance and degradation processes [14]. Following childbirth, it undergoes swift clearance from the maternal circulation, with a half-life measured in minutes. This phenomenon of rapid elimination distinguishes it from the clearance kinetics of fetal cells, which have been shown to persist for extended periods. Understanding the mechanisms behind cffDNA clearance and degradation is essential for accurate interpretation of prenatal genetic testing results. One of the primary mechanisms involved in cffDNA clearance is the maternal immune system. Research suggests that cffDNA stimulates an immune response in pregnant women, leading to its subsequent clearance. In earlier investigations, researchers explored the clearance of cell-free DNA (cfDNA) in plasma through observations of the therapeutic outcomes of administering venous DNase I in systemic lupus erythematosus mouse models and patients [15, 16]. These studies revealed a correlation between the rate of cfDNA clearance and the concentration of DNase I. Furthermore, they discovered heightened levels of DNase I in the blood plasma of patients who developed sepsis following significant trauma. The precise mechanisms by which the immune system recognizes and clears cffDNA are still under investigation, but it is believed to involve a combination of cellular and humoral immune responses.

The dynamics of cffDNA clearance over gestation are complex. Studies have demonstrated that cffDNA levels decrease as pregnancy progresses, indicating increased efficiency in clearance mechanisms [17]. It is believed that the combination of immune clearance, hepatic and renal clearance, and enzymatic degradation collectively contributes to its reduction in maternal plasma. Several factors influence its clearance and degradation, although not all are fully understood. Gestational age appears to be a significant factor, with higher clearance rates observed in later stages of pregnancy [18]. Maternal factors, such as age, body mass index (BMI), and overall health, may also influence its clearance [19]. Additionally, the characteristics of cffDNA itself, such as size and fragmentation, can affect its susceptibility to clearance and degradation processes.

3.2.4 Factors Influencing cffDNA Levels in Maternal Plasma

The stage of pregnancy plays a pivotal role in determining cffDNA levels. Pregnancy stages demonstrate a consistent pattern of increasing cffDNA concentrations as the pregnancy advances. Although the exact mechanism behind this rise remains somewhat elusive, it is widely speculated that the escalation is linked to the gradual release of cffDNA from the placenta during the course of pregnancy [13]. The maternal-fetal interface plays a crucial role in its release into maternal plasma. Factors such as placental function and integrity can affect the amount of cffDNA that enters the maternal circulation. Placental abnormalities or pathologies, such as preeclampsia or placental insufficiency, may result in its altered levels in maternal plasma [20]. Its levels are not solely determined by fetal factors, but maternal factors also play a role. Various aspects of maternal characteristics, such as maternal age, BMI, and overall health, have been identified as factors that can influence cffDNA concentrations. Studies have revealed that advanced maternal age is associated with higher levels of cffDNA, suggesting a potential correlation between age-related changes in placental function and increased release of fetal DNA [21, 22]. Additionally, higher maternal BMI has been linked to elevated cffDNA levels, although the precise mechanisms driving this association are still being explored. It is evident that maternal factors contribute to the intricate dynamics of cffDNA levels, shedding light on the multifaceted nature of this biomarker in noninvasive prenatal screening (NIPS). Fetal factors can also affect cffDNA levels in maternal plasma. Fetal sex has been reported to play a role, with studies showing that pregnancies carrying male fetuses tend to have higher cffDNA levels compared to pregnancies with female fetuses [23, 24]. Additionally, fetal genetic abnormalities or aneuploidies can lead to alterations in cffDNA levels [25].

Sampling and technical factors can influence cffDNA measurements as well. The timing and method of sample collection, as well as the quality of the sample, can impact cffDNA levels. Factors such as the interval between sample collection and processing, the storage conditions, and the use of specific isolation methods can affect the stability and yield of cffDNA [26].

3.3 Extraction and Analysis of cffDNA

3.3.1 Techniques for Isolating and Enriching cffDNA from Maternal Plasma

Isolating cffDNA from maternal plasma is a critical step in prenatal screening and research. The ability to extract and analyze cffDNA accurately and efficiently is essential for reliable results. Various techniques are in place to isolate cffDNA from maternal plasma, each with its advantages and considerations. One widely used method is differential centrifugation. This technique involves a series of centrifugation steps to separate different components of the plasma based on their density [27]. By carefully selecting the centrifugation parameters, including speed and time, it is possible to isolate cffDNA from the plasma fraction [28]. The advantage of this technique is its simplicity and cost-effectiveness, making it accessible in many laboratory settings. Another commonly employed technique is based on size selection using membrane filtration or gel electrophoresis. In this approach, cffDNA is separated from other components of maternal plasma based on its size [29]. Membrane filtration involves passing the plasma through a filter with specific pore size, allowing cffDNA to pass while retaining larger molecules [30]. Size-based techniques provide high specificity in isolating cffDNA and are particularly useful when analyzing specific DNA fragments or regions. In addition to manual methods, the market offers various commercial kits designed for the isolation of cffDNA. These kits commonly employ spin-column-based, magnetic bead-based, or automated magnetic bead-based methodologies.

Commercially Available Kits

- *QIAGEN's QIAamp Circulating Nucleic Acid Kit*: Designed for the effective extraction of cffDNA from maternal plasma, this kit employs a silica membrane-based column purification method, ensuring the isolation of high-quality cffDNA. In addition to this kit, QIAGEN offers several others such as the QIAamp MinElute ccfDNA Kit, QIAamp DSP Circulating NA Kit, QIAamp ccfDNA/RNA Kit, and QIAseq cfDNA All-in-One Kits.
- *cfPure® Cell-Free DNA Extraction Kit*: AMSBIO's cfPure® Cell-Free DNA Extraction Kit, designed for scalable and automation-friendly isolation of circulating cfDNA from serum or plasma samples. This kit utilizes silica-coated paramagnetic particles, enabling the purification of cfDNA from samples of less than 1 mL.
- *MagMAX Cell-Free DNA Isolation Kit*: This kit from the Thermo Fisher Scientific is a cutting-edge solution for the scalable extraction of circulating cfDNA. Based on magnetic bead technology, this kit is specifically designed for the enrichment of small (<300 bp) circulating DNA. It is ideal for use with blood, plasma, and serum samples. It ensures increased recovery and facilitates processing with lower starting volumes.
- *NucleoSnap cfDNA Kit*: This kit from Macherey-Nagel is based on silica membrane technology, ensuring efficient and high-yield extraction of cfDNA.
- *MagNA Pure 24 Total NA Isolation Kit*: The MagNA Pure 24 Total NA Isolation Kit by Roche provides a reliable and streamlined process for obtaining high-quality cfDNA from plasma samples.

3.3.2 Quantification and Quality Assessment of cffDNA

Quantification and quality assessment are integral aspects of prenatal screening process. The accurate measurement of cffDNA concentration and the assessment of its quality play a crucial role in ensuring reliable results. Quality assessment includes the evaluation of the integrity and purity of cffDNA [31]. This involves examining factors such as DNA fragment size, which is critical for downstream applications. Agarose gel electrophoresis and capillary electrophoresis are commonly utilized to assess the size distribution of DNA fragments [32]. Additionally, the assessment of cffDNA quality involves checking for potential contaminants or inhibitors that might affect downstream analyses. Quality control steps are crucial to ensure that the isolated cffDNA is suitable for subsequent molecular analyses, such as sequencing or PCR [33]. Various techniques, some of them mentioned below, are employed to assess the quality of cffDNA (Table 3.2).

- *Fragment Size Analysis*: Agarose gel electrophoresis or capillary electrophoresis can be used to analyze the size distribution of cffDNA fragments. Fragment sizes are compared to DNA markers to determine its integrity [34]. Lower molecular weight fragments are indicative of intact cffDNA entities.
- *Fragment Ratio Analysis*: Assessing the ratio of short to long fragments of cffDNA provides insights into its quality. Techniques such as quantitative (qPCR) or next-generation sequencing (NGS) are used to measure the relative abundance of short and long DNA fragments [35]. An increased ratio of short fragments suggests DNA degradation.
- *DNA Integrity Index*: The DNA Integrity Index is a quantitative measure of cffDNA quality. It is calculated by comparing the amplification of short and long DNA fragments using qPCR or other methods [36]. A higher Index value indicates better DNA integrity and quality.

Quantification methods involve the precise determination of the amount of DNA present in maternal plasma. Besides others, various qPCR methods or techniques are commonly used for this purpose. These methods enable the precise measurement of the cffDNA fraction relative to the total cfDNA or maternal DNA in the sample.

Table 3.2 Quantification and quality assessment of cffDNA

Technique	Description	Applications	Insights
Fragment size analysis	Agarose gel electrophoresis or capillary electrophoresis used to analyze the size distribution of cffDNA fragments	Indicates integrity, with lower molecular weight fragments suggesting intact cffDNA	Provides insights into fragment size distribution
Fragment ratio analysis	Techniques such as qPCR or NGS employed to measure the relative abundance of short and long DNA fragments	Provides insights into quality, with an increased ratio of short fragments suggesting DNA degradation	Measures the ratio of short to long fragments
DNA Integrity Index	Quantitative measure of cffDNA quality calculated by comparing the amplification of short and long DNA fragments using qPCR or other methods	Reflects DNA integrity and quality, with a higher DNA Integrity Index indicating better integrity	Quantifies the quality of cffDNA using a numerical index

qPCR quantitative PCR, *cffDNA* cell-free fetal DNA

3.3.3 Challenges in cffDNA Extraction

The total DNA isolation from maternal plasma is a simple process; however, isolating and quantifying cffDNA from the total DNA is challenging and difficult. As the preformed cffDNA is crucial for the downstream process, it is important to mitigate these challenges. Some of the challenges include the following:

1. *Low Concentration of cffDNA*. One of the primary challenges is the low fetal fraction in maternal plasma [37]. The amount of cffDNA is relatively small compared to the total amount of cfDNA in the plasma, which mainly consists of maternal DNA. This low concentration makes it difficult to isolate sufficient cffDNA for accurate analysis.
2. *Contamination with Maternal DNA*. Another challenge is the contamination with maternal DNA. The maternal DNA can overshadow the presence of cffDNA, leading to false results or reduced sensitivity in the test. Efficient separation and removal of maternal DNA during extraction are necessary to ensure accurate analysis.
3. *Fragmentation of cffDNA*. During release into maternal plasma, cffDNA undergoes fragmentation [38]. The extent of fragmentation can vary, and highly fragmented cffDNA may be challenging in terms of obtaining intact genetic information.
4. *Inhibitors in Maternal Plasma*. Maternal plasma contains various inhibitors that interfere with DNA extraction and downstream analysis. Substances such as hemoglobin, heparin, and other proteins can inhibit the enzymatic reactions involved in extraction process [39], affecting the yield and quality of extracted cffDNA. Proper sample processing and purification steps are crucial to minimize the impact of these inhibitors.
5. *Variability in Extraction Efficiency*. The efficiency of DNA extraction can vary among different samples and extraction methods. Variability in extraction efficiency can lead to inconsistent results and hinder the reliability and reproducibility of genetic testing. Standardization of extraction protocols and quality control measures is essential to minimize this variability.
6. *Gestational Age Dependency*. The concentration and quality of cffDNA in maternal plasma can vary depending on the gestational age [39, 40]. In early pregnancy, the amount of cffDNA may be lower, making extraction more challenging. Conversely, in later stages of pregnancy, the concentration increases, but so does the presence of maternal DNA, which can complicate extraction and analysis.

3.4 Clinical Applications of cffDNA

3.4.1 Screening for Common Chromosomal Abnormalities

NIPS method offers a safe and precise approach for identifying common chromosomal abnormalities in the developing fetus [41]. Detecting chromosomal abnormalities using cffDNA is one of the most dominant applications in NIPS to date

(Fig. 3.1). Some biotechnology companies in prenatal genetic testing business have launched tests specifically for detecting fetal chromosomal abnormalities, and these tests are based on cffDNA.

By analyzing the cffDNA's presence in maternal blood, NIPT enables the detection of conditions like trisomy 21 (Down syndrome), trisomy 18 (Edwards syndrome), and trisomy 13 (Patau syndrome) [42]. Enormous literature has been published regarding this, and some countries have formally introduced cffDNA-based tests for the detection of chromosomal abnormalities [43–45]. Numerous studies have investigated the effectiveness and reliability of NIPT for the screening of these chromosomal abnormalities. A meta-analysis conducted by Gil et al. evaluated the performance of NIPT in screening for trisomy 21, 18, and 13 based on the analysis of cffDNA [46]. The analysis included data from multiple studies and

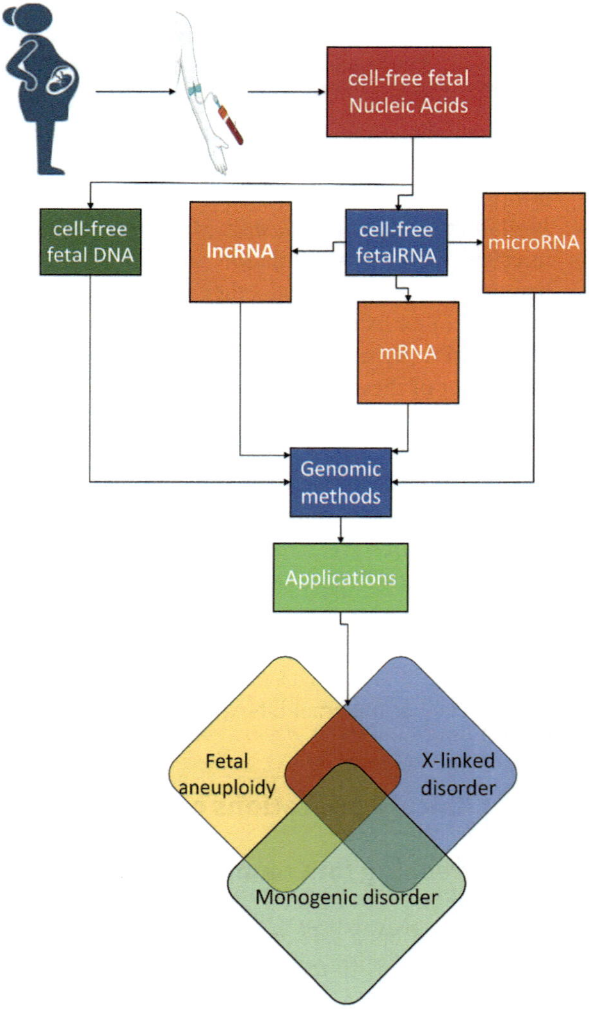

Fig. 3.1 The schematic diagram illustrates the isolation and utilization of cell-free fetal DNA (cffDNA) in diverse applications within prenatal testing. Initially, a blood sample is obtained from pregnant women, and this blood undergoes a series of steps for DNA isolation. The resulting isolated sample comprises cell-free nucleic acids (cfNAs), such as cffDNA or cell-free RNA, in addition to maternal cell-free DNA. The fetal cfNAs are subjected to various genomic processes aimed at screening for specific fetal conditions, including chromosomal aneuploidies, monogenic disorders, and X-linked disorders. It is important to note that the list of fetal conditions depicted in this diagram is not exhaustive; rather, it serves as a representation

demonstrated high sensitivity and specificity for these conditions. The results of this meta-analysis support the use of NIPT as a reliable method for screening common chromosomal abnormalities. Another study assessed the fetal aneuploidy in singleton pregnancies. The study utilized the analysis of cffDNA to examine these chromosomal abnormalities. The results showed that NIPT had high sensitivity and specificity for the detection of these conditions, providing valuable information for prenatal screening and management [47].

It is crucial to acknowledge that NIPT functions as a screening test rather than a diagnostic one. In instances where a positive result is obtained, it is imperative to conduct confirmatory diagnostic testing to establish a definitive diagnosis. Commonly employed follow-up diagnostic procedures include amniocentesis or chorionic villus sampling.

3.4.2 Fetal Sex and RhD Status Detection

Traditionally, the determination of fetal sex is reliant on invasive procedures or through visual observation of the genitals during ultrasound scans or by applying nub theory. However, the advent of cffDNA analysis has provided a noninvasive alternative. Detecting fetal sex using cffDNA has emerged as a valuable tool in prenatal screening. Detecting fetal sex may stem from parental enthusiasm to know the gender of the fetus, or its identification may be driven by medical considerations [48]. Several studies have investigated the efficacy and reliability of NIPT in detecting fetal sex [49–51]. A review analysis carried by Bowman et al. demonstrated the feasibility and accuracy of determining fetal sex using this cffDNA-based NIPT method [52]. The analysis of cffDNA facilitated the identification of Y-chromosome sequences, signifying a male fetus, whereas the absence of these sequences indicated a female fetus. The review documented consistently high sensitivity and specificity for fetal sex determination through cffDNA across various studies.

Beyond fetal sex determination, evaluating the RhD status of the fetus holds significant importance in pregnancy management, especially for RhD-negative mothers. RhD incompatibility between the mother and fetus can result in maternal alloimmunization and subsequent complications in subsequent pregnancies [53]. The noninvasive assessment of fetal RhD status using cffDNA has demonstrated reliability as a method. Two distinct studies carried out by Rather et al., involving diverse cohorts with varying ethnicities, have shown a satisfactory level of accuracy in determining fetal RhD status using cffDNA [54, 55]. The research successfully identified RhD status by targeting different exon regions of the *RHD* gene, unveiling a high level of concordance between the analysis of cffDNA and the determination of fetal RhD status. Likewise, numerous other studies have effectively determined fetal RhD status by utilizing cffDNA extracted from maternal plasma [56–58].

The utilization of cffDNA technology for detecting fetal sex and RhD status presents numerous advantages. It obviates the requirement for invasive procedures, thereby reducing the risk of complications like miscarriage. Additionally, it furnishes early and precise information, facilitating appropriate prenatal management and counseling. The noninvasive characteristic of this approach renders it a favored choice for many expectant parents. Despite these advantages, it is crucial to recognize the limitations of cffDNA analysis for fetal sex and RhD determination. Variables such as maternal obesity, fetal fraction, and technical constraints can impact result accuracy. Consequently, healthcare providers should meticulously interpret the findings and contemplate confirmatory diagnostic testing if deemed necessary.

3.4.3 Assessment of Aneuploidies and Microdeletion Syndromes

Microdeletion syndromes are characterized by small deletions of genetic material and can lead to a range of developmental and health issues in affected individuals [59]. Microdeletion syndromes, characterized by small deletions of genetic material, can result in a spectrum of developmental and health issues in affected individuals [60]. The utilization of cffDNA has emerged as a valuable tool for assessing microdeletion syndromes through NGS. For example, one of the most well-known microdeletion syndromes, 22q11.2 deletion syndrome, also recognized as DiGeorge syndrome, has been subject to evaluation in various studies. Bevilacqua et al. conducted a study with the aim of evaluating the efficacy of cffDNA analysis in detecting 22q11.2 deletion syndrome [61]. The research focused on pregnant women identified as high risk for this syndrome based on prenatal screening results. Maternal blood samples were subjected to cffDNA analysis, and the outcomes were compared with results obtained from invasive diagnostic testing. The study revealed a notably high sensitivity and specificity of cffDNA analysis in detecting 22q11.2 deletion syndrome, indicating its potential as a reliable screening method. Beyond 22q11.2 deletion syndrome, cffDNA technology offers a means to assess various other microdeletion syndromes. These encompass syndromes associated with chromosomes 1p36, 5p, 7q11.23 (Williams syndrome), and 15q11-q13 (Prader-Willi and Angelman syndromes) [62]. The effectiveness of cffDNA analysis for these conditions has been scrutinized in numerous studies. Similarly, Straver et al. conducted a study to evaluate the efficacy of cffDNA analysis in detecting multiple microdeletion syndromes. The researchers analyzed cffDNA extracted from maternal blood samples and compared the results with those derived from invasive diagnostic testing. The study unveiled a satisfactory level of sensitivity and specificity for detecting the microdeletion syndromes under consideration [63]. These outcomes emphasize the credibility of cffDNA as a screening marker for microdeletion syndromes.

3.4.4 Evaluation of Fetal Growth and Placental Function

Fetal conditions associated with placental dysfunction encompass a spectrum of health issues that arise due to impaired functioning of the placenta during pregnancy [64]. Placental dysfunction can lead to inadequate nutrient and oxygen supply to the developing fetus, impacting its growth and overall well-being. Conditions linked to placental dysfunction include fetal growth restriction, where the fetus doesn't attain its expected size, and preeclampsia, a hypertensive disorder affecting both maternal and fetal health [65]. Placental insufficiency, marked by an insufficient exchange of nutrients and waste between the mother and fetus, can result in various complications. These may include preterm birth, as the fetus responds to the compromised environment by initiating an early delivery. Additionally, conditions like fetal distress, characterized by abnormalities in fetal heart rate, may arise due to inadequate oxygen supply from the malfunctioning placenta [66]. Monitoring and timely intervention are crucial in managing fetal conditions related to placental dysfunction, often involving close antenatal surveillance and, in severe cases, delivery to mitigate potential risks to both the mother and the baby [67]. A study conducted by Rolnik et al. examined the potential association between the fetal fraction and first-trimester markers in predicting adverse pregnancy outcomes, specifically for conditions like preeclampsia and fetal growth restriction [68]. They analyzed cffDNA from maternal plasma samples and identified specific genetic markers associated with fetal growth. The study suggests that low fetal fraction might have the potential to act as an independent marker in the first trimester for screening preeclampsia and fetal growth restriction.

Placental function is another crucial aspect of fetal well-being, and its evaluation is essential for identifying potential complications during pregnancy. A study conducted by Tsui et al. explored the correlation between circulating RNA species in maternal plasma and placental function, employing massively parallel sequencing for plasma transcriptome profiling across the first, second, and third trimesters of pregnancy [34]. The researchers identified a panel of pregnancy-associated genes through this comprehensive analysis. The utilization of plasma RNA sequencing offered a holistic perspective on the maternal plasma transcriptomic repertoire. Furthermore, cffDNA analysis has also been explored in the evaluation of preeclampsia, a pregnancy complication characterized by high blood pressure and organ damage [20, 69].

3.4.5 Assessment of Fetal Development, Preterm Birth, and Organ Maturity

The assessment of fetal development throughout pregnancy is of paramount importance to ensure the well-being and appropriate management of pregnancies. Customary methods of assessing fetal development relied on ultrasound measurements and other invasive procedures. However, cffDNA analysis provides an alternative approach that can provide valuable information about the genetic and molecular aspects of fetal development. In a study conducted by Lo et al., the researchers elucidated the genomic and mutational profile of the fetus utilizing cffDNA [70]. This analysis led to the identification of multiple molecular markers with the potential to detect fetal anomalies or assess the overall well-being of the developing fetus.

Preterm birth is a significant concern in obstetrics, as it can lead to numerous complications and adverse outcomes for both the fetus and the mother. Early identification of the risk of preterm birth allows for timely interventions and management strategies. Evidence suggests that preterm birth is linked to elevated levels of cffDNA after 20 weeks of gestation or upon the onset of spontaneous preterm birth symptoms [71]. Several studies have measured cffDNA in maternal blood and explored its association with preterm birth. For instance, a study conducted by Darghahi et al. demonstrated that cffDNA biomarkers can serve as a direct and noninvasive method for predicting preterm labor. The study reported a specificity of 46%, a sensitivity of 58%, a positive predictive value of 51%, and a negative predictive value of 52% [72]. Furthermore, cffDNA analysis has been explored in assessing the maturity of fetal organs, particularly the lungs [29]. Lung maturity is a critical factor in determining the readiness of a fetus for extrauterine life. Normal methods for assessing lung maturity involve invasive procedures; however, cffDNA analysis offers a noninvasive alternative [73, 74].

3.5 Clinical Inferences

3.5.1 Advantages of cffDNA-Based Testing over Invasive Procedures

One of the primary advantages of cffDNA-based testing is its noninvasive nature. Instead of relying on invasive procedures that require the insertion of a needle into the amniotic sac or placenta, cffDNA-based testing utilizes a simple maternal blood draw [75]. This significantly reduces the risk of procedure-related complications, such as infection or injury to the fetus or mother [76]. A study conducted by Mokhtar et al. compared the safety and accuracy of cffDNA-based testing for the detection of trisomy 21 with invasive procedures and found that cffDNA-based testing had a lower risk of procedure-related fetal loss [77]. Another advantage is its high accuracy and sensitivity. Several studies have demonstrated the excellent performance of this testing in detecting common chromosomal abnormalities. A meta-analysis conducted by Taylor et al. concluded that NIPT utilizing cffDNA exhibits very high sensitivity and specificity in the detection of fetal aneuploidy [78]. A large-scale clinical trial by Norton et al. evaluated the accuracy of cffDNA-based testing for the detection of Down syndrome and reported a sensitivity of over 99% [79]. Similarly, other studies have shown high sensitivity and specificity for the detection of Edwards syndrome and Patau syndrome [78, 80].

Its noninvasive nature and high accuracy have additional advantages in terms of patient acceptance and reduced psychological stress. Invasive procedures often cause anxiety and stress in pregnant individuals due to their invasive nature and associated risks [81]. cffDNA-based testing provides a less stressful alternative, resulting in improved patient satisfaction and adherence to prenatal screening recommendations. A study by Benn et al. assessed patient preferences for prenatal testing methods and found that the majority of participants preferred cffDNA-based testing over invasive procedures [82]. It offers the potential for comprehensive screening of multiple conditions simultaneously, providing a more comprehensive assessment of the fetal genetic health.

3.5.2 Impact on Patient Management and Counseling

Testing based on cffDNA provides valuable insights into the genetic condition of the fetus, enabling healthcare providers to tailor patient management strategies effectively. Accurate detection of fetal anomalies using cffDNA allows for timely interventions and appropriate counseling. When a high-risk result for chromosomal abnormalities is revealed, further diagnostic testing or patient counseling becomes essential. The precision and early detection afforded by cffDNA-based testing empower healthcare providers to deliver comprehensive and personalized counseling on the condition, available treatment options, and potential outcomes [83]. Another advantageous aspect is its significant impact on patient counseling, providing accurate and reliable information about the genetic condition of the fetus. This enhanced knowledge empowers medical practitioners to engage in comprehensive discussions with patients, considering factors such as the test's limitations, potential false-positive or false-negative results, and the need for further diagnostic testing, if required. A study by Allyse et al. underscores the importance of comprehensive counseling to ensure patient understanding and informed decision-making regarding cffDNA testing [84].

3.5.3 Ethical Considerations and Informed Consent

The utilization of cffDNA testing for prenatal screening is intricately tied to ethical considerations and the imperative need for informed consent. As this technology advances, addressing the ethical implications becomes crucial to ensure that patients are fully informed and engaged in decision-making. One primary ethical concern associated with this testing is the potential for incidental findings—unexpected genetic information unrelated to the original purpose of the test. Scientists emphasize the importance of implementing suitable counseling and consent procedures to effectively address the potential occurrence of incidental findings in molecular diagnosis [85]. Another vital ethical consideration is the appropriate use and interpretation of cffDNA testing results. Given that cffDNA testing can detect a wide range of genetic information, healthcare providers must ensure accurate interpretation and communication of results to patients. Ethical guidelines, such as those proposed by the Institute of Medicine (USA), Committee on Assessing Genetic Risks, underscore the importance of appropriately trained professionals and clear communication in providing accurate and meaningful information to patients [86].

The issue of test limitations and the possibility of false-positive/false-negative results hold ethical significance. It is crucial for patients to grasp the limitations and potential uncertainties associated with cffDNA testing. Transparent communication about the potential for test inaccuracies is vital for managing patient expectations and facilitating informed decision-making. The American College of Obstetricians and Gynecologists offers guidelines addressing these ethical considerations, advocating for comprehensive pretest counseling [87].

Informed consent is a crucial aspect of ethical practice in any form of medical testing. Patients must fully comprehend the benefits, risks, and limitations of the test before deciding. This ensures active patient engagement in decision-making, allowing them to discuss their values, preferences, and concerns [88]. Effective communication of accurate information, including the benefits and risks of medical testing, is paramount. Providing comprehensive details and obtaining informed consent not only ensures ethical practice but also offers practical benefits for all involved parties. Privacy and data security are additional ethical concerns associated with cffDNA testing [89]. The highly sensitive genetic information obtained requires strict measures to protect patient confidentiality and prevent unauthorized access or misuse of data. Ethical guidelines, such as those by the European Society of Human Genetics, emphasize safeguarding patient privacy and ensuring data security [90]. Furthermore, ensuring equitable access to cffDNA testing is an ethical consideration. As the technology evolves, addressing potential disparities in access and affordability is crucial. Efforts should be made to ensure that all pregnant individuals, regardless of their socioeconomic background, have equal opportunities to benefit from cffDNA testing [91].

3.6 Challenges and Limitations

While cffDNA has gained extensive use in NIPS, it is not without its share of limitations and challenges, like any other molecule or biomarker. Here, we discuss a few challenges associated with the use of cffDNA in NIPS.

Fetal Fraction and Sample Quality: One of the main challenges in cffDNA-based testing is the fetal fraction. As mentioned in earlier section(s) of the chapter, various factors like maternal BMI, gestational age, and placental health, can affect the fetal fraction. Low fetal fraction leads to reduced sensitivity and accuracy of the test. Efforts are in place to optimize sample collection and processing techniques to improve the fetal fraction and ensure reliable results [92].

False-Positive and False-Negative Results: False-positive and false-negative results pose a risk of potential misdiagnosis, prompting unwarranted anxiety or interventions. Factors contributing to such inaccuracies include maternal mosaicism, confined placental mosaicism, and rare genetic variants. Continuous research endeavors are underway to enhance the accuracy and reliability of cffDNA-based testing, with the ultimate goal of minimizing the occurrence of false results [93].

Restricted Range of Genetic Conditions and Copy Number Variations: cffDNA-based testing exhibits a limitation in its focus on specific genetic conditions, primarily targeting common chromosomal abnormalities like trisomy 21, trisomy 18, and trisomy 13. While these conditions bear significant clinical implications, it's noteworthy that this testing might not encompass the detection of other genetic disorders or structural abnormalities [79]. To evaluate a broader spectrum of conditions, complementary screening methods or diagnostic procedures may be essential. Furthermore, the identification of copy number variations encounters challenges due to the low fetal fraction and dependence on sequencing depth. Current technologies are more adept at detecting larger copy number variations, potentially overlooking smaller ones. Researchers are actively exploring methods and leveraging bioinformatics tools to improve the sensitivity of copy number variation detection in cffDNA-based testing [94].

Cost, Validation, and Standards: The cost of implementing cffDNA-based testing presents a potential barrier to widespread adoption. Currently, it is more expensive compared to traditional screening methods or invasive procedures. Moreover, accessibility to this noninvasive test may be limited in specific regions or healthcare settings [95, 96]. Ongoing efforts are focused on enhancing cost-effectiveness and broadening access to cffDNA-based testing, with the aim of making it more readily available to a diverse population. As it undergoes continued development, it is imperative to validate and standardize the methodologies and interpretation algorithms. Consistency and reproducibility are paramount to ensuring accurate and reliable results across different laboratories and platforms.

3.7 Current and Future Directions

3.7.1 Advancements in cffDNA Technology and Methodology

Over the years, significant advancements have been made in NIPS, enhancing its accuracy, reliability, and clinical utility. Here, we discuss some of the key advancements in cffDNA technology, drawing upon previous literature.

Improved Sequencing Techniques: The advent of NGS has been pivotal in propelling cffDNA technology forward. NGS allows for high-throughput sequencing of cffDNA fragments, facilitating a comprehensive analysis of the fetal genome [97]. The development of more efficient and cost-effective sequencing platforms has considerably contributed to the widespread adoption of cffDNA-based testing.

Enhanced Fetal Fraction Estimation: The accurate calculation of fetal fraction, representing the proportion of cffDNA in maternal plasma, is crucial for reliable test outcomes. Various approaches, including the use of informative single nucleotide polymorphs and advanced bioinformatics algorithms, have been introduced to enhance fetal fraction estimation [19, 98, 99].

Single-Molecule Analysis: Techniques like droplet digital PCR and single-molecule real-time sequencing, falling under single-molecule analysis, have emerged as potent tools in cffDNA analysis. These techniques enable the detection and quantification of individual DNA molecules, thereby enhancing the sensitivity and precision of the test [100].

Expanded Range of Conditions: Initially concentrating on common chromosomal abnormalities, technological progress has broadened the testing panel to encompass additional conditions. This expansion includes screening for specific genetic disorders such as cystic fibrosis and hemoglobinopathies, along with the detection of fetal microdeletion and microduplication syndromes [101]. Beyond fetal genetic information, cffDNA analysis offers insights into placental health and function. Research explores the potential of cffDNA analysis in predicting conditions related to placental dysfunction, such as preeclampsia and intrauterine growth restriction [102].

Integration of Artificial Intelligence: Although artificial intelligence (AI) has transformed medical diagnosis, its full application in prenatal diagnosis is still evolving. Incorporating AI-based approaches into NIPS holds promise for enhancing result accuracy and reliability. AI can optimize data interpretation, aiding in the precise identification of rare genetic variants, and improve quality control processes, thereby enhancing overall NIPS performance [103, 104].

Standardization and Quality Control: With the growing prevalence of cffDNA-based testing, standardization and quality control are paramount. Efforts to establish guidelines and ensure consistency across laboratories and platforms, such as initiatives like the Sequencing Quality Control consortium and external proficiency testing programs, contribute to the standardization of cffDNA-based testing.

Technological Challenges: Despite the advancements, several technological challenges persist. These include the detection of low-frequency variants, accurate measurement of fragment size, and the need for cost-effective platforms for widespread implementation. Ongoing research aims to address these challenges and further refine cffDNA technology.

3.7.2 Integration of cffDNA Testing into Routine Prenatal Screening

The integration of cffDNA testing into routine prenatal care has garnered significant interest due to its array of compelling advantages. Foremost among these is its non-invasive nature, eliminating the necessity for invasive procedures [5]. This integration minimizes associated risks, discomfort, and potential complications related to invasive procedures. Moreover, cffDNA testing has showcased superior accuracy and detection rates for common chromosomal abnormalities compared to traditional screening methods [105, 106]. By seamlessly incorporating this testing into routine prenatal care, healthcare providers can enhance the detection of conditions such as Down syndrome, Edwards syndrome, and Patau syndrome, thereby facilitating more accurate risk assessment and informed decision-making.

Another compelling reason for the integration lies in the ability of cffDNA testing to allow early detection of genetic conditions, offering a window for timely intervention and personalized management strategies [107]. This early identification during routine prenatal care enables to develop tailored care plans, optimizing outcomes for both mother and baby. Furthermore, it provides a more precise risk assessment for various genetic conditions, enabling healthcare providers to offer tailored counseling and support. The integration into routine care allows for better stratification of patient risk and the provision of accurate information about the likelihood of specific genetic conditions, empowering patients to make informed decisions about their pregnancy.

The integration of cffDNA testing into routine prenatal care also significantly reduces the need for invasive procedures. This reduction not only minimizes potential risks and complications associated with invasive procedures but also alleviates patient anxiety and stress related to these interventions. Furthermore, cffDNA testing, being noninvasive, contributes to an enhanced patient experience by offering a relatively stress-free approach to pregnant individuals. This patient-centered strategy prioritizes safety, comfort, and convenience, thus improving overall satisfaction and engagement in prenatal care. Lastly, cffDNA testing's ability to provide comprehensive genetic information about the fetus facilitates earlier identification of potential genetic conditions, allowing stakeholders to offer detailed counseling and support. Integrating this into routine care ensures comprehensive counseling sessions that address patient concerns and guide decision-making about pregnancy management.

3.7.3 Areas of Future Research and Development

Current cffDNA testing focuses on common chromosomal abnormalities and specific genetic conditions, but there is ongoing research to expand its scope to include a broader range of genetic disorders. Researchers are working to enhance its sensitivity and specificity by refining testing algorithms, optimizing sequencing techniques, and developing advanced bioinformatics approaches [108]. cffDNA-based testing is being explored for evaluating fetal growth, organ maturity, and placental function by analyzing specific fetal-derived DNA markers [109–111]. Combining cffDNA testing with other omics technologies could provide a more comprehensive understanding of fetal health and development, improving the accuracy and predictive power of testing. Its potential to analyze fetal epigenetic markers, offering insights into epigenetic patterns and their implications for health and disease [112], is extensively under consideration. Efforts are underway to establish standardized protocols and quality control measures to ensure accurate and reproducible results across different laboratories and platforms.

3.8 Conclusion

cffDNA stands out as one of the most promising molecules in prenatal testing, opening new avenues in NIPS by its presence in maternal circulation. Nowadays, the majority of NIPS tests are cffDNA-based, with proven efficacy in screening for chromosomal aneuploidies, monogenic disorders, and other fetal anomalies, as demonstrated by numerous studies. Some countries have formally incorporated NIPS into routine prenatal screening. Beyond chromosomal anomalies, cffDNA has been utilized for determining fetal gender, Rh status, congenital heart disease, and even maternal health parameters such as gestational diabetes and preeclampsia. Despite its burgeoning applications in prenatal screening, cffDNA testing comes with certain limitations, including the challenge of low amounts of circulating fetal DNA in maternal circulation, ethical considerations, issues of equitable access to healthcare resources, and cost-related challenges. Ongoing technological advances are addressing these challenges. In its entirety, this molecule has opened new doors in prenatal screening, mitigating the risks associated with earlier invasive procedures. Realizing the full potential of cffDNA necessitates essential collaboration among researchers, clinicians, and policymakers. This collaborative effort is pivotal for translating research findings into responsible and ethical clinical practices, ensuring the conscientious implementation of cffDNA testing.

References

1. Invernizzi P, Biondi ML, Battezzati PM, Perego F, Selmi C, Cecchini F, et al. Presence of fetal DNA in maternal plasma decades after pregnancy. Hum Genet. 2002;110(6):587–91.
2. Taglauer ES, Wilkins-Haug L, Bianchi DW. Review: cell-free fetal DNA in the maternal circulation as an indication of placental health and disease. Placenta. 2014;35 Suppl(Suppl):S64–8.
3. Flori E, Doray B, Gautier E, Kohler M, Ernault P, Flori J, et al. Circulating cell-free fetal DNA in maternal serum appears to originate from cyto- and syncytio-trophoblastic cells. Case report. Hum Reprod. 2004;19(3):723–4.
4. Wang E, Batey A, Struble C, Musci T, Song K, Oliphant A. Gestational age and maternal weight effects on fetal cell-free DNA in maternal plasma. Prenat Diagn. 2013;33(7):662–6.
5. Lo YM, Corbetta N, Chamberlain PF, Rai V, Sargent IL, Redman CW, et al. Presence of fetal DNA in maternal plasma and serum. Lancet. 1997;350(9076):485–7.
6. Tabor A, Alfirevic Z. Update on procedure-related risks for prenatal diagnosis techniques. Fetal Diagn Ther. 2010;27(1):1–7.
7. Lo YMD, Tein MS, Lau TK, Haines CJ, Leung TN, Poon PM, et al. Quantitative analysis of fetal DNA in maternal plasma and serum: implications for noninvasive prenatal diagnosis. Am J Hum Genet. 1998;62(4):768–75.
8. Chiu RWK, Chan KC, Gao Y, Lau VY, Zheng W, Leung TY, et al. Noninvasive prenatal diagnosis of fetal chromosomal aneuploidy by massively parallel genomic sequencing of DNA in maternal plasma. Proc Natl Acad Sci U S A. 2008;105(51):20458–63.
9. Mahdi Mortazavipour M, Mahdian R, Shahbazi S. The current applications of cell-free fetal DNA in prenatal diagnosis of single-gene diseases: a review. Int J Reprod Biomed. 2022;20(8):613–26.
10. Alberry M, Maddocks D, Jones M, Abdel Hadi M, Abdel-Fattah S, Avent N, et al. Free fetal DNA in maternal plasma in anembryonic pregnancies: confirmation that the origin is the trophoblast. Prenat Diagn. 2007;27(5):415–8.
11. Rather RA, Saha SC. Reappraisal of evolving methods in non-invasive prenatal screening: discovery, biology and clinical utility. Heliyon. 2023;9(3):e13923.
12. Bianchi DW. Circulating fetal DNA: its origin and diagnostic potential—a review. Placenta. 2004;25(Suppl A):S93–101.
13. Bischoff FZ, Lewis DE, Simpson JL. Cell-free fetal DNA in maternal blood: kinetics, source and structure. Hum Reprod Update. 2005;11(1):59–67.
14. Lo YM, Zhang J, Leung TN, Lau TK, Chang AM, Hjelm NM. Rapid clearance of fetal DNA from maternal plasma. Am J Hum Genet. 1999;64(1):218–24.
15. Yu SC, Lee SW, Jiang P, Leung TY, Chan KC, Chiu RW, et al. High-resolution profiling of fetal DNA clearance from maternal plasma by massively parallel sequencing. Clin Chem. 2013;59:1228–37.
16. Garcia Moreira V, de la Cera Martinez T, Gago Gonzalez E, Prieto Garcia B, Alvarez Menendez FV. Increase in and clearance of cell-free plasma DNA in hemodialysis quantified by real-time PCR. Clin Chem Lab Med. 2006;44:1410–5.
17. Lo YM. Fetal DNA in maternal plasma. Ann N Y Acad Sci. 2000;906:141–7.
18. Ashoor G, Syngelaki A, Poon LC, Rezende JC, Nicolaides KH. Fetal fraction in maternal plasma cell-free DNA at 11-13 weeks' gestation: relation to maternal and fetal characteristics. Ultrasound Obstet Gynecol. 2013;41(1):26–32.

19. Deng C, Liu S. Factors affecting the fetal fraction in noninvasive prenatal screening: a review. Front Pediatr. 2022;10:812781.
20. Levine RJ, Maynard SE, Qian C, Lim KH, England LJ, Yu KF, et al. Circulating angiogenic factors and the risk of preeclampsia. N Engl J Med. 2004;350(7):672–83.
21. Lo YM. The quest for accurate measurement of fetal DNA in maternal plasma. Clin Chem. 2011;57(3):522–3.
22. Deng C, Liu J, Liu S, Liu H, Bai T, et al. Maternal and fetal factors influencing fetal fraction: a retrospective analysis of 153,306 pregnant women undergoing noninvasive prenatal screening. Front Pediatr. 2023;11:1066178.
23. Picchiassi E, Coata G, Fanetti A, Centra M, Pennacchi L, Di Renzo GC. The best approach for early prediction of fetal gender by using free fetal DNA from maternal plasma. Prenat Diagn. 2008;28(6):525–30.
24. Liu FM, Wang XY, Feng X, Wang W, Ye YX, Chen H. Feasibility study of using fetal DNA in maternal plasma for non-invasive prenatal diagnosis. Acta Obstet Gynecol Scand. 2007;86(5):535–41.
25. Fan HC, Blumenfeld YJ, Chitkara U, Hudgins L, Quake SR. Noninvasive diagnosis of fetal aneuploidy by shotgun sequencing DNA from maternal blood. Proc Natl Acad Sci U S A. 2008;105(42):16266–71.
26. Risberg B, Tsui DWY, Biggs H, Martin R-V, de Almagro A, Dawson SJ, Hodgkin C, et al. Effects of collection and processing procedures on plasma circulating cell-free DNA from cancer patients. J Mol Diagn. 2018;20(6):883–92.
27. Pös Z, Pös O, Styk J, Mocova A, Strieskova L, Budis J, et al. Technical and methodological aspects of cell-free nucleic acids analyzes. Int J Mol Sci. 2020;21(22):8634.
28. Martignano F. Cell-free DNA: an overview of sample types and isolation procedures. Methods Mol Biol. 2019;1909:13–27.
29. Edlow AG, Bianchi DW. Tracking fetal development through molecular analysis of maternal biofluids. Biochim Biophys Acta. 2012;1822(12):1970–80.
30. Campos CDM, Gamage SST, Jackson JM, Witek MA, Park DS, Murphy MC, et al. Microfluidic-based solid phase extraction of cell free DNA. Lab Chip. 2018;18(22):3459–70.
31. Meddeb R, Pisareva E, Thierry AR. Guidelines for the preanalytical conditions for analyzing circulating cell-free DNA. Clin Chem. 2019;65(5):623–33.
32. Lee PY, Costumbrado J, Hsu CY, Kim YH. Agarose gel electrophoresis for the separation of DNA fragments. J Vis Exp. 2012;(62):3923.
33. Tang DL, Li Y, Zhou X, Li X, Zheng F. Multiplex fluorescent PCR for noninvasive prenatal detection of fetal-derived paternally inherited diseases using circulatory fetal DNA in maternal plasma. Eur J Obstet Gynecol Reprod Biol. 2009;144(1):35–9.
34. Tsui NB, Jiang P, Wong YF, Leung TY, Chan KC, Chiu RW, Sun H, Lo YM. Maternal plasma RNA sequencing for genome-wide transcriptomic profiling and identification of pregnancy-associated transcripts. Clin Chem. 2014;60(7):954–62.
35. Fan HC, Blumenfeld YJ, Chitkara U, Hudgins L, Quake SR. Analysis of the size distributions of fetal and maternal cell-free DNA by paired-end sequencing. Clin Chem. 2010;56(8):1279–86.
36. Chudasama DY, Aladag Z, Felicien MI, Hall M, Beeson J, Asadi N, et al. Prognostic value of the DNA integrity index in patients with malignant lung tumors. Oncotarget. 2018;9(30):21281–8.
37. Karami F, Noori Daloii MR, Hantooshzadeh S, Modarressi MH. Comparing the efficiency of three protocols in isolation of cell free fetal DNA from maternal blood. J Family Reprod Health. 2017;11(3):146–51.
38. Tounta G, Kolialexi A, Papantoniou N, Tsangaris GT, Kanavakis E, Mavrou A. Non-invasive prenatal diagnosis using cell-free fetal nucleic acids in maternal plasma: progress overview beyond predictive and personalized diagnosis. EPMA J. 2011;2(2):163–71.
39. Hou Y, Yang J, Deng F, Wang F, Peng H, Guo F, et al. Association between cell-free DNA fetal fraction and pregnant character: a retrospective cohort study of 27,793 maternal plasmas. Sci Rep. 2023;13(1):11420.

40. Xing L, Bai T, Liu S, Liu J, Jing X, Deng C, et al. Maternal, neonatal, pregnancy outcome characteristics of pregnant women with high plasma cell-free DNA concentration in noninvasive prenatal screening: a retrospective analysis. Front Pediatr. 2023;11:1195818.
41. Wapner RJ, Babiarz JE, Levy B, Stosic M, Zimmermann B, Sigurjonsson S, et al. Expanding the scope of noninvasive prenatal testing: detection of fetal microdeletion syndromes. Am J Obstet Gynecol. 2015;212(332):e331–9.
42. Tian M, Feng L, Li J, Zhang R. Focus on the frontier issue: progress in noninvasive prenatal screening for fetal trisomy from clinical perspectives. Crit Rev Clin Lab Sci. 2023;60(4):248–69.
43. Gadsbøll K, Petersen OB, Gatinois V, Strange H, Jacobsson B, Wapner R, et al. Current use of noninvasive prenatal testing in Europe, Australia and the USA: a graphical presentation. Acta Obstet Gynecol Scand. 2020;99(6):722–30.
44. Takahashi M, Linh LK, Sayed AM, Imoto A, Sato M, Dila KAS, et al. Non-invasive prenatal testing (NIPT) implementation in Japan: a comparison with the United Kingdom, Germany, Italy, Sweden, and Taiwan. Int J Environ Res Public Health. 2022;19(24):16404.
45. Ravitsky V, Roy MC, Haidar H, Henneman L, Marshall J, Newson AJ, et al. The emergence and global spread of noninvasive prenatal testing. Annu Rev Genomics Hum Genet. 2021;22:309–38.
46. Gil MM, Accurti V, Santacruz B, Plana MN, Nicolaides KH. Analysis of cell-free DNA in maternal blood in screening for aneuploidies: updated meta-analysis. Ultrasound Obstet Gynecol. 2017;50(3):302–14. Update in: Ultrasound Obstet Gynecol 2019;53(6):734–42.
47. Chitayat D, Langlois S, Wilson RD. No. 261-prenatal screening for fetal aneuploidy in singleton pregnancies. J Obstet Gynaecol Can. 2017;39(9):e380–94.
48. Wu D, Hou Q, Li T, Chu Y, Guo Q, Kang B, et al. The use of cffDNA in fetal sex determination during the first trimester of pregnancy of female DMD carriers. Intractable Rare Dis Res. 2012;1(4):157–60.
49. Hill M, Finning K, Martin P, Hogg J, Meaney C, Norbury G, et al. Non-invasive prenatal determination of fetal sex: translating research into clinical practice. Clin Genet. 2011;80(1):68–75.
50. Fernández-Martínez FJ, Galindo A, Garcia-Burguillo A, Vargas-Gallego C, Nogués N, Moreno-García M, Moreno-Izquierdo A. Noninvasive fetal sex determination in maternal plasma: a prospective feasibility study. Genet Med. 2012;14(1):101–6.
51. Koumbaris G, Kypri E, Tsangaras K, Achilleos A, Mina P, Neofytou M, et al. Cell-free DNA analysis of targeted genomic regions in maternal plasma for non-invasive prenatal testing of trisomy 21, trisomy 18, trisomy 13, and fetal sex. Clin Chem. 2016;62(6):848–55.
52. Bowman-Smart H, Savulescu J, Gyngell C, Mand C, Delatycki MB. Sex selection and noninvasive prenatal testing: a review of current practices, evidence, and ethical issues. Prenat Diagn. 2020;40(4):398–407.
53. Tyndall C, Cuzzilla R, Kane SC. The rhesus incompatible pregnancy and its consequences for affected fetuses and neonates. Transfus Apher Sci. 2020;59(5):102948.
54. Rather RA, Dhawan V, Saha SC. Non-invasive prenatal rhesus D genotyping using cell-free foetal DNA. Indian J Med Res. 2019;150(1):62–6.
55. Niguse B, Ermias M, Berhanu S, Abayneh L, Chakiso B, Rather RA. RHD exon 5, 7 and 10 targeted non-invasive prenatal screening of fetal rhesus-D (RhD) in selected RhD negative pregnant women in Ethiopia. PLoS One. 2022;17(3):e0265583.
56. Finning K, Martin P, Summers J, Daniels G. Fetal genotyping for the K (Kell) and Rh C, c, and E blood groups on cell-free fetal DNA in maternal plasma. Transfusion. 2007;47:2126–33.
57. Gutensohn K, Müller SP, Thomann K, Stein W, Suren A, Körtge-Jung S, et al. Diagnostic accuracy of noninvasive polymerase chain reaction testing for the determination of fetal rhesus C, c and E status in early pregnancy. BJOG. 2010;117:722–9.
58. O'Brien H, Hyland C, Schoeman E, Flower R, Daly J, Gardener G. Non-invasive prenatal testing (NIPT) for fetal Kell, Duffy and Rh blood group antigen prediction in alloimmunised pregnant women: power of droplet digital PCR. Br J Haematol. 2020;189:e90–4.

59. Watson CT, Marques-Bonet T, Sharp AJ, Mefford HC. The genetics of microdeletion and microduplication syndromes: an update. Annu Rev Genomics Hum Genet. 2014;15:215–44.
60. Bevilacqua E, Jani JC, Chaoui R, Suk EA, Palma-Dias R, Ko TM, Warsof S, Stokowski R, Jones KJ, Grati FR, Schmid M. Performance of a targeted cell-free DNA prenatal test for 22q11.2 deletion in a large clinical cohort. Ultrasound Obstet Gynecol. 2021;58(4):597–602.
61. McDonald-McGinn DM, Sullivan KE, Marino B, Philip N, Swillen A, Vorstman JA, et al. 22q11.2 deletion syndrome. Nat Rev Dis Primers. 2015;1:15071.
62. Williams CA, Beaudet AL, Clayton-Smith J, Knoll JH, Kyllerman M, Laan LA, et al. Angelman syndrome 2005: updated consensus for diagnostic criteria. Am J Med Genet A. 2006;140(5):413–8.
63. Straver R, Sistermans EA, Holstege H, Visser A, Oudejans CB, Reinders MJ. WISECONDOR: detection of fetal aberrations from shallow sequencing maternal plasma based on a within-sample comparison scheme. Nucleic Acids Res. 2014;42(5):e31.
64. Lee SA, Ding C. The dysfunctional placenta epigenome: causes and consequences. Epigenomics. 2012;4(5):561–9.
65. Audette MC, Kingdom JC. Screening for fetal growth restriction and placental insufficiency. Semin Fetal Neonatal Med. 2018;23(2):119–25.
66. Tharmaratnam S. Fetal distress. Baillieres Best Pract Res Clin Obstet Gynaecol. 2000;14(1):155–72.
67. Molina LCG, Odibo L, Zientara S, Običan SG, Rodriguez A, Stout M, et al. Validation of Delphi procedure consensus criteria for defining fetal growth restriction. Ultrasound Obstet Gynecol. 2020;56(1):61–6.
68. Rolnik DL, da SilvaCosta F, Lee TJ, Schmid M, McLennan AC. Association between fetal fraction on cell-free DNA testing and first trimester markers for pre-eclampsia. Ultrasound Obstet Gynecol. 2018;52:722–7.
69. Tsatsaris V, Goffin F, Foidart JM. Circulating angiogenic factors and preeclampsia. N Engl J Med. 2004;350(19):2003–4; author reply 2003–4.
70. Lo YM, Chan KC, Sun H, Chen EZ, Jiang P, Lun FM, et al. Maternal plasma DNA sequencing reveals the genome-wide genetic and mutational profile of the fetus. Sci Transl Med. 2010;2(61):61ra91.
71. van Boeckel SR, Davidson DJ, Norman JE, Stock SJ. Cell-free fetal DNA and spontaneous preterm birth. Reproduction. 2018;155(3):R137–45.
72. Darghahi R, Mobaraki-Asl N, Ghavami Z, Pourfarzi F, Hosseini-Asl S, Jalilvand F. Effect of cell-free fetal DNA on spontaneous preterm labor. J Adv Pharm Technol Res. 2019;10(3):117–20.
73. Ahmed B, Konje JC. Fetal lung maturity assessment: a historic perspective and non-invasive assessment using an automatic quantitative ultrasound analysis (a potentially useful clinical tool). Eur J Obstet Gynecol Reprod Biol. 2021;258:343–7.
74. Bronkhorst AJ, Ungerer V, Diehl F, Anker P, Dor Y, Fleischhacker M, et al. Towards systematic nomenclature for cell-free DNA. Hum Genet. 2021;140(4):565–78.
75. Harraway J. Non-invasive prenatal testing. Aust Fam Physician. 2017;46(10):735–9.
76. Geifman-Holtzman O, Berman JO. Prenatal diagnosis: update on invasive versus noninvasive fetal diagnostic testing from maternal blood. Expert Rev Mol Diagn. 2008;8(6):727–51.
77. Mokhtar R, Hans P, Sinha A. Comparing non-invasive prenatal testing with invasive testing for the detection of trisomy 21. Cureus. 2022;14(11):e31252.
78. Taylor-Phillips S, Freeman K, Geppert J, Agbebiyi A, Uthman OA, Madan J, et al. Accuracy of non-invasive prenatal testing using cell-free DNA for detection of Down, Edwards and Patau syndromes: a systematic review and meta-analysis. BMJ Open. 2016;6(1):e010002.
79. Norton ME, Brar H, Weiss J, Karimi A, Laurent LC, Caughey AB, et al. Non-invasive chromosomal evaluation (NICE) study: results of a multicenter prospective cohort study for detection of fetal trisomy 21 and trisomy 18. Am J Obstet Gynecol. 2012;207(2):137.e1–8.
80. Iwarsson E, Jacobsson B, Dagerhamn J, Davidson T, Bernabé E, Arnlind MH. Analysis of cell-free fetal DNA in maternal blood for detection of trisomy 21, 18 and 13 in a general

pregnant population and in a high risk population—a systematic review and meta-analysis. Acta Obstet Gynecol Scand. 2017;96(1):7–18.
81. Green JM, Hewison J, Bekker HL, Bryant LD, Cuckle HS. Psychosocial aspects of genetic screening of pregnant women and newborns: a systematic review. Health Technol Assess. 2004;8(33):iii, ix–x, 1–109.
82. Dubois ML, Winters PD, Rodrigue MA, Gekas J. Patient attitudes and preferences about expanded noninvasive prenatal testing. Front Genet. 2023;14:976051.
83. Singh ON. Clinical experience and follow-up with large-scale single-nucleotide polymorphism-based noninvasive prenatal aneuploidy testing. Am J Obstet Gynecol. 2015;213(2):253.
84. Allyse M, Minear MA, Berson E, Sridhar S, Rote M, Hung A, et al. Non-invasive prenatal testing: a review of international implementation and challenges. Int J Womens Health. 2015;7:113–26.
85. Krier JB, Green RC. Management of incidental findings in clinical genomic sequencing. Curr Protoc Hum Genet. 2015;87:9.23.1–9.23.16.
86. Institute of Medicine (US). Committee on assessing genetic risks. In: Andrews LB, et al., editors. Washington, DC: National Academies Press (US); 1994.
87. American College of Obstetricians and Gynecologists. ACOG committee opinion no. 545: noninvasive prenatal testing for fetal aneuploidy. Obstet Gynecol. 2012;120(6):1532–4.
88. Manti S, Licari A. How to obtain informed consent for research. Breathe (Sheff). 2018;14(2):145–52.
89. Majumder MA, Guerrini CJ, McGuire AL. Direct-to-consumer genetic testing: value and risk. Annu Rev Med. 2021;72:151–66.
90. European Society of Human Genetics. Statement of the ESHG on direct-to-consumer genetic testing for health-related purposes. Eur J Hum Genet. 2010;18(12):1271–3.
91. Skirton H, Lewis C, Kent A, Coviello DA, Members of Eurogentest Unit 6 and ESHG Education Committee. Genetic education and the challenge of genomic medicine: development of core competences to support preparation of health professionals in Europe. Eur J Hum Genet. 2010;18(9):972–7.
92. Hou Y, Yang J, Qi Y, Guo F, Peng H, Wang D, et al. Factors affecting cell-free DNA fetal fraction: statistical analysis of 13,661 maternal plasmas for non-invasive prenatal screening. Hum Genomics. 2019;13(1):62.
93. Bianchi DW. From prenatal genomic diagnosis to fetal personalized medicine: progress and challenges. Nat Med. 2012;18(7):1041–51.
94. Hyblova M, Gnip A, Kucharik M, Budis J, Sekelska M, Minarik G. Maternal copy number imbalances in non-invasive prenatal testing: do they matter? Diagnostics (Basel). 2022;12(12):3056.
95. Carbone L, Cariati F, Sarno L, Conforti A, Bagnulo F, et al. Non-invasive prenatal testing: current perspectives and future challenges. Genes (Basel). 2020;12(1):15.
96. de Sousa MJ, Albuquerque M, Cruz G. International guidelines for implementation of NIPT/cffDNA testing. In: Prenatal diagnostic testing for genetic disorders: the revolution of the non-invasive prenatal test. Cham: Springer International Publishing; 2023. p. 389–428.
97. Wienzek-Lischka S, Bachmann S, Froehner V, Bein G. Potential of next-generation sequencing in noninvasive fetal molecular blood group genotyping. Transfus Med Hemother. 2020;47(1):14–22.
98. Welker NC, Lee AK, Kjolby RAS, Wan HY, Theilmann MR, Jeon D, et al. High-throughput fetal fraction amplification increases analytical performance of noninvasive prenatal screening. Genet Med. 2021;23(3):443–50.
99. Hui L, Bianchi DW. Fetal fraction and noninvasive prenatal testing: what clinicians need to know. Prenat Diagn. 2020;40:155–63.
100. Sawakwongpra K, Tangmansakulchai K, Ngonsawan W, Promwan S, Chanchamroen S, Quangkananurug W, et al. Droplet-based digital PCR for non-invasive prenatal genetic diagnosis of α and β-thalassemia. Biomed Rep. 2021;15(4):82.

101. Jeppesen LD, Lildballe DL, Hatt L, Hedegaard J, Singh R, Toft CLF, et al. Noninvasive prenatal screening for cystic fibrosis using circulating trophoblasts: detection of the 50 most common disease-causing variants. Prenat Diagn. 2023;43(1):3–13.
102. Poon LC, Maiz N, Valencia C, Plasencia W, Nicolaides KH. First-trimester maternal serum pregnancy-associated plasma protein-A and pre-eclampsia. Ultrasound Obstet Gynecol. 2009;33(1):23–33.
103. Boddupally K, Thuraka ER. Artificial intelligence for prenatal chromosome analysis. Clin Chim Acta. 2024;552:117669.
104. Al-Antari MA. Artificial intelligence for medical diagnostics-existing and future AI technology! Diagnostics (Basel). 2023;13(4):688.
105. Strom CM, Anderson B, Tsao D, Zhang K, Liu Y, Livingston K, et al. Improving the positive predictive value of non-invasive prenatal screening (NIPS). PLoS One. 2017;12(3):e0167130.
106. Ashoor G, Syngelaki A, Wang E, Stuble C, Oliphant A, Song K, et al. Trisomy 13 detection in the first trimester of pregnancy using a chromosome-selective cell-free DNA analysis method. Ultrasound Obstet Gynecol. 2013;41:21–2.
107. Andari MVC, Bussamra SLC, Tedesco TGD, Peixoto PAB, Pares PDBS, Braga A, et al. Noninvasive prenatal testing: benefits and limitations of the available tests. Ceska Gynekol. 2020;85(1):41–8.
108. Zhang J, Li J, Saucier JB, Feng Y, Jiang Y, Sinson J, et al. Non-invasive prenatal sequencing for multiple Mendelian monogenic disorders using circulating cell-free fetal DNA. Nat Med. 2019;25(3):439–47.
109. Chang L, Jiao H, Chen J, Wu G, Liu P, et al. Single-cell whole-genome sequencing, haplotype analysis in prenatal diagnosis of monogenic diseases. Life Sci Alliance. 2023;6(5):e202201761.
110. Hayward J, Chitty LS. Beyond screening for chromosomal abnormalities: advances in non-invasive diagnosis of single gene disorders and fetal exome sequencing. Semin Fetal Neonatal Med. 2018;23(2):94–101.
111. Bedei I, Wolter A, Weber A, Signore F, Axt-Fliedner R. Chances and challenges of new genetic screening technologies (NIPT) in prenatal medicine from a clinical perspective: a narrative review. Genes (Basel). 2021;12(4):501.
112. Tsui DW, Chiu RW, Lo YD. Epigenetic approaches for the detection of fetal DNA in maternal plasma. Chimerism. 2010;1(1):30–5.

Methods and Protocols for Extraction of Cell-Free Fetal DNA from the Maternal Circulation

4

Shahnawaz Majeed, Safiya Mehraj, and Mohammad Iqbal Banday

4.1 Introduction

Cell-free DNA (cfDNA) refers to the DNA released from cells into the circulatory system, making it detectable in peripheral blood, cerebral spinal fluid, pleural fluid, urine, and saliva [1–4]. Mandel and Metais first discovered cfDNA in peripheral blood in 1948 [5]. In healthy individuals, cfDNA occurs in plasma, originating from the hematopoietic system [6]. However, changes in physiology or the development of pathology leads to the release of additional and mutated DNA into circulation, providing potential for noninvasive detection of underlying conditions. During pregnancy, the mother harbors fetal cfDNA (cffDNA) in her circulation. These are short DNA fragments released into the bloodstream due to apoptosis. The total cfDNA content in maternal blood during pregnancy includes both maternal and fetal DNA, with fetal content varying from 2 to 20% of total maternal cfDNA. Placental cfDNA can be detected around 2 months into gestation and disappears shortly after birth. Quantitative analysis of cffDNA may correlate with maternal risk for various pregnancy-related complications. However, challenges exist in recovering sufficient quantity and quality of cffDNA from maternal circulation, impacting the accuracy of prenatal diagnostics. Some studies suggest that the majority of fetal DNA circulates in membrane-bound vesicles, such as apoptotic bodies [7].

S. Majeed (✉)
Clinical Microbiology and PK/PD Division, CSIR-Indian Institute of Integrative Medicine, Srinagar, Jammu and Kashmir, India

S. Mehraj · M. I. Banday
Clinical Microbiology and PK/PD Division, CSIR-Indian Institute of Integrative Medicine, Srinagar, Jammu and Kashmir, India

Academy of Scientific and Innovative Research (AcSIR), Srinagar, Jammu and Kashmir, India

R. A. Rather, S. C. Saha (eds.), *Non-invasive Prenatal Screening (NIPS) in Clinical Practice*, https://doi.org/10.1007/978-981-97-6402-0_4

4.2 General Steps of cffDNA Isolation

1. Collection of maternal blood
2. Isolation of plasma from blood
3. Separation of cell-free DNA from maternal plasma

Collection of maternal blood: While blood collection is a routine procedure in the field of diagnostics, there is a pressing need for improved quality and quantitative yield of cffDNA for diagnostic purposes. Numerous efforts are currently underway to address this requirement. A study details a method aimed at enhancing the stability of cffDNA in maternal blood over extended periods and across varying temperatures [8]. The study utilizes Blood Exo DNA ProTeck® (ProTeck) tubes specifically designed for the collection of blood for cfDNA extraction and amplification. The ProTeck reagent employed in the study effectively inhibits blood cell metabolism during sample storage, resulting in increased stability of cfDNA at temperatures of 4, 22, and 30 °C for 21, 28, and 7 days, respectively. Additionally, it has been noted that treating blood samples with formaldehyde contributes to the stabilization of cffDNA, as it primarily exists within apoptotic bodies.

Isolation of plasma from blood: The general procedure for plasma isolation typically involves centrifugation of blood followed by the harvesting of the supernatant. This step, illustrated in Fig. 4.1, is of paramount significance as it helps prevent the contamination of cellular material during the separation of plasma from other blood components, regardless of the technique employed.

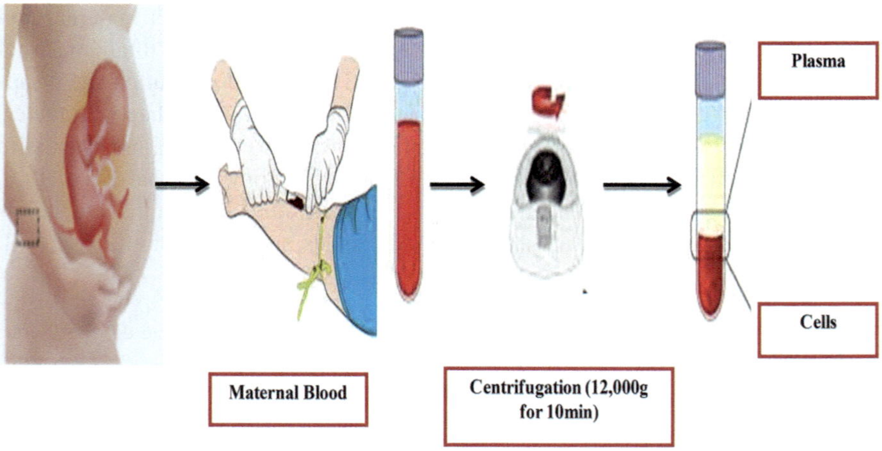

Fig. 4.1 The isolation of plasma from maternal blood is a critical procedure in the field of obstetrics and prenatal care, as it allows for the assessment of maternal health and the monitoring of specific biomarkers during pregnancy. This process involves centrifugation, which capitalizes on the varying densities of blood components. A maternal blood sample is collected using a specialized tube or container. This blood specimen is then carefully placed in a centrifuge. During centrifugation, the blood sample undergoes rapid rotation, subjecting it to gravitational forces

As a result of this spinning action, the heavier cellular components, including red blood cells and white blood cells, are driven toward the bottom of the container due to their higher density. Meanwhile, the lighter plasma, which contains various proteins, hormones, and other soluble substances, rises to the top. This process effectively separates these two major components, allowing for the collection of isolated plasma. Subsequently, this plasma can be subjected to a wide range of laboratory tests, enabling healthcare professionals to evaluate the mother's health status, screen for specific markers, and monitor the well-being of the developing fetus. The method of isolating plasma from maternal blood through centrifugation is a fundamental diagnostic tool in prenatal care, providing valuable insights to ensure a healthy pregnancy.

4.3 Separation of Cell-Free DNA from Maternal Plasma

Numerous protocols have emerged to enhance the efficiency of plasma separation. The established procedure outlined by Hyuang et al. [9] for plasma isolation employs a two-step centrifugation process. The initial step utilizes high-speed centrifugation to ensure the separation of cellular material from the plasma layer. Following plasma separation, various commercial kits have been utilized to extract cfDNA from this isolated plasma layer.

Anticoagulants and enzyme inhibitors play a crucial role in inhibiting 90% of deoxyribonuclease activity in blood plasma. The choice of different anticoagulants (e.g., EDTA, citrate, or heparin) and cross-linking reagents in blood collection tubes (BCTs) can significantly impact the quality and integrity of cfDNA isolated from blood plasma. This is primarily attributed to cell lysis, particularly from leukocytes, and DNA damage caused by traditional cell-stabilizing reagents like formaldehyde and glutaraldehyde. Interestingly, certain studies have highlighted the stability of cfDNA in BCTs, particularly favoring K3EDTA vials. BCT Streck tubes have been demonstrated to stabilize cell membranes in whole blood, thereby minimizing cell lysis and the subsequent release of cellular DNA into plasma, thus proving to be a valuable preanalytical factor in optimizing cfDNA studies for clinical research. Further studies confirm that cfDNA in serum is susceptible to contamination by blood cells, with contamination increasing with the time between blood withdrawal and serum separation. The cfDNA concentration is significantly higher in serum (3- to 24-fold) compared to plasma, likely due to background maternal DNA and DNA liberation during clotting. Delaying serum processing significantly increases cfDNA concentration, while plasma shows minimal changes within 24 h. However, conflicting observations have been made regarding DNA concentrations in serum and plasma, with some studies reporting higher mean DNA concentrations in serum than in plasma. Plasma generally contains lower cfDNA levels compared to serum, highlighting the disparities between the two. Furthermore, serum samples tend to contain significantly more cell-free genomic DNA than plasma samples, and cfDNA levels in serum increase with storage at 4 °C. Storage of collection tubes at room temperature for 4–25 h post-venipuncture results in a 1.6-fold increase in cfDNA concentration, potentially leading to false-positive results. Real-time polymerase chain reaction (PCR) data suggests that plasma cfDNA is more reliable than serum for detecting mutations. Considering clinical evidence and research findings from PubMed, blood plasma emerges as the preferred source of cfDNA in studies [10]. There are mainly two types of protocols for separation of DNA from plasma, namely, manual and kit based.

4.4 Manual Methods

Manual methods involve DNA extraction from plasma or serum using a combination of reagents and procedures, with the Triton-Heat-Phenol extraction method being a common approach [11]. Here's a step-by-step breakdown of the method:

1. Sample preparation
   - Take 500 μL of plasma or serum and mix it with 5 μL of Triton X-100.
2. Heat denaturation
   - Heat the mixture at 98 °C for 5 min.
3. Cooling and extraction
   - After heat denaturation, place the samples on ice for 5 min.
   - Extract the samples by adding an equal volume of phenol-chloroform-isoamyl alcohol (25:24:1, v:v:v) to the mixture.
4. Centrifugation
   - Centrifuge the mixture for 10 min at 14,000 *g*.
5. Aqueous phase separation
   - Collect the aqueous (upper) phase, leaving behind the organic (lower) phase and any interphase.
6. Precipitation
   - Precipitate the DNA by adding 1/10 volume of 3 M sodium acetate and 2.5 volumes of 100% ethanol to the aqueous phase.
   - Incubate the mixture at −20 °C overnight for precipitation.
7. DNA pellet collection
   - Centrifuge the precipitated DNA at high speed (e.g., 14,000 *g*) to pellet the DNA.
8. Washing and resuspension
   - Wash the DNA pellet with ethanol.
   - Allow the DNA pellet to air-dry.
   - Resuspend the DNA pellet in 50 μL of ddH_2O (double-distilled water).
9. Quantification: PicoGreen assay and SYBR-Green real-time PCR

Another method also involves Triton X-100 and heat, followed by DNA extraction using a phenol-chloroform-isoamyl alcohol mixture but with little modifications. Hufnagl et al. have modified the method as described under:

4.4.1 Materials

- Plasma samples (1 mL per sample)
- EDTA-NaCl solution
- 100 μL of 250 mmol/L EDTA
- 100 μL of 750 mmol/L NaCl
- 100 g/L sodium dodecyl sulfate (SDS) solution
- Proteinase K solution (20 mg/mL)
- Saturated 6 M NaCl solution
- Phenol-chloroform mixture (1:1, v:v)
- Absolute ethanol
- 70% ethanol
- Double-distilled water (ddH_2O)

4.5 Procedure

Protocol 1: Standard DNA extraction method

1. Sample preparation
   - Mix 1 mL of plasma with 100 μL of EDTA-NaCl solution.
   - Add 100 μL of 100 g/L SDS solution.
   - Thoroughly mix the solution.
   - Add 20 μL of proteinase K (final concentration 20 mg/mL).
2. Incubation
   - Incubate the samples for 2 h at 56 °C.
3. Protein precipitation
   - Add 200 μL of saturated 6 M NaCl solution to the samples (final concentration, 0.86 mol/L).
   - Mix thoroughly and centrifuge the samples for 15 min at 14,000 *g*.
   - Transfer the cleared supernatant into a new tube.
4. DNA extraction
   - Add an equal volume of phenol-chloroform mixture to the supernatant at room temperature.
   - Incubate for 5 min at room temperature.
   - Centrifuge the mixture for 15 min at 14,000 *g*.
   - Transfer the supernatant to a new tube.
5. DNA precipitation
   - Add the same volume of absolute ethanol to the supernatant.
   - Mix gently and incubate overnight at −20 °C.
   - Centrifuge for 15 min at 14,000 *g*.
6. DNA pellet washing and resuspension
   - Discard the supernatant.
   - Wash the DNA pellet with 70% ethanol.
   - Centrifuge for 15 min at 14,000 *g*.

- Carefully remove the ethanol.
- Allow the DNA pellet to air-dry.
- Resuspend the DNA pellet in 50 μL of ddH_2O.

7. Quantification
   - Use 5 μL of template DNA for quantification using quantitative PCR (qPCR).

Protocol 2: Plasma DNA extraction using MPC method

1. Sample preparation
   - Mix 1 mL of plasma with 50 μL of 25% SDS.
   - Add 30 μL of 20 mg/mL proteinase K.
   - Incubate the mixture at 55 °C for 16 h.
2. Digestion and extraction
   - After digestion, add an equal volume of water-saturated phenol to the sample and mix well.
   - Transfer the mixture into a phase lock gel tube (MaXtract low density tube, Qiagen, Hilden, Germany).
   - Centrifuge at 16,000 *g* at room temperature for 5 min.
3. Supernatant preparation
   - Transfer the supernatant into a new 2 mL Eppendorf tube.
4. Chloroform-isoamyl alcohol extraction
   - Mix the supernatant with an equal volume of chloroform-isoamyl alcohol (24:1) mixture.
   - For PC method
     - Centrifuge the mixture at 16,000 *g* for 15 min.
   - For MPC method
     - Transfer the mixture into a phase lock gel tube (MaXtract high density tube, Qiagen, Hilden, Germany).
     - Centrifuge at 16,000 *g* at room temperature for 5 min.
5. DNA precipitation
   - Add 3 mM sodium acetate with a 1:10 ratio of the supernatant and an equal volume of isopropanol.
   - Incubate for DNA precipitation overnight at −20 °C.
6. DNA dissolution and storage:
   - Dissolve the DNA in 50 μL of DNA hydration solution (Qiagen, Hilden, Germany).
   - Store the DNA at −20 °C until further use.

Protocol 3: Sodium N-lauryl sarcosine with glycogen method.

1. Sample preparation
   - Obtain 500 μL of peripheral blood plasma for DNA isolation.
2. Preparation of Solution 1
   - Combine the following components to create Solution 1:
     - 6 M NaCl

- 13 mM EDTA
- 10 µg glycogen (used as a carrier)
- 0.5% sodium N-lauryl sarcosine
- 26 mM Tris-HCl pH 8.0

3. Adding Solution 1 and incubation
   - Add 500 µL of Solution 1 to the 500 µL of plasma in a suitable tube.
   - Mix the contents thoroughly by vortexing or gentle inversion.
   - Incubate the mixture at 55 °C for 1 h.
4. Isopropanol precipitation
   - After incubation, add 1 mL of isopropanol to the tube containing the incubated mixture.
   - Vortex the tube for 30 s to promote DNA precipitation.
5. First centrifugation
   - Centrifuge the tube at 16,000 *g* for 5 min to pellet the DNA.
6. Transferring supernatant
   - Carefully transfer the supernatant (containing impurities) to a new tube.
7. Second isopropanol precipitation
   - Add 1 mL of 40% isopropanol to the tube containing the supernatant.
   - Vortex the tube for 10 s to precipitate any remaining DNA.
8. Second centrifugation
   - Centrifuge the tube at 16,000 *g* for 5 min.
9. Supernatant removal and drying
   - Carefully remove the supernatant without disturbing the DNA pellet.
   - Allow the tube to air-dry to remove any residual isopropanol.
10. Resuspending DNA
    - Resuspend the DNA pellet in 100 µL of TE buffer by gently pipetting or vortexing.
11. DNA quantification and storage
    - Quantify the isolated DNA using appropriate methods such as ultraviolet (UV) spectroscopy or fluorometry.
    - Store the resuspended DNA at an appropriate temperature.

The progression of science and technology undeniably rests upon the bedrock of conventional methodologies and principles. These traditional approaches form a sturdy framework from which innovation and advancement can flourish. As we venture into developing more sophisticated methods, we often draw upon these foundational principles to refine and optimize processes. A notable outcome of this evolution is the emergence of commercially available kits, which encapsulate years of scientific knowledge and techniques into convenient packages. These kits have revolutionized numerous fields by simplifying procedures, saving time, and reducing errors. Researchers and practitioners now have access to these ready-made solutions that embody the collective expertise of the scientific community. However, it's essential to recognize that the use of these kits doesn't diminish the importance of understanding the underlying principles; rather, they enhance efficiency and facilitate broader participation in scientific endeavors.

This accessibility can catalyze further innovations and discoveries, democratizing access to advanced technologies and knowledge. Furthermore, these advancements underscore the harmonious relationship between traditional wisdom and cutting-edge technology, demonstrating how building upon the past can pave the way for a more efficient and promising future. As science and technology forge ahead, it's imperative to acknowledge and integrate the insights gleaned from conventional methodologies while embracing the opportunities presented by modern advancements.

Various kit-based methods have been developed for cfDNA isolation, including the following:

- NucleoSpin plasma XS kit
- FitAmpTM plasma/serum DNA isolation kit
- Polymer Mediated Enrichment (PME)-free circulating DNA extraction kit
- QIAamp circulating nucleic acid kit
- Chemagic circulating nucleic acid kit
- Quick-cfDNATM serum/plasma kit
- Norgen cfDNA kit

Typically, these kits are utilized following the manufacturer's protocol. Here, we outline a protocol for cfDNA extraction using one such commercial kit.

Protocol for cfDNA isolation with NucleoSpin plasma XS kit

Materials needed for the DNA isolation protocol include fresh or frozen human EDTA blood samples, a DNA isolation kit with a specialized column type, binding buffer, wash buffer, and elution buffer. Here's the step-by-step procedure:

1. Preparation of plasma
   - Centrifuge fresh or frozen human EDTA blood samples for 10 min at 2000 ×*g* to obtain plasma.
   - Carefully separate plasma from the cell pellet, ensuring no leukocyte carryover.
   - Optionally, store some plasma samples at −80 °C for several days to weeks.
2. Preprocessing of plasma
   - Before DNA isolation, recentrifuge plasma samples for 3 min at 11,000 ×*g* to remove any particulate matter.
   - Transfer the cleared supernatant for further processing.
3. DNA isolation
   - Perform DNA isolation using the specialized DNA isolation kit following the manufacturer's instructions.
   - Utilize a new column type within the kit, directing the sample to a very small silica membrane, enabling DNA elution in 5–30 μL.
4. Rapid protocol
   - Mix 200 μL of plasma sample with 1.5 times the volume of binding buffer.
   - Apply the mixture to the spin column and spin for 30 s at 11,000 ×*g*.
   - Wash the column twice with 500 μL of wash buffer each time:

  - First wash: Spin for 30 s at 11,000 ×*g*.
  - Second wash: Spin for 3 min at 11,000 ×*g*.
- Elute the DNA with 20 μL of elution buffer.

5. High-sensitivity protocol
   - Mix 240 μL of plasma sample with 1.5 times the volume of binding buffer.
   - Apply the mixture to the spin column and spin for 30 s at 11,000 ×*g*.
   - Wash the column twice with 500 μL of wash buffer each time:

- First wash: Spin for 30 s at 11,000 ×*g*.
- Second wash: Spin for 3 min at 11,000 ×*g*.
  - Elute the DNA with 20 μL of elution buffer.

Optionally, for some experiments, heat-treat the eluted DNA for 8 min at 90 °C.

The inconsistency observed in research involving cfDNA arises from its low concentration and the quantification methods used before downstream applications. Quality assessment typically involves one of the following methods: (1) measuring band intensity on an agarose gel, (2) fluorescence measurements using DNA-binding dyes, or (3) estimating UV absorbance at 260 nm. However, these methods often fail to cover the entire physiological cfDNA concentration range.

UV absorbance, in particular, has limitations due to significant contributions from nucleotides, single-stranded nucleic acids, and proteins to the elevated absorbance values. It struggles to distinguish between DNA and RNA and has relatively low sensitivity (requiring 2 mg/mL dsDNA). Due to these limitations, alternative techniques for reliable quantification with improved efficacy and reduced variance need to be explored for cfDNA quantification.

In addition to the spectrophotometric method, other commonly used cfDNA quantification methods (Table 4.1) include those utilizing fluorescent dyes such as

Table 4.1 Methods of quantifying cfDNA for further downstream applications

Quantification method	Detection limit (cfDNA concentration)	Special features
Agarose gel intensity	N/A	Measures band intensity on an agarose gel
Fluorescence (DNA-binding dyes)	N/A	Utilizes fluorescence for quantification
UV absorbance (260 nm)	2 mg/mL dsDNA	Estimates absorbance at 260 nm
Hoechst 33258 fluorescence	10 pg/mL	Detects and quantifies DNA at low concentrations
PicoGreen fluorescence	25 pg/mL	Allows detection of dsDNA at low concentrations
qPCR (quantitative PCR)	N/A	High sensitivity and specificity for DNA quantification
Qubit fluorometer	10 pg/mL	High analytical sensitivity; improved tolerance to contaminants
Bioanalyzer 2100	N/A	Chip-based nucleic acid analysis system; provides fragment size information

Hoechst 33258, PicoGreen, and qPCR analysis of multi-copy genes. Hoechst 33258 enables the detection and quantification of DNA at concentrations as low as 10 pg/mL. PicoGreen, a popular method, allows detection of dsDNA at concentrations as low as 25 pg/mL. However, it underestimates the concentration of double-stranded DNA with a size less than 23 kbps when using fluorescent dyes to assess nucleic acids.

Another effective method for cfDNA quantification is qPCR, known for its high sensitivity and specificity in molecular applications. It proves to be an excellent choice for qualitative and quantitative DNA assessments. qPCR targets such as ribosomal DNA, Alu repeats of varying lengths (83–115 bp for smaller/longer fragments and 244–247 bp for longer fragments), and glyceraldehyde 3-phosphate dehydrogenase (GAPDH) of 125 bp can enhance qPCR sensitivity compared to conventional PCR. Furthermore, two advanced options for cfDNA quantification are the Qubit fluorometer and the Bioanalyzer 2100. The Qubit fluorometer offers quantification with cfDNA sample concentrations as low as 10 pg/mL, providing high analytical sensitivity, throughput, and improved tolerance to contaminants. This approach is increasingly being adopted as a best practice for next-generation sequencing applications. The second option is the Bioanalyzer 2100 system by Agilent Technologies, which utilizes microfluidic technology to pass 1 mL of the sample through a microchannel. The result is presented as an electropherogram, offering information about the size range of fragments in the nucleic acid.

4.6 Comparison of Kits and Manual Methods

Three spin column-based kits—DNA Blood Mini Kit (DBM), DSP Virus Kit (DSP), and Circulating Nucleic Acid (CNA) Kit—were compared for isolating cffDNA from the plasma of pregnant women by [12]. The comparison involved original and optimized protocols in two study phases, assessing genomic equivalents per mL (GEq/mL) values. Results revealed a significant difference in kit performance ($p < 0.0001$), with CNA and DSP Kits yielding significantly higher GEq/mL compared to the DBM Kit. The CNA Kit consistently demonstrated higher GEq/mL values compared to the DSP Kit. The commonly used DBM Kit could be substituted with CNA or DSP Kits, proving advantageous in qualitative and quantitative tests such as gender determination and aneuploidy detection, where isolation yield impacts downstream analyses. However, the decision to switch from the DBM Kit to DSP or CNA Kits should consider cost implications, as the DBM Kit is notably more cost-effective.

Recent research has extensively evaluated cffDNA extraction methods, including the conventional Triton-Heat-Phenol (cTHP) method, DSP, and NEXTprep-Mag cfDNA Isolation Kit (NPM), from plasma [11–14]. Xue et al. highlighted cTHP's efficacy in extracting short DNA fragments (<100 bp). Keshavarz et al. also affirmed cTHP's efficiency in extracting cffDNA and cfDNA from plasma, showcasing improved DNA quantity [15]. In an international workshop comparing fetal DNA extraction methods, the DSP Kit stood out, yielding higher cffDNA from maternal plasma. The DSP Kit's use of carrier RNA significantly improved

DNA recovery by enhancing short DNA fragment affinity to the membrane column [14, 16]. Similarly, it was found that using glycogen as a co-precipitant significantly increased DNA yield during extraction, akin to the Triton-Heat-Phenol-Glycogen (THPG) method employing glycogen as a carrier RNA replacement [17]. Contrastingly, the NPM Kit exhibited lower methylated DNA quantity compared to other protocols, aligning with findings that the magnetic bead kit is inadequate for low methylated cffDNA extraction for downstream analyses. Intramethod comparison highlighted that the THPG method produced significantly higher GEq/mL values for methylated cffDNA and methylated cfDNA compared to other protocols, indicating the potential for improved methylated cfDNA enrichment. The THPG method proves simple, swift, and cost-effective, with an estimated cost per case markedly lower than that of tested commercial kits, making it a viable choice for enhancing NIPT accuracy.

The challenges posed by the low quantity of cffDNA in maternal plasma and its high molecular similarity to maternal DNA are significant hurdles to ideal extraction [18, 19]. Several studies have investigated the biochemical characteristics of circulatory fetal DNA, revealing that fetal DNA molecules are generally smaller than their maternal counterparts. This unique property of fetal DNA was utilized to precisely delineate relatively small DNA species [20]. They employed a combinatorial approach involving agarose gel electrophoresis, Southern blotting, and real-time PCR to analyze circulatory DNA and selectively enrich circulatory fetal DNA sequences. These enriched sequences can be instrumental in identifying DNA polymorphisms that might otherwise be obscured by maternal sequences within the original plasma sample.

4.7 Semiautomated and Automated Methods of cfDNA Isolation

4.7.1 Southern Blot Analysis

1. **Sample collection**
   - Collect blood from pregnant woman in EDTA coated tubes.
   - Centrifuge at 1600 *g* for 10 min.
   - Perform a second centrifugation at 16,000 *g* for 10 min. Use plasma immediately for DNA extraction.
2. **DNA extraction**
   - Extract circulatory DNA from maternal plasma.
   - Use a phenol-chloroform procedure with modification (chaotropic guanidinium isothiocyanate solution).
3. **Gel electrophoresis**
   - Separate extracted DNA on a 1.0% agarose gel.
   - Use a 100-bp ladder and HindIII-digested Lambda phage DNA for size estimation.
   - Transfer DNA to nylon membranes using capillary transfer.

4. **Southern hybridization**
   - Detect transferred DNA using the Roche® DIG labeling and detection system.
   - Utilize the Alu sequence as a hybridization probe.
   - Prehybridization at 42 °C, followed by overnight hybridization.
   - Wash at high stringency and expose to X-ray film.

4.7.2 Determination of Size Distribution Flowchart

1. Preparation of circulatory DNA
   - Collect plasma samples from pregnant women (early and third trimester) and controls.
   - Use Roche High Pure Template DNA Purification Kit with a custom-made vacuum pump for DNA isolation.
   - Elute DNA in 40 μL of elution buffer.
2. Gel electrophoresis and DNA isolation
   - Subject total circulatory DNA to agarose gel electrophoresis.
   - Cut each lane into six discrete sections based on size markers.
   - Sizes of fragments: 0.09–0.3, 0.3–0.5, 0.5–1.0, 1.0–1.5, 1.5–23, and >23 kb.
   - Extract DNA from gel sections using QIAEX®II Gel Extraction Kit.
3. Determination of proportion of DNA
   - Use TaqMan® real-time PCR assay for SRY gene (Y chromosome) and GAPDH gene (total DNA).
   - Run PCR reactions with eluted DNA.
   - Express the relative proportions of each gene as percentages

Due to the challenge of low fetal DNA presence in maternal plasma (typically <10%) during early gestation, effective extraction methods are crucial for successful downstream detection in noninvasive prenatal diagnostic testing. The Akonni Biosystems TruTip technology employs a binding matrix within a pipette tip, while Qiagen's Circulating Nucleic Acids Kit uses a spin column approach. The TruTip method incorporates an additional step to reduce the recovery of DNA fragments exceeding 600 bp from the sample, resulting in a higher proportion of smaller molecular weight DNA and effectively enriching for fetal DNA. Three distinct extraction comparison studies have been conducted, including a dilution series of fragmented DNA in plasma, a set of clinical maternal samples, and a study involving maternal samples at various time points using blood collection tubes. Both extraction methods demonstrated efficient extraction of small fragment DNA from large plasma volumes. In amended samples, the TruTip extraction method exhibited approximately 15% lower overall DNA recovery efficiency, but it yielded an 87% increase in the percentage of fetal DNA compared to the Qiagen method. Across all blinded clinical sample sets, the average increase in fetal DNA using the TruTip method was 55%. A study involving whole blood samples incubated for up to 48 h prior to processing into plasma indicated more consistent percentages of fetal DNA recoveries with the TruTip method. The extracted products were evaluated using two detection platforms, quantitative real-time PCR, and droplet digital PCR, yielding similar results for both extraction methods.

4.8 Conclusion

The discussion surrounding the methods of isolating fetal DNA from maternal circulation is crucial in understanding the practical implications and the potential challenges associated with these techniques. However, challenges persist. One primary concern is the quality of fetal DNA obtained, which raises the possibility of false positives or false negatives, having profound consequences. The sensitivity and specificity of the methods are of paramount importance, and ongoing research aims to improve these aspects to reduce the risk of misdiagnosis. Ethical considerations also play a vital role in the discussion. The ability to obtain fetal genetic information from maternal blood samples raises questions about consent, privacy, and the potential misuse of this information. Cost issues further complicate matters, as making these diagnostic tests available at minimal charges to parents is essential for accessibility. It is crucial for healthcare providers and policymakers to establish ethical guidelines and regulations to ensure the responsible and secure handling of this sensitive data and to make these diagnoses available to the common masses at affordable costs.

In conclusion, the methods discussed in this chapter represent a significant breakthrough in the field of prenatal diagnostics, offering great promise for safer and more accessible prenatal testing. However, the ongoing evolution and the need for careful consideration of ethical and clinical implications emphasize the importance of continued research and thoughtful implementation of these methodologies. The future of noninvasive prenatal testing is bright, and its continued advancement will undoubtedly benefit both patients and the medical community.

References

1. Wang Y, Springer S, Zhang M, McMahon KW, Kinde I, Dobbyn L, et al. Detection of tumor-derived DNA in cerebrospinal fluid of patients with primary tumors of the brain and spinal cord. Proc Natl Acad Sci U S A. 2015;112(31):9704–9.
2. Sriram KB, Relan V, Clarke BE, Duhig EE, Windsor MN, Matar KS, et al. Pleural fluid cell-free DNA integrity index to identify cytologically negative malignant pleural effusions including mesotheliomas. BMC Cancer. 2012;12:428.
3. Sidransky D, Von Eschenbach A, Tsai YC, Jones P, Summerhayes I, Marshall F, et al. Identification of p53 gene mutations in bladder cancers and urine samples. Science. 1991;252(5006):706–9.
4. Mithani SK, Smith IM, Zhou S, Gray A, Koch WM, Maitra A, et al. Mitochondrial resequencing arrays detect tumor-specific mutations in salivary rinses of patients with head and neck cancer. Clin Cancer Res. 2007;13(24):7335–40.
5. Mandel P, Metais P. Nuclear acids in human blood plasma. C R Seances Soc Biol Fil. 1948;142(3–4):241–3.
6. Sun K, Jiang P, Chan KCA, Wong J, Cheng YKY, Liang RHS, et al. Plasma DNA tissue mapping by genome-wide methylation sequencing for noninvasive prenatal, cancer, and transplantation assessments. Proc Natl Acad Sci U S A. 2015;112(40):E5503–12.

7. Bischoff FZ, Lewis DE, Simpson JL. Cell-free fetal DNA in maternal blood: kinetics, source and structure. Hum Reprod Update. 2005;11(1):59–67.
8. Fernando MR, Jiang C, Krzyzanowski GD, Somer-Shely T, Ryan WL. A novel approach to stabilize fetal cell-free DNA fraction in maternal blood samples for extended period of time. PLoS One. 2018;13(12):e0208508.
9. Huang DJ, Mergenthaler-Gatfield S, Hahn S, Holzgreve W, Zhong XY. Isolation of cell-free DNA from maternal plasma using manual and automated systems. Methods Mol Biol. 2008;444:203–8.
10. Bronkhorst AJ, Aucamp J, Pretorius PJ. Cell-free DNA: preanalytical variables. Clin Chim Acta. 2015;450:243–53.
11. Xue X, Teare MD, Holen I, Zhu YM, Woll PJ. Optimizing the yield and utility of circulating cell-free DNA from plasma and serum. Clin Chim Acta. 2009;404(2):100–4.
12. Repiská G, Sedláčková T, Szemes T, Celec P, Minárik G. Selection of the optimal manual method of cell free fetal DNA isolation from maternal plasma. Clin Chem Lab Med. 2013;51(6):1185–9.
13. Lim JH, Lee BY, Kim JW, Han YJ, Chung JH, Kim MH, et al. Evaluation of extraction methods for methylated cell-free fetal DNA from maternal plasma. J Assist Reprod Genet. 2018;35(4):637–41.
14. Legler TJ, Liu Z, Mavrou A, Finning K, Hromadnikova I, Galbiati S, et al. Workshop report on the extraction of foetal DNA from maternal plasma. Prenat Diagn. 2007;27(9):824–9.
15. Keshavarz Z, Moezzi L, Ranjbaran R, Aboualizadeh F, Behzad-Behbahani A, Abdullahi M, et al. Evaluation of a modified DNA extraction method for isolation of cell-free fetal DNA from maternal serum. Avicenna J Med Biotechnol. 2015;7(2):85–8.
16. Warton K, Graham LJ, Yuwono N, Samimi G. Comparison of 4 commercial kits for the extraction of circulating DNA from plasma. Cancer Genet. 2018;228–229:143–50.
17. Akbariqomi M, Heidari R, Gargari SS, Omrani MD, Rigi G, Sanikhani NS, et al. Evaluation and statistical optimization of a method for methylated cell-free fetal DNA extraction from maternal plasma. J Assist Reprod Genet. 2019;36(5):1029–38.
18. Houfflin-Debarge V, O'Donnell H, Overton T, Bennett PR, Fisk NM. High sensitivity of fetal DNA in plasma compared to serum and nucleated cells using unnested PCR in maternal blood. Fetal Diagn Ther. 2000;15(2):102–7.
19. Chan KCA, Zhang J, Hui ABY, Wong N, Lau TK, Leung TN, et al. Size distributions of maternal and fetal DNA in maternal plasma. Clin Chem. 2004;50(1):88–92.
20. Li Y, Zimmermann B, Rusterholz C, Kang A, Holzgreve W, Hahn S. Size separation of circulatory DNA in maternal plasma permits ready detection of fetal DNA polymorphisms. Clin Chem. 2004;50(6):1002–11.

Part II

Detection Strategies and Clinical Implementation

Scope of Cell- Versus Cell-Free-Based Detection Approach in Noninvasive Prenatal Screening

5

Subhas Chandra Saha and Riyaz Ahmad Rather

5.1 Introduction

Noninvasive prenatal screening (NIPS) is a screening technique that enables the detection of certain genetic conditions in the fetus. The test is conducted by utilizing a blood sample from a pregnant woman [1, 2]. Unlike invasive prenatal testing methods, NIPS poses no direct harm to the fetus or to the expectant mother [3]. This noninvasive approach has changed prenatal care by providing a safer alternative for prenatal screening.

The significance of NIPS in prenatal care is important. It furnishes crucial information about the fetus's health early in pregnancy, enabling parents and healthcare professionals to make well-informed decisions regarding additional diagnostic testing or possible interventions. Its effectiveness is particularly notable in screening for common chromosomal abnormalities [4–7]. Early detection of these conditions can help parents prepare for the birth of a child with special needs. Invasive prenatal testing methods, while widely used, are not without limitations. Procedures like amniocentesis and chorionic villus sampling involve the insertion of a needle into the uterus or to the placental tissue, respectively, to obtain fetal cells or DNA for analysis [8]. Although screened by invasive methods, the early identification of chromosomal abnormalities conditions plays a crucial

S. C. Saha
Department of Obstetrics and Gynecology, Post Graduate Institute of Medical Education and Research, Chandigarh, India

R. A. Rather (✉)
Department of Biotechnology, College of Natural and Computational Science, Wachemo University, Hossana, Ethiopia

R. A. Rather, S. C. Saha (eds.), *Non-invasive Prenatal Screening (NIPS) in Clinical Practice*, https://doi.org/10.1007/978-981-97-6402-0_5

role in helping parents prepare for the birth of a child with special needs, and it enables to prepare appropriate medical care and support. While invasive prenatal testing methods are widely used, they come with inherent limitations. They carry a small but notable risk of complications, including miscarriage or injury to the fetus [9]. The invasive nature of these tests also restricts their availability and acceptance among pregnant women, as potential risks may outweigh perceived benefits in some cases. To overcome these limitations, noninvasive approaches have been developed, primarily relying on two detection methods (Fig. 5.1): cell-based and cell-free-based approaches. Cell-based approaches involve isolating and analyzing fetal cells circulating in the maternal bloodstream [10, 11]. These cells, known as circulating fetal cells (CFCs), are relatively rare, posing technical challenges for their isolation and analysis [12]. In contrast, cell-free-based detection focuses on analyzing cell-free nucleic acids like cell-free fetal DNA (cffDNA) and cell-free fetal RNA (cffRNA) present in maternal blood [13]. This chapter aims to provide a comprehensive overview of the cell- versus cell-free-based detection approach in NIPS.

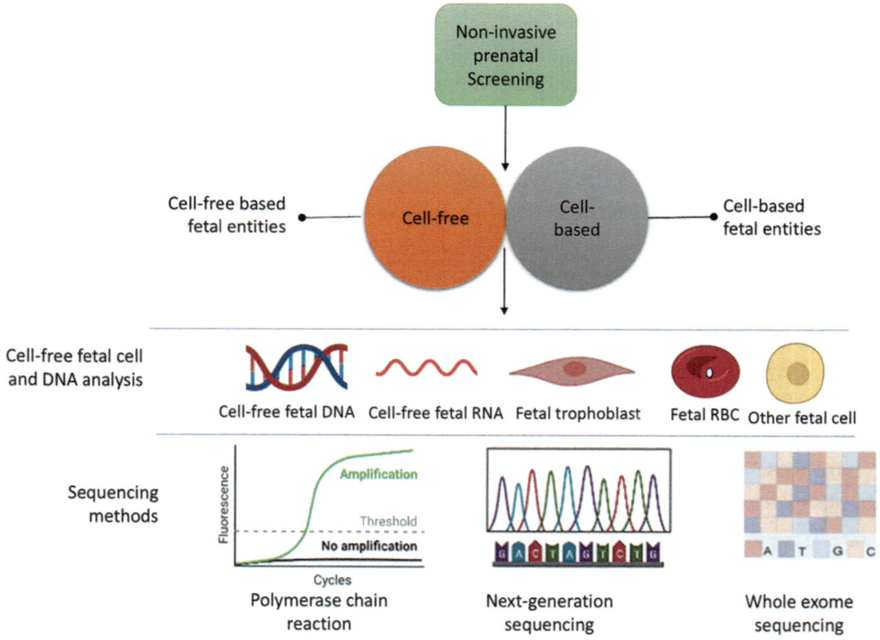

Fig. 5.1 Schematic flow of noninvasive prenatal screening (NIPS) employing both cell-free- and cell-based detection approaches. NIPS is conducted by adopting either a cell-free- or cell-based methodology, where cell-free fetal nucleic acids and fetal-originated cells are utilized. These components undergo analysis/screening through genomic techniques such as polymerase chain reaction or sequencing to interpret results and draw conclusions

5.2 Cell-Based Detection Approach

5.2.1 Cell-Based Detection: Principles and Techniques

Understanding the principles and techniques of cell-based detection in NIPS is crucial for appreciating the potential of this approach. This technique involves isolating and analyzing fetal cells to identify genetic abnormalities or other chromosomal disorders, presenting a noninvasive alternative to conventional prenatal testing methods [14]. The fundamental concept behind cell-based detection lies in the fact that fetal cells circulating in the maternal bloodstream harbor the same genetic information as the developing fetus [15]. Analyzing these cells provides valuable insights into the fetal genome and its condition. Different techniques are employed to isolate and analyze these cells, ensuring the accuracy and reliability of the results [16].

5.2.2 Sample Collection and Processing

Sample collection and processing play pivotal roles in the efficacy of cell-based detection methods, ensuring the isolation and preservation of fetal cells in maternal blood for subsequent analysis [17]. Maternal blood sample collection is a relatively simple and minimally invasive procedure. In this process, a blood sample is drawn from the expectant mother using a syringe or needle. Specialized tubes with anticoagulants are utilized to prevent clotting and maintain cell integrity during transportation and processing [18, 19]. Following collection, the sample undergoes processing steps to extract fetal cells from blood components. Subsequently, a density gradient centrifugation is performed, wherein the blood sample is placed on a density gradient medium and subjected to centrifugation. This process segregates blood components according to density, streamlining the isolation of the targeted fetal cells [20, 21]. Methods such as magnetic-activated cell sorting and fluorescence-activated cell sorting have also found application [22, 23]. The effectiveness and purity of fetal cell isolation are pivotal as they directly influence the subsequent processes. Following isolation, the next crucial step involves the preservation of cells. Preserving the integrity of isolated fetal cells is important for accurate analysis, necessitating careful handling and adherence to proper storage and transportation conditions. Specialized vials are used to protect the cells from temperature fluctuations, light exposure, and mechanical damage during the handling of the cell sample in the laboratory.

5.2.3 Cell-Based Detection Methods in NIPS

5.2.3.1 Fetal Nucleated Red Blood Cells

Fetal nucleated red blood cells (fnRBCs) are a unique type of blood cell found specifically in the fetal circulation. In adults, red blood cells don't have a nucleus, but fnRBCs do [24]. This unique characteristic allows for the analysis of fetal genetic material within these cells. By examining the genetic content of fnRBCs, it is possible to detect chromosomal abnormalities like aneuploidies and single gene disorders [25–27]. fnRBCs are considered rare in maternal peripheral blood, with only a few to several tens per milliliter [28]. They become detectable as early as 6 weeks of gestation and progressively increase in number as pregnancy advances, peaking around 17–18 weeks [22, 29]. The level of fnRBCs is also associated with the fetus's gender, with males generally having more than females [30]. Additionally, the fnRBC count might be linked to certain diseases. Conditions like premature birth, chronic lack of oxygen, ABO hemolysis, maternal diabetes, maternal smoking, and congenital TORCH (toxoplasma, other pathogens, rubella, cytomegalovirus, herpes simplex virus) infections have been connected to fetal hypoxia and anemia [31, 32]. These factors can lead to increased erythropoietin, causing an abnormal rise in fnRBC numbers.

Several techniques are employed for identifying and analyzing fnRBCs in NIPS. A widely used method is fluorescence in situ hybridization (FISH), which uses fluorescently labeled probes to target specific chromosome regions [33]. This method allows for the detection of aneuploidies or genetic abnormalities linked to specific chromosomal areas [34]. Polymerase chain reaction (PCR) is another approach utilized to amplify specific DNA sequences within fnRBCs, facilitating targeted genetic analysis [26, 35].

Despite the potential significance of fnRBCs in NIPS, their rarity presents a notable challenge in isolating and detecting them. fnRBCs make up only a small fraction of the total fetal cells in maternal blood, posing technical difficulties in their identification and analysis [36]. Moreover, their delicate nature and susceptibility to degradation during sample processing add complexity to their isolation and characterization [37]. To overcome these challenges, ongoing research focuses on improving the sensitivity and efficiency of techniques for fnRBC isolation and analysis. Genomic advances, including microfluidic platforms and high-resolution imaging, show promise in improving the detection and characterization of fnRBCs.

5.2.3.2 Trophoblast Cells

Trophoblast cells are specialized cells that form the outer layer of the blastocyst, which later develops into the placenta [38]. They play a vital role in establishing and maintaining the connection between the maternal and fetal circulations. Trophoblast cells are found in maternal blood as a result of their natural shedding from the placenta [39, 40]. Examining trophoblast cells allows the identification of fetal genetic material without the need for invasive techniques. Various methods are in place for the isolation and analysis of trophoblast cells. For instance, immunomagnetic separation is a common method used for their separation. Immunomagnetic technique relies on the use of specific antibodies attached to magnetic beads [41]. These antibodies target antigens expressed on the surface of trophoblast cells, allowing for their selective isolation. This technique offers a high level of purity and enables the subsequent analysis of the isolated trophoblast cells. Once isolated, trophoblast cells are subjected to various genetic analysis methods like FISH. In addition to

FISH, methods such digital PCR and next-generation sequencing are applied to analyze trophoblast cells [42–44].

While trophoblast cells hold great promise, there are challenges associated with their isolation and analysis. The heterogeneity of trophoblast cells and the variability in their surface antigen expression pose difficulties in their selective isolation. Additionally, the fragility and vulnerability of these cells during sample processing can affect the quality and integrity of the genetic material obtained. Further advancements in isolation techniques, as well as the standardization of protocols, are necessary to enhance the efficiency and reliability.

5.2.3.3 Exosomes

Exosomes are nanosized vesicles released by cells into the extracellular space [45]. They are generated through the inward budding of endosomal membranes, resulting in the formation of multivesicular bodies that eventually fuse with the plasma membrane, releasing their contents into the extracellular environment. These vesicles are found in various biological fluids, including blood, urine, and amniotic fluid, making them accessible for analysis [46]. The cargo carried by exosomes, such as microRNAs (miRNAs), messenger RNAs (mRNAs), lipids, and cell-specific proteins, reflects the molecular profile of their originating cells [47, 48]. These small extracellular vesicles play a crucial role in intercellular communication. This characteristic makes exosomes an attractive source for the detection of fetal genetic material. Exosomes have emerged as a fascinating aspect of NIPS, holding great potential for the detection of fetal genetic conditions [49]. Understanding their role is essential for harnessing their diagnostic capabilities and advancing noninvasive prenatal test methods. By analyzing the genetic content within exosomes, it is possible to gain insights into the fetal genome and detect potential chromosomal abnormalities or genetic disorders. Numerous studies have highlighted the diagnostic potential of exosomes. In a study by Salomon et al., the presence of fetal exosomes in maternal plasma was investigated, revealing specific miRNAs from placental exosomes that could serve as biomarkers for detecting pregnancy-associated conditions, such as gestational diabetes [50]. Another study by Lu et al. explored the use of exosomal RNA in maternal blood for detecting preeclampsia. Differentially expressed mRNAs in exosomes from pregnant women with preeclampsia symptoms were identified, highlighting the potential of exosomal RNA analysis for preeclampsia screening [51].

The benefits of employing exosomes for NIPS are apparent. These vesicles exhibit stability in diverse biological fluids, safeguarding the enclosed cargo from degradation. Additionally, exosomes present the advantage of accessibility through minimally invasive procedures like blood sampling. Despite their promising potential, several challenges must be tackled. The technical demands of isolating and purifying exosomes from complex biological fluids arise from their heterogeneity and the existence of other extracellular vesicles. Ensuring the accuracy and reliability, exosomal analysis necessitates the standardization of isolation protocols and the development of efficient purification techniques.

5.2.3.4 Circulating Fetal Cells

CFCs are fetal cells that enter the maternal circulation during pregnancy. These cells are isolated from maternal blood and analyzed for their genetic content, enabling the detection of chromosomal anomalies [22]. They include cell types, such as stem cells, lymphocytes, or progenitor cells, besides nucleated red blood cells and trophoblast cells. These cells originate from different tissues of the developing fetus. Several studies have explored their diagnostic potential [52, 53]. In certain studies, the detection of aneuploidy through intact fetal cell analysis in maternal blood has demonstrated a sensitivity of approximately 75%, with an estimated false-positive rate ranging from 0.6 to 4.1% [54]. The National Institute of Child Health and Human Development conducted a comprehensive national study in 1994 to assess the feasibility of using fetal cells in maternal blood for the diagnosis of fetal chromosomal abnormalities. Interim data analysis on 2744 maternal samples was performed after the first 5 years of the study [55]. When the fetus had a normal chromosomal count (euploid), fetal male cells were correctly identified in 41.4% of cases ($n = 1292$). The detection rate increased to 74.4% in cases where aneuploidy was confirmed. These findings imply that free fetal cells are relatively more abundant and can be isolated more easily after enrichment methods. Furthermore, a case involving a mother with a balanced translocation (46,XX,t(1;6)(p31;q14)) was examined in a report, where FISH was employed on retrieved fetal cells to rule out fetal aneuploidy [56]. The results illustrated the potential of utilizing CFCs for the noninvasive detection of these genetic abnormalities.

In another retrospective study conducted by Liao et al., on 55 subjects, all 55 cases of fetal trisomy 21 were successfully identified using the semiconductor sequencing platform (SSP) method, with a remarkable sensitivity of 99.94% and a specificity of 99.46%. The SSP technique also exhibited excellent performance in detecting cases of trisomy 18, with a sensitivity of 100% and a specificity of 99.24% for the 16 identified cases. Additionally, the method achieved perfect sensitivity and specificity in identifying the three cases of trisomy 13. Moreover, the SSP method successfully detected 15 cases of sex chromosome aneuploidies, including 10 cases of 45,X, 2 cases of 47,XYY, 2 cases of 47,XXX, and 1 case of 47,XXY [57]. Additionally, placenta plays a vital role in supporting the growth and development of the fetus. In addition to being a valuable source of regenerative progenitor cells, recent studies have shown that placenta originated cells have the potential to promote regenerative and reparative processes. The placenta originated cells are composed of progenitor cells, and emerging research suggests that these components possess regenerative properties. This newfound understanding highlights the multifaceted nature of CFCs, as both its cellular and cell-free components contribute to its regenerative potential [58].

Despite the potential of CFCs in NIPS, their rarity and technical challenges with their isolation and analysis present obstacles. The development of robust and standardized techniques for CFC isolation, such as immunomagnetic separation and fluorescence-activated cell sorting, is crucial for efficient and reliable analysis. Further research and technological advancements are necessary to optimize their detection and utilization.

5.2.4 Advantages and Challenges of Cell-Based Detection Approach

The use of cell-based detection presents various advantages along with certain challenges. It is imperative to understand these aspects to evaluate the feasibility and effectiveness of this approach in prenatal care. Exploring both the benefits and difficulties associated with employing cell-based detection is essential.

Cell-based detection enables the direct analysis of fetal cells found in maternal circulation or other biological samples. By examining fnRBCs, trophoblast cells, and other CFCs, it becomes feasible to directly access and analyze the fetal genetic makeup. This offers a comprehensive view of the fetal genome, facilitating the detection of chromosomal abnormalities. Vital issues like low abundance of fetal cells, heterogeneity of cell populations, sample processing and cell fragility, and technical expertise and resources are the key challenges in using cell-based detection approaches [34]. For instance, fetal cells are rare compared to maternal cells, making their isolation and analysis technically challenging. The low fetal cell-to-maternal cell ratio requires sensitive and efficient isolation methods to ensure reliable analysis. Additionally, the variability in the number of fetal cells among different pregnancies poses further challenges in obtaining sufficient genetic material for accurate testing. Further, the heterogeneity of cell populations in maternal circulation poses challenges. Fetal cells can vary in their origin, developmental stage, and expression of surface markers [59]. This heterogeneity requires the development of specialized isolation techniques that can selectively capture the desired fetal cell populations. The variability in cell types also necessitates the use of multiple analysis methods to cover different cell populations of interest.

The processing of samples requires careful handling to preserve the integrity and quality of fetal cells. Maintaining the viability and integrity of these cells is crucial for accurate analysis and reliable results. Standardization of sample processing protocols is essential to ensure consistency and reproducibility across different laboratories. Major challenges potentially limit the widespread implementation of cell-based detection approaches in NIPS.

5.3 Cell-Free-Based Detection Approach in NIPS

5.3.1 Cell-Free-Based Detection: Principles and Techniques

Cell-free-based detection relies on the release of fetal nucleic acids into the maternal bloodstream during pregnancy. These nucleic acids, originating from the placenta, contain genetic information from the fetus. By isolating and analyzing these fragments, it becomes possible to identify genetic variations in developing fetus [13]. Techniques in the cell-free-based detection approach, such as DNA extraction, targeted sequencing, single nucleotide polymorphism analysis, methylation analysis, and bioinformatics, are employed to process and derive insights about the genetic makeup of the fetus.

5.3.2 Cell-Free Fetal DNA: Application

cffDNA has emerged as a game-changing component of NIPS, revolutionizing prenatal diagnosis. Small pieces of fetal DNA are released into the maternal bloodstream during pregnancy, and these fragments, known as cffDNA, provide a noninvasive alternative to typical invasive testing procedures [60]. The premise behind cffDNA analysis is that fetal DNA can be discovered and examined in maternal plasma. This groundbreaking discovery has paved the way for advances in prenatal diagnosis. By isolating and analyzing cffDNA, chromosomal abnormalities, genetic mutations, and other genetic variations associated with fetus can be identified [61]. Seminal research conducted by Lo et al., who demonstrated the presence of fetal DNA in maternal plasma, laid the foundation for its analysis in prenatal screening [60]. The researchers discovered its potential for diagnosing fetal chromosomal aneuploidies such as trisomy 21 by quantifying the fraction of cffDNA in total cell-free DNA in maternal plasma. Building on this foundation, further investigations have investigated the applications of cffDNA in prenatal diagnosis. Rather et al. investigated the use of cffDNA to predict fetal RhD status. The group successfully determined the RhD status of fetuses by examining maternal plasma of RhD-negative mothers, minimizing the requirement for invasive treatments. Other research have also shown the potential of cffDNA in personalized medicine, emphasizing its importance in improving prenatal care [62].

cffDNA analysis has displayed promise in identifying specific genetic mutations and variations. Chen et al. conducted a study focusing on the detection of paternally inherited mutations [63]. Their analysis successfully identified paternally inherited mutations associated with β-thalassemia. This research highlights the potential of cffDNA in identifying inherited genetic disorders and facilitating early intervention.

In another study, Wang et al. utilized cffDNA in chromosomal microarray analysis for the detection of microdeletions and microduplications linked to genetic syndromes. Out of 24,613 pregnant women undergoing NIPS, a suspected fetal copy number variation condition was observed in 124 cases, constituting approximately 0.50% of the sample. Subsequent prenatal diagnosis with microarray analysis was conducted for 66 of these women, leading to the confirmation of true-positive results in 13 cases. The detection rate and positive predictive were estimated to be approximately 51.1% and 19.7%, respectively [25]. Furthermore, cffDNA analysis has extended beyond genetic disorders to encompass pregnancy-related complications. For instance, the use of cffDNA in predicting preeclampsia, a common pregnancy disorder characterized by high blood pressure, is evident from previous studies [64, 65]. This research underscores its potential in prenatal risk assessment and improving maternal-fetal health outcomes. Despite its immense potential, the analysis of cffDNA in NIPS does have certain limitations. The proportion of cffDNA in maternal plasma is relatively low, necessitating the use of sensitive techniques to accurately distinguish fetal DNA from maternal DNA. Continued advancements in genomic methods hold the potential to further enhance the accuracy and reliability of cffDNA-based testing.

5.3.3 Cell-Free RNA Analysis

Similar to cffDNA, cffRNA is found in the maternal bloodstream, originating from fetal-derived molecules released from the placenta [66, 67]. The genesis of cffRNA analysis lies in the recognition that maternal plasma contains a diverse repertoire of RNA molecules, including both coding and noncoding RNA [68]. These RNA molecules originate from the placenta and reflect the gene expression patterns and molecular profiles of the developing fetus. By isolating and analyzing cffRNA, insights into the health status of the fetus can be obtained. There are multiple types of cffRNA commonly studied as cell-free-based molecules in NIPS and their roles are discussed below:

mRNA: The analysis of fetal mRNA in maternal plasma has shown promise in detecting specific genetic disorders [69, 70]. For example, the detection of abnormal levels of mRNA associated with neural tube defects, such as spina bifida, can aid in the diagnosis and management of these conditions.

miRNAs: Aberrant expression of miRNAs in maternal plasma has been associated with various fetal anomalies. For instance, altered levels of specific miRNAs have been linked to congenital heart defects, neural tube defects, and chromosomal abnormalities [71]. Abnormal expression of certain RNA molecules in maternal plasma has been associated with conditions such as preeclampsia, gestational diabetes, and preterm birth. A study carried by Luizon et al. identified differentially expressed miRNAs in maternal plasma, providing potential biomarkers for the prediction and diagnosis of preeclampsia [72].

Long noncoding RNAs (lncRNAs): Dysregulated expression of lncRNAs has been observed in several fetal abnormalities. For instance, expression of certain lncRNAs has been associated with neurodevelopmental disorders, such as autism spectrum disorder [73, 74]. Luo et al. conducted a study that investigated the connection between lncRNAs and preeclampsia. The findings of the study suggest that lncRNAs NR_027457, AF085938, G36948, and AK002210 have the potential to be used as diagnostic biomarkers for preeclampsia. Additionally, the study suggests that NR_027457 and G36948 might play a role in the development and progression of preeclampsia [75].

Circular RNAs (circRNAs): circRNAs have been implicated in various diseases, including fetal diseases. In the context of fetal development, circRNAs have shown potential as biomarkers for conditions such as placenta accreta spectrum and trophoblast invasion. Their detection and characterization aids in the early diagnosis and monitoring of fetal conditions [76].

Piwi-interacting RNAs (piRNAs): While the application of piRNAs is still emerging, studies have suggested their involvement in congenital disorders and cell development. Exploration of piRNAs analysis may provide insights into the diagnosis and understanding of congenital malformations like nonsyndromic cleft lip and palate [77].

The analysis of different types of cffRNA in maternal circulation offers the potential for noninvasive and early detection of abnormalities. By identifying specific RNA biomarkers, medical practitioner can enhance prenatal care and provide appropriate interventions for fetal conditions. Continued advancements in cfRNA research will further contribute to the development of diagnostic tools for improved fetal disease management.

5.3.4 Other Cell-Free Biomarkers in NIPS

Ongoing research has uncovered other promising biomarkers with significant potential for noninvasive fetal diagnosis. For instance, biomarkers such as cell-free placental proteins and cell-free placental DNA methylation markers [78] are emerging ones. These biomarkers offer new avenues for understanding placental function, predicting pregnancy complications, and detecting fetal conditions. Further investigation is needed to validate their effectiveness and establish standardized protocols for their clinical relevance.

5.3.5 Advantages and Challenges of Cell-Free-Based Detection Approach

The cell-free-based detection approach offers several advantages but also presents unique challenges. Understanding these aspects is crucial for evaluating the feasibility and effectiveness of this approach. Here, we describe the advantages and challenges associated with cell-free-based detection in NIPS.

Advantages

1. *Noninvasive*: One of the primary advantages is its noninvasive nature. It eliminates the need for invasive procedures, which carry a significant risk, including miscarriage. Instead, cell-free-based detection utilizes maternal blood samples, making it a safer and more comfortable option for pregnant women [2].
2. *Early detection*: Cell-free-based detection allows for early detection of fetal abnormalities [5]. By analyzing cell-free biomarkers such as cffDNA, cffRNA, or other emerging biomarkers, it becomes possible to detect genetic disorders, chromosomal abnormalities, and other fetal conditions at an early stage of pregnancy. Detecting issues early allows for well-informed decisions about the pregnancy.
3. *High accuracy*: The cell-free-based detection approach has demonstrated high accuracy in detecting fetal abnormalities [7]. Advances in sequencing technologies and bioinformatics algorithms have significantly improved the sensitivity and specificity of NIPS. By utilizing techniques, such as next-generation sequencing and targeted sequencing, cell-free-based detections provide reliable and precise results.
4. *Wide range of applications*: Cell-free-based detection has a wide range of applications in prenatal screening. It can detect specific genetic mutations associated with inherited disorders, providing valuable information about the fetus's health and potential medical interventions.

Challenges

1. *False-positive and false-negative results*: Despite its high accuracy, cell-free-based detection is not foolproof. False results can occur, leading to potential anxiety or unnecessary medical interventions for expectant parents [79]. The presence of maternal genetic variations, mosaicism, or technical limitations in sample processing and analysis contribute to these errors. It is essential to provide proper counseling and follow-up testing to confirm positive results and address any concerns or uncertainties.
2. *Limited diagnostic scope*: While cell-free-based detection is effective in detecting common chromosomal abnormalities and genetic mutations, it has limitations in detecting rare genetic disorders or structural abnormalities [80]. Some fetal conditions may require additional diagnostic methods, such as ultrasound imaging or invasive procedures, to provide a comprehensive assessment.
3. *Cost and accessibility*: The cost can be a significant challenge for widespread implementation. Advanced sequencing technologies and analysis methods contribute to the overall cost of the testing. Additionally, the availability and accessibility of NIPS may vary across different regions or healthcare systems, limiting its widespread use.
4. *Ethical considerations*: The use of cell-free-based detection raises ethical considerations, particularly regarding the potential for incidental findings [81].

5.4 Comparative Analysis of Cell- Versus Cell-Free-Based Detection Approaches

Both methods have their advantages and limitations, making it essential to understand their similarities and differences (Table 5.1). The two methods vary in sensitivity and specificity, clinical utility, cost-effectiveness, diagnostic accuracy, and potential for false-positive/false-negative results. Here, we provide a comparison of these approaches to shed light on their respective strengths and challenges.

Cell-based detection and cell-free-based detection share the common goal of identifying genetic abnormalities and fetal conditions during pregnancy [13]. However, their methodologies and sample sources differ significantly. Cell-based detection involves the isolation and analysis of fetal cells from maternal blood. In contrast, cell-free-based detection focuses on analyzing cell-free fetal nucleic acids present in maternal blood. This fundamental difference in sample sources sets the stage for various similarities and differences between the approaches.

In terms of sensitivity and specificity, both cell- and cell-free-based approaches have shown remarkable improvements over traditional screening methods. Cell-based detection methods, such as analysis of fnRBCs, offer high sensitivity due to the direct presence of fetal cells. However, they often suffer from low specificity due to the presence of maternal cells that can confound the analysis. On the other hand, cell-free-based detection approaches provide acceptable specificity but can have slightly lower sensitivity due to their relatively low abundance in maternal blood [82].

Table 5.1 Comparison of cell-based and cell-free-based detection approaches in noninvasive prenatal screening (NIPS)

Property	Cell-based detection approach	Cell-free-based detection approach
Similarities	1. Aims to detect fetal genetic abnormalities without invasive procedures	1. Contributes to the field of noninvasive prenatal screening
	2. Can provide valuable information for prenatal diagnosis and screening	2. Reduces the need for invasive procedures such as amniocentesis or chorionic villus sampling
	3. Has the potential to identify genetic disorders and chromosomal abnormalities	3. Can detect common chromosomal abnormalities, such as Down syndrome (trisomy 21)
	4. Can aid in identifying structural abnormalities in the fetus	4. Enables monitoring of fetal health and assessment of treatment response
	5. Helps in making informed decisions regarding pregnancy management	5. Allows for early detection and intervention in certain fetal conditions
Differences	1. Requires isolation and analysis of intact fetal cells	1. Analyzes cell-free fetal DNA and RNA in maternal blood or other biofluids
	2. Relies on techniques such as fetal nucleated red blood cell (fnRBC) analysis	2. Utilizes DNA extraction and sequencing methods, such as polymerase chain reaction and next-generation sequencing
	3. Faces challenges related to sample availability and potential contamination from maternal cells	3. Relies on the presence of cell-free fetal DNA and RNA in maternal blood
	4. Requires specialized equipment and expertise for isolating and analyzing intact fetal cells	4. Requires advanced sequencing technologies and bioinformatics analysis for analyzing cell-free fetal DNA and RNA
	5. Higher costs compared to the cell-free-based approach	5. Broader applicability in screening for various fetal genetic abnormalities
	6. Requires isolation and analysis of intact fetal cells	6. Analyzes cell-free fetal DNA and RNA in maternal blood
Advantages	1. Direct access to intact fetal cells for genetic analysis	1. Broader applicability in screening for common chromosomal abnormalities and genetic mutations
	2. Can provide valuable information for diagnosing specific genetic disorders or structural abnormalities	2. Allows for monitoring of fetal health and prediction of pregnancy complications
	3. Offers the potential for higher sensitivity in detecting certain fetal abnormalities	3. Can provide higher sensitivity and specificity and low false results
	4. Can aid in personalized treatment planning and informed decision-making for expectant parents	4. Allows for noninvasive screening, reducing the risks associated with invasive procedures

(continued)

Table 5.1 (continued)

Property	Cell-based detection approach	Cell-free-based detection approach
Disadvantages	1. Lower specificity compared to the cell-free-based approach	1. Slightly lower sensitivity in determining certain genetic conditions like X-linked syndromes
	2. Challenges in sample availability, isolation, and potential contamination from maternal cells	2. Potential for false-positive or false-negative results
	3. Higher costs associated with isolation and analysis of intact fetal cells	3. Requires advanced sequencing technologies and bioinformatics analysis
	4. Requires specialized equipment and expertise for cell isolation and analysis	4. Limited by the availability and quality of cell-free fetal DNA and RNA in maternal blood
	5. May not be applicable or effective in certain clinical scenarios or conditions	5. Potential limitations in detecting rare genetic mutations or structural abnormalities

Note: This table provides a general overview, and the specific advantages and disadvantages may vary based on the techniques, technologies, and advancements within each approach

The clinical utility and applicability of cell- and cell-free-based approaches vary depending on the specific scenario and the condition being targeted. Cell-based detection methods hold promise in cases where the isolation and characterization of intact fetal cells are crucial, such as the diagnosis of specific genetic disorders or the analysis of structural abnormalities [83]. Cell-free-based approaches, on the other hand, offer broader applicability and are particularly valuable for the screening of common chromosomal abnormalities, such as trisomy 21, 18, and 13 [74]. They also enable the detection of specific genetic mutations associated with inherited disorders. The choice of approach depends on the clinical context, the specific condition being targeted, and the available resources.

Cost-effectiveness is a critical consideration in healthcare decision-making. Cell-based detection methods may involve more complex and labor-intensive procedures, including cell isolation and characterization, leading to higher costs [84]. In contrast, cell-free-based approaches benefit from advances in automation, which have reduced costs over time [85]. However, it is essential to consider not only the direct costs of the testing but also the potential downstream costs associated with false-positive or false-negative results and the need for confirmatory testing. Cost-effectiveness analyses specific to each approach and condition are necessary to make informed decisions.

Both cell- and cell-free-based approaches offer improved diagnostic accuracy compared to traditional screening methods. Cell-based detection methods can provide definitive genetic information due to the direct analysis of intact fetal cells. However, false-positive results occur due to the presence of maternal cells or technical limitations in cell isolation and analysis [86]. Cell-free-based approaches, while highly accurate, can also generate false-positive or false-negative results. Factors such as the gestational age, sample quality, analytical techniques, and the presence of maternal genetic variations can influence the accuracy and potential for false results [87]. Refinement of analytical methods aims to mitigate these limitations.

5.5 Conclusion

The scope of both cell-based and cell-free-based detection approaches in NIPS is significant and holds immense promise in improving feto-maternal health. These approaches offer complementary advantages and face distinct challenges. Cell-based detection methods provide direct access to fetal genetic material and can offer valuable information in diagnosing specific genetic disorders or structural abnormalities. However, they may encounter challenges related to sample availability, isolation, and potential contamination from maternal cells. On the other hand, cell-free-based detection approaches offer broader applicability and have demonstrated remarkable utility in screening for common chromosomal abnormalities and genetic mutations associated with inherited disorders. These approaches benefit from advancements in genomic technologies, which have improved their sensitivity and specificity over time. The scope of both approaches extends beyond primary screening to include additional applications such as monitoring fetal health, assessing treatment response, and predicting pregnancy complications. The growing body of research in this field continues to explore the potential of cell- and cell-free-based detection approaches, paving the way for future advancements in prenatal screening. While these approaches have shown great promise, it is important to consider the limitations and challenges associated with each method. Factors such as cost-effectiveness, diagnostic accuracy, potential for false-positive/false-negative results, and clinical utility in different scenarios should be carefully evaluated to ensure optimal use and implementation. Continued research and development in this field will further enhance the accuracy, accessibility, and clinical value of these approaches, ultimately benefiting expectant mothers and improving prenatal care outcomes.

References

1. Lo YM, Tein MS, Lau TK, Haines CJ, Leung TN, Poon PM, et al. Quantitative analysis of fetal DNA in maternal plasma and serum: implications for noninvasive prenatal diagnosis. Am J Hum Genet. 1998;62(4):768–75.
2. Lo YM. Fetal DNA in maternal plasma: biology and diagnostic applications. Clin Chem. 2000;46(12):1903–6.
3. Pertl B, Bianchi DW. Fetal DNA in maternal plasma: emerging clinical applications. Obstet Gynecol. 2001;98(3):483–90.
4. Porreco RP, Garite TJ, Maurel K, Marusiak B, Obstetrix Collaborative Research Network, Ehrich M, et al. Noninvasive prenatal screening for fetal trisomies 21, 18, 13 and the common sex chromosome aneuploidies from maternal blood using massively parallel genomic sequencing of DNA. Am J Obstet Gynecol. 2014;211(4):365.e1–365.e12.
5. Gil MM, Quezada MS, Revello R, Akolekar R, Nicolaides KH. Analysis of cell-free DNA in maternal blood in screening for fetal aneuploidies: updated meta-analysis. Ultrasound Obstet Gynecol. 2015;45(3):249–66.
6. Deng C, Liu J, Liu S, Liu H, Bai T, Jing X, et al. Maternal and fetal factors influencing fetal fraction: a retrospective analysis of 153,306 pregnant women undergoing noninvasive prenatal screening. Front Pediatr. 2023;11:1066178.
7. Lau TK, Cheung SW, Lo PS, Pursley AN, Chan MK, Jiang F, et al. Non-invasive prenatal testing for fetal chromosomal abnormalities by low-coverage whole-genome sequencing of

maternal plasma DNA: review of 1982 consecutive cases in a single center. Ultrasound Obstet Gynecol. 2014;43(3):254–64.

8. Alfirevic Z, Navaratnam K, Mujezinovic F. Amniocentesis and chorionic villus sampling for prenatal diagnosis. Cochrane Database Syst Rev. 2017;9(9):CD003252.
9. Salomon LJ, Sotiriadis A, Wulff CB, Odibo A, Akolekar R. Risk of miscarriage following amniocentesis or chorionic villus sampling: systematic review of literature and updated meta-analysis. Ultrasound Obstet Gynecol. 2019;54(4):442–51.
10. Sekizawa A, Purwosunu Y, Matsuoka R, Koide K, Okazaki S, Farina A, et al. Recent advances in non-invasive prenatal DNA diagnosis through analysis of maternal blood. J Obstet Gynaecol Res. 2007;33(6):747–64.
11. Holzgreve W, Garritsen HS, Ganshirt-Ahlert D. Fetal cells in the maternal circulation. J Reprod Med. 1992;37(5):410–8.
12. Rezaei M, Winter M, Zander-Fox D, Whitehead C, Liebelt J, Warkiani ME, et al. A reappraisal of circulating fetal cell noninvasive prenatal testing. Trends Biotechnol. 2019;37(6):632–44.
13. Rather RA, Saha SC. Reappraisal of evolving methods in non-invasive prenatal screening: discovery, biology and clinical utility. Heliyon. 2023;9(3):e13923.
14. Sadeghi S, Rahaie M, Ostad-Hasanzadeh B. Nanostructures in non-invasive prenatal genetic screening. Biomed Eng Lett. 2021;12(1):3–18.
15. Posey JE, O'Donnell-Luria AH, Chong JX, Harel T, Jhangiani SN, Coban Akdemir ZH, et al. Insights into genetics, human biology and disease gleaned from family based genomic studies. Genet Med. 2019;21(4):798–812.
16. Vona G, Béroud C, Benachi A, Quenette A, Bonnefont JP, Romana S, et al. Enrichment, immunomorphological, and genetic characterization of fetal cells circulating in maternal blood. Am J Pathol. 2002;160(1):51–8.
17. Bohmer RM, Stroh HP, Johnson KL, LeShane ES, Bianchi DW. Fetal cell isolation from maternal blood cultures by flow cytometric hemoglobin profiles. Results of a preliminary clinical trial. Fetal Diagn Ther. 2002;17(2):83–9.
18. Ordoñez E, Rueda L, Cañadas MP, Fuster C, Cirigliano V. Evaluation of sample stability and automated DNA extraction for fetal sex determination using cell-free fetal DNA in maternal plasma. Biomed Res Int. 2013;2013:195363.
19. Prieto B, Alonso R, Paz A, Cándenas M, Venta R, Ladenson JH, et al. Optimization of nucleated red blood cell (NRBC) recovery from maternal blood collected using both layers of a double density gradient. Prenat Diagn. 2001;21(3):187–93.
20. Al-Mufti R, Hambley H, Farzaneh F, Nicolaides KH. Assessment of efficacy of cell separation techniques used in the enrichment of foetal erythroblasts from maternal blood: triple density gradient vs. single density gradient. Clin Lab Haematol. 2004;26(2):123–8.
21. Sekizawa A, Farina A, Zhen DK, Wang JY, Falco VM, Elmes S, et al. Improvement of fetal cell recovery from maternal blood: suitable density gradient for FACS separation. Fetal Diagn Ther. 1999;14(4):229–33.
22. Sabbatinelli G, Fantasia D, Palka C, Morizio E, Alfonsi M, Calabrese G. Isolation and enrichment of circulating fetal cells for NIPD: an overview. Diagnostics (Basel). 2021;11(12):2239.
23. Emadi E, Bordbar AK, Nadri H, Shams A, Taheri-Kafrani A, Kalantar SM. Isolation of HLA-G+ cells using MEM-G/9 antibody-conjugated magnetic nanoparticles for prenatal screening: a reliable, fast and efficient method. RSC Adv. 2021;11(49):30990–1001.
24. He Z, Guo F, Feng C, Cai B, Lata JP, He R, et al. Fetal nucleated red blood cell analysis for non-invasive prenatal diagnostics using a nanostructure microchip. J Mater Chem B. 2017;5(2):226–35.
25. Wang Z, Cheng L, Sun Y, Wei X, Cai B, Liao L, et al. Enhanced isolation of fetal nucleated red blood cells by enythrocyte-leukocyte hybrid membrane-coated magnetic nanoparticles for noninvasive pregnant diagnostics. Anal Chem. 2021;93(2):1033–42.
26. Chen Y, Wu Z, Sutlive J, Wu K, Mao L, Nie J, et al. Noninvasive prenatal diagnosis targeting fetal nucleated red blood cells. J Nanobiotechnol. 2022;20(1):546. Erratum in: J Nanobiotechnology. 2023;21(1):67.

27. Wang Z, Cheng L, Wei X, Cai B, Sun Y, Zhang Y, et al. High-throughput isolation of fetal nucleated red blood cells by multifunctional microsphere-assisted inertial microfluidics. Biomed Microdevices. 2020;22(4):75.
28. Feng C, He Z, Cai B, Peng J, Song J, Yu X, et al. Non-invasive prenatal diagnosis of chromosomal aneuploidies and microdeletion syndrome using fetal nucleated red blood cells isolated by nanostructure microchips. Theranostics. 2018;8:1301–11.
29. Kuo PL. Frequencies of fetal nucleated red blood cells in maternal blood during different stages of gestation. Fetal Diagn Ther. 1998;13:375–9.
30. Alhur NF, Al Qahtani NH, AlSuhaibani ES, Alsulmi E, Almandil NB, AbdulAzeez S, et al. Fetal nucleated red blood cells preferable than cell-free fetal DNA for early determination of gender among invasive and non-invasive source using novel four genes multiplex PCR. Int J Gen Med. 2021;14:9697–5.
31. Hermansen MC. Nucleated red blood cells in the fetus and newborn. Arch Dis Child Fetal Neonatal Ed. 2001;84:F211–5.
32. Cheng L, Wei X, Wang Z, Feng C, Gong Q, Fu Y, et al. Silica microbeads capture fetal nucleated red blood cells for noninvasive prenatal testing of fetal ABO genotype. Electrophoresis. 2020;41:966–72.
33. Howe B, Umrigar A, Tsien F. Chromosome preparation from cultured cells. J Vis Exp. 2014;83:e50203.
34. Vossaert L, Chakchouk I, Zemet R, Van den Veyver IB. Overview and recent developments in cell-based noninvasive prenatal testing. Prenat Diagn. 2021;41(10):1202–14.
35. Tang Y, Tang Q, Luo H, Zhang X, Chen Q, Tang W, et al. Research progress in isolation and enrichment of fetal cells from maternal blood. J Chem. 2022;2022:1–8.
36. Kølvraa S, Christensen B, Lykke-Hansen L, Philip J. The fetal erythroblast is not the optimal target for non-invasive prenatal diagnosis: preliminary results. J Histochem Cytochem. 2005;53(3):331–6.
37. Schlütter JM, Kirkegaard I, Petersen OB, Larsen N, Christensen B, Hougaard DM, et al. Fetal gender and several cytokines are associated with the number of fetal cells in maternal blood—an observational study. PLoS One. 2014;9(9):e106934.
38. Deng Q, Chen Y, Yin N, Shan N, Luo X, Tong C, et al. N-acetylglucosaminyltransferase V inhibits the invasion of trophoblast cells by attenuating MMP2/9 activity in early human pregnancy. Placenta. 2015;36(11):1291–9.
39. Knöfler M, Haider S, Saleh L, Pollheimer J, Gamage TKJB, James J. Human placenta and trophoblast development: key molecular mechanisms and model systems. Cell Mol Life Sci. 2019;76(18):3479–96.
40. Burton GJ, Charnock-Jones DS, Jauniaux E. Regulation of vascular growth and function in the human placenta. Reproduction. 2009;138:895–902.
41. Mahajan KD, Nabar GM, Xue W, Anghelina M, Moldovan NI, Chalmers JJ, et al. Mechanotransduction effects on endothelial cell proliferation via CD31 and VEGFR2: implications for immunomagnetic separation. Biotechnol J. 2017;12(9). https://doi.org/10.1002/biot.201600750.
42. Hawes CS, Suskin HA, Petropoulos A, Latham SE, Mueller UW. A morphologic study of trophoblast isolated from peripheral blood of pregnant women. Am J Obstet Gynecol. 1994;170(5 Pt 1):1297–300.
43. Papageorgiou EA, Patsalis PC. Non-invasive prenatal diagnosis of aneuploidies: new technologies and clinical applications. Genome Med. 2012;4(5):46.
44. Mao X, Liu C, Tong H, Chen Y, Liu K. Principles of digital PCR and its applications in current obstetrical and gynecological diseases. Am J Transl Res. 2019;11(12):7209–22.
45. Pegtel DM, Gould SJ. Exosomes. Annu Rev Biochem. 2019;88:487–514.
46. Kalluri R, LeBleu VS. The biology, function, and biomedical applications of exosomes. Science. 2020;367(6478):eaau6977.
47. Rashed M, Bayraktar E, Helal G, Abd-Ellah MF, Amero P, Chavez-Reyes A, et al. Exosomes: from garbage bins to promising therapeutic targets. Int J Mol Sci. 2017;18(3):538.
48. Qin J, Xu Q. Functions and application of exosomes. Acta Pol Pharm. 2014;71(4):537–43.

49. Guo XR, Ma Y, Ma ZM, Dai TS, Wei SH, Chu YK, et al. Exosomes: the role in mammalian reproductive regulation and pregnancy-related diseases. Front Physiol. 2023;14:1056905.
50. Salomon C, Scholz-Romero K, Sarker S, Sweeney E, Kobayashi M, Correa P, et al. Gestational diabetes mellitus is associated with changes in the concentration and bioactivity of placenta-derived exosomes in maternal circulation across gestation. Diabetes. 2016;65(3):598–609.
51. Lu J, Sun Y, Cao Y, Zhang Y. Small RNA sequencing reveals placenta-derived exosomal microRNAs associated with preeclampsia. J Hypertens. 2022;40(5):1030–41.
52. Huang Z, Fong CY, Gauthaman K, Sukumar P, Choolani M, Bongso A. Novel approaches to manipulating foetal cells in the maternal circulation for non-invasive prenatal diagnosis of the unborn child. J Cell Biochem. 2011;112(6):1475–85.
53. Daley R, Hill M, Chitty LS. Non-invasive prenatal diagnosis: progress and potential. Arch Dis Child Fetal Neonatal Ed. 2014;99(5):F426–30.
54. de la Cruz F, Shifrin H, Elias S, Bianchi DW, Jackson L, Evans MI, et al. Low false-positive rate of aneuploidy detection using fetal cells isolated from maternal blood. Fetal Diagn Ther. 1998;13(6):380.
55. Bianchi DW, Simpson JL, Jackson LG, Elias S, Holzgreve W, et al. Fetal gender and aneuploidy detection using fetal cells in maternal blood: analysis of NIFTY I data. National Institute of Child Health and Development Fetal Cell Isolation Study. Prenat Diagn. 2002;22(7):609–15.
56. Wang JY, Zhen DK, Zilberstein ME, Falco VM, Bianchi DW. Non-invasive exclusion of fetal aneuploidy in an at-risk couple with a balanced translocation. Mol Hum Reprod. 2000;6(2):103–6.
57. Liao C, Yin AH, Peng CF, Fu F, Yang JX, Li R, et al. Noninvasive prenatal diagnosis of common aneuploidies by semiconductor sequencing. Proc Natl Acad Sci U S A. 2014;111(20):7415–20.
58. Bowen CM, Ditmars FS, Gupta A, Reems JA, Fagg WS. Cell-free amniotic fluid and regenerative medicine: current applications and future opportunities. Biomedicines. 2022;10(11):2960.
59. Bianchi DW, Khosrotehrani K, Way SS, MacKenzie TC, Bajema I, O'Donoghue K. Forever connected: the lifelong biological consequences of fetomaternal and maternofetal microchimerism. Clin Chem. 2021;67(2):351–62.
60. Lo YM, Corbetta N, Chamberlain PF, Rai V, Sargent IL, Redman CW, et al. Presence of fetal DNA in maternal plasma and serum. Lancet. 1997;350(9076):485–7.
61. Invernizzi P, Biondi ML, Battezzati PM, Perego F, Selmi C, Cecchini F, et al. Presence of fetal DNA in maternal plasma decades after pregnancy. Hum Genet. 2002;110(6):587–91.
62. Clausen FB, Steffensen R, Christiansen M, Rudby M, Jakobsen MA, Jakobsen TR, et al. Routine noninvasive prenatal screening for fetal RHD in plasma of RhD-negative pregnant women-2 years of screening experience from Denmark. Prenat Diagn. 2014;34(10):1000–5.
63. Chen JJ, Tan JA, Chua KH, Tan PC, George E. Non-invasive prenatal diagnosis using fetal DNA in maternal plasma: a preliminary study for identification of paternally-inherited alleles using single nucleotide polymorphisms. BMJ Open. 2015;5(7):e007648.
64. Sarzynska-Nowacka U, Kosinski P, Wielgos M. Is there a future for cell-free fetal DNA tests in screening for preeclampsia? Ginekol Pol. 2019;90(1):55–60.
65. Papantoniou N, Bagiokos V, Agiannitopoulos K, Kolialexi A, Destouni A, Tounta G, et al. RASSF1A in maternal plasma as a molecular marker of preeclampsia. Prenat Diagn. 2013;33(7):682–7.
66. Jin H, Zhang Y, Fan Z, Wang X, Rui C, Xing S, et al. Identification of novel cell-free RNAs in maternal plasma as preterm biomarkers in combination with placental RNA profiles. J Transl Med. 2023;21(1):256.
67. Zhong XY, Holzgreve W, Huang DJ. Isolation of cell-free RNA from maternal plasma. Methods Mol Biol. 2008;444:269–73.
68. Tsui NB, Lo YM. A microarray approach for systematic identification of placental-derived RNA markers in maternal plasma. Methods Mol Biol. 2008;444:275–89.
69. Ng EK, Tsui NB, Lau TK, Leung TN, Chiu RW, Panesar NS, et al. mRNA of placental origin is readily detectable in maternal plasma. Proc Natl Acad Sci U S A. 2003;100(8):4748–53.
70. Ferguson-Smith MA. Placental mRNA in maternal plasma: prospects for fetal screening. Proc Natl Acad Sci U S A. 2003;100(8):4360–2.

71. Cai M, Kolluru GK, Ahmed A. Small molecule, big prospects: microRNA in pregnancy and its complications. J Pregnancy. 2017;2017:6972732.
72. Luizon MR, Conceição IMCA, Viana-Mattioli S, Caldeira-Dias M, Cavalli RC, Sandrim VC. Circulating microRNAs in the second trimester from pregnant women who subsequently developed preeclampsia: potential candidates as predictive biomarkers and pathway analysis for target genes of miR-204-5p. Front Physiol. 2021;12:678184.
73. van de Vondervoort II, Gordebeke PM, Khoshab N, Tiesinga PH, Buitelaar JK, Kozicz T, et al. Long non-coding RNAs in neurodevelopmental disorders. Front Mol Neurosci. 2013;6:53.
74. Nagy B. Cell-free nucleic acids in prenatal diagnosis and pregnancy-associated diseases. EJIFCC. 2019;30(2):215–23.
75. Luo X, Li X. Long non-coding RNAs serve as diagnostic biomarkers of preeclampsia and modulate migration and invasiveness of trophoblast cells. Med Sci Monit. 2018;24:84–91.
76. Shi J, Yang M, Cao X, Huang Q, He F, Peng Y, et al. Identification of circular RNAs as regulators and noninvasive biomarkers for placenta accreta spectrum. Reproduction. 2023;165(4):417–30.
77. Jia S, Zhang Q, Wang Y, Wang Y, Liu D, He Y, et al. PIWI-interacting RNA sequencing profiles in maternal plasma-derived exosomes reveal novel non-invasive prenatal biomarkers for the early diagnosis of nonsyndromic cleft lip and palate. EBioMedicine. 2021;65:103253.
78. Tsui DW, Chiu RW, Lo YD. Epigenetic approaches for the detection of fetal DNA in maternal plasma. Chimerism. 2010;1(1):30–5.
79. Lin Y, Liang D, Wang Y, Li H, Liu A, Hu P, et al. Analyzing false-negative results detected in low-risk non-invasive prenatal screening cases. Mol Genet Genomic Med. 2020;8(4):e1185.
80. Deng C, Liu S. Factors affecting the fetal fraction in noninvasive prenatal screening: a review. Front Pediatr. 2022;10:812781.
81. Zaami S, Orrico A, Signore F, Cavaliere AF, Mazzi M, Marinelli E. Ethical, legal and social issues (ELSI) associated with non-invasive prenatal testing: reflections on the evolution of prenatal diagnosis and procreative choices. Genes (Basel). 2021;12(2):204.
82. Li X, Zhang D, Zhao X, Huang S, Han M, Wang G, et al. Exploration of a novel noninvasive prenatal testing approach for monogenic disorders based on fetal nucleated red blood cells. Clin Chem. 2023;69(12):1396–408.
83. Doffini A, Forcato C, Mangano C, Lattuada D, Aversa R, Maranta C, et al. Isolation of single circulating trophoblasts from maternal circulation for noninvasive fetal copy number variant profiling. Prenat Diagn. 2023;43(1):14–27.
84. Pensold D, Zimmer-Bensch G. Methods for single-cell isolation and preparation. Adv Exp Med Biol. 2020;1255:7–27.
85. Hess JF, Kohl TA, Kotrová M, Rönsch K, Paprotka T, Mohr V, et al. Library preparation for next generation sequencing: a review of automation strategies. Biotechnol Adv. 2020;41:107537.
86. Lüthgens K, Häbig K, Sonek J, Kagan KO. Screen-positive rate in cell free DNA screening for trisomy 21. Prenat Diagn. 2023;43(12):1536–43.
87. Wei L, Zhang J, Shi N, Luo C, Bo L, Lu X, et al. Association of maternal risk factors with fetal aneuploidy and the accuracy of prenatal aneuploidy screening: a correlation analysis based on 12,186 karyotype reports. BMC Pregnancy Childbirth. 2023;23(1):136.

6 Advanced Genomic Approaches in Noninvasive Prenatal Screening: The Road from Conventional Karyotyping to Next-Generation Sequencing

Anja Weise and Thomas Liehr

6.1 Introduction

6.1.1 Definition of Noninvasive Prenatal Screening

Noninvasive prenatal screening (NIPS) serves to assess the likelihood of a developing fetus having a chromosomal disorder. As its name suggests, it functions as a screening rather than a diagnostic test, offering insight into whether the baby is likely affected or unaffected by the tested genetic conditions. Should NIPS indicate an elevated risk for a specific chromosomal abnormality, further diagnostic testing of fetal tissue, such as amniocentesis or umbilical cord blood testing, is recommended. This is necessary because NIPS analyzes placenta-derived DNA, which may differ from the fetal genetic makeup in 2–5% of cases [1–3].

6.1.2 Importance of NIPS in Prenatal Care

NIPS assesses placenta-derived cell-free fetal DNA (cffDNA), which circulates in maternal blood exclusively during pregnancy, along with maternal cell-free DNA (cfDNA). Typically accessible from the tenth week of pregnancy onward, NIPS can be offered as a primary screening option, preceding other measures. It is particularly recommended for assessing autosomal trisomies associated with viable pregnancies, namely, trisomies 13, 18, and 21. However, in some countries like Germany, fetal gender determination before the 12th week of gestation is restricted, necessitating separate consideration for NIPS regarding gonosomal trisomies in each case.

A. Weise (✉) · T. Liehr
Jena University Hospital, Friedrich Schiller University, Institute of Human Genetics, Jena, Germany
e-mail: Anja.Weise@med.uni-jena.de

R. A. Rather, S. C. Saha (eds.), *Non-invasive Prenatal Screening (NIPS) in Clinical Practice*, https://doi.org/10.1007/978-981-97-6402-0_6

Additionally, the false-positive rate for sex chromosomal aberrations exceeds 50%. Guidelines advise against using NIPS to screen for rare autosomal trisomies or microdeletion/duplication syndromes due to the less reliable nature of the results. In regions lacking comprehensive prenatal sonographic screening for the entire pregnant population, NIPS holds greater significance for prenatal care compared to Western countries with more extensive prenatal services [1–3].

6.1.3 A Brief History of NIPS Development

One hundred thirty years ago, Walter Flemming made the pioneering discovery of chromatin, laying the foundation for the field of cytogenetics. However, it was not until 1956 that standard cytogenetic diagnostics, involving the examination of human chromosomes for numerical and structural abnormalities, became feasible after the accurate determination of the modal number of human chromosomes as 46 [4, 5]. This breakthrough was particularly significant for prenatal diagnostics, as common numerical chromosomal aberrations such as trisomies were detected shortly thereafter, between 1956 and the early 1960s [5].

The limitations of established genetic diagnostics have been a key driver for the development of new methodologies. Consequently, subsequent technical advancements such as fluorescence in situ hybridization (FISH), molecular karyotyping (chromosomal microarray; CMA), and next-generation sequencing (NGS) have played pivotal roles in addressing these limitations of banding cytogenetics [6]. These approaches, to varying degrees, have been instrumental in enhancing the diagnostic yields of prenatal diagnostics [7].

As the association between invasive prenatal diagnostics and a minimal risk of fetal loss persisted, the demand for noninvasive prenatal screening methods remained. Efforts to mitigate this risk included advancements in sonographic techniques for studying fetal phenotype. Additionally, noninvasive tests based on hormone levels in maternal blood were introduced. The latest breakthrough in this domain is NIPS. Utilizing readily accessible maternal blood samples, it has emerged as a relatively cost-effective and straightforward screening tool for contemporary gynecologists [1–3, 7].

Initial concerns arose due to the lack of understanding regarding the mechanisms behind false-positive and false-negative results in NIPS [8, 9]. These uncertainties, coupled with unresolved ethical dilemmas, have yet to be adequately addressed through open societal discourse. Consequently, these unresolved ethical issues are now passed on to every individual pregnant woman undergoing NIPS [3].

Despite the prevalence of high-resolution DNA-focused techniques in current human genetic diagnostics, performing banding cytogenetics remains a fundamental prerequisite and complement for the application of all other mentioned higher resolution methods, even today [6, 8]. This necessity extends to NIPS as well.

6.2 Conventional Karyotyping

6.2.1 Explanation of Conventional Karyotyping

The first numerical and structural aberrations in constitutional and acquired human syndromes and diseases were identified through classical cytogenetics. It is worth noting that neither chromosome preparation nor staining methods were well established to a high standard before the beginning of the 1970s [5]. Particularly, structural aberrations were (almost) undetectable until the introduction of banding cytogenetics in 1970 by Lore Zech [10]. Banding cytogenetics, which allows for the differentiation of distinct black-gray-white patterns along the chromosome, laid the foundation for what we now know as conventional karyotyping. This method continues to be recommended as a standard first-tier diagnostic genetic test, including in prenatal care, even today [6, 8].

Conventional karyotyping can be performed using samples obtained from the blood, bone marrow, or prenatal tissues. The involvement of dividing cells is crucial, as metaphase chromosomes cannot be obtained without them. Typically, cells are cultured before chromosomal preparation is possible. This culturing process usually lasts between 1 and 3 days for blood samples and 1–3 weeks for prenatal tissues. Once metaphases have been prepared, they can be spread onto slides using the "air-drying method." This method ensures well-spread chromosomes, and subsequent treatment with trypsin and Giemsa stain results in reproducible black and white banding [5, 11].

6.2.2 Advantages and Limitations of Conventional Karyotyping

Conventional karyotyping resolves microscopically visible numerical and large scale genomic alterations (>5–10 Mb) on a single cell level. Such chromosomal aberrations can be causative for a patients' constitutional or an acquired somatic phenotype by affecting gene dosage (aneuploidies, deletions, duplications), disrupting genes, leading to gene fusion, deregulating of gene expression by position effects, or predisposing for unbalances in offspring (translocations, inversions, insertions) [6, 11]. Besides that, unequal parental contribution of chromosomes leading to uniparental disomy (UPD) can contribute to clinical features if this involves chromosomes inheriting imprinted genes [12]. UPD, small cell mosaics, as well as submicroscopic rearrangements (microdeletions, microduplications, DNA sequence mutations, epigenetic changes) cannot be resolved by conventional cytogenetics and need subsequent stepwise diagnostics including FISH, CMA, and/or NGS. Cases with mosaic conditions should also be tested cytogenetically in other tissues [11, 13, 14].

Conventional karyotyping is a single cell directed genome-wide diagnostic tool providing information about the aberration type. Only by this approach, e.g., free trisomy 21, mosaic, and translocation trisomy 21 can be distinguished [6]. This differentiation is essential to counsel the recurrence risk in parents of a Down syndrome patient. However, as a limitation, conventional karyotyping depends on living and dividing cells. In a prenatal setting, this requires an invasive procedure like acquisition of amniotic fluid, placenta, or cord blood, which goes together with an abortion risk (as low as 0.1%) and should be offered only by specialized gynecologists [1, 15].

6.2.3 Advantages and Disadvantages of Conventional Karyotyping in Connection with NIPS

Although NIPS is based on high throughput NGS, the main difference and at the same time the main limitation compared to the invasive diagnostic methods above are the origin of the DNA. NIPS is performed on cffDNA, and results are obtained in comparison with maternal cfDNA. As already mentioned, the so-called cffDNA circulating in maternal bloodstream originates from the extra embryonal cytotrophoblast = placenta and not from fetus-derived, as the name implies falsely [1–3].

This placental origin is one of several reasons why NIPS can only be a screening and no diagnostic tool: Discordant mosaics between the placenta and fetus mainly cause false-positive and false-negative NIPS results. In addition, NIPS results can be masked by several other instances, e.g., a subclinical maternal mosaic, an acquired maternal somatic mutation, or a vanishing twin pregnancy. Furthermore, in ~3–5% of NIPS, the fetal fraction is too low due to maternal obesity or insufficient placental function, which might but must not be a marker for adverse pregnancy outcomes [1–3, 16].

Therefore, NIPS providers recommend invasive diagnostic to confirm high-risk NIPS results, which uncovers false-positive cases but misses most if not all false negative ones [1]. By focusing NIPS on trisomies 13, 18, and 21, the test misses at least 50% of aberrations that would be accessed by first trimester screening and invasive procedure based prenatal karyotyping, leaving pregnant women with a wrong sense of security [1, 17, 18].

This means conventional karyotyping and NIPS test are two completely different things—the one is an invasive diagnostic, the other a noninvasive screening test. It is not about the question if one test is better than the other. It must be decided for each case what is the question, which test is most helpful and what needs to be done. The tests rather complement each other, than they are in concurrence [8].

6.3 Advances in Genomic Approaches

Since ~50 years, there was only one major sequencing approach—the Sanger sequencing technique. It is reliable but can only analyze one relatively small DNA stretch at a time. In addition, it is laborious, time-consuming, and expensive [19].

NGS techniques dramatically reduced sequencing costs and time. When entering routine diagnostics and it thereby replaced in most diagnostic setting time intensive and low throughput sequencing of single genes by Sanger sequencing. The main reasons for this progress and triumphant march of NGS are that in one experimental setting, (1) many sequences and (2) several patients can be "tagged" and studied in parallel [19].

The most straightforward strategy in order to find causative sequence variants by NGS is to run whole-genome (WGS) or whole-exome (WES) sequencing (ideally in family trios). Since the cost efficiency of NGS got better and better within the last years, even targeted gene panels were replaced in routine diagnostics by genome-wide NGS approaches [19].

NIPS can be based on different NGS approaches—detailed in Sect. 6.5. Also some other ideas on how to combine NGS with other approaches in NIPS were published [1–3]. However, NIPS is yet technically, strictly NGS based. The common principle of all NIPS approaches is that they do not use NGS for characterizing any DNA sequences in detail. NGS in NIPS is applied to quantify DNA [19] in a chromosome-wise manner and cannot be applied for causal clarification of monogenic diseases (currently).

6.4 The Role of NGS in NIPS

NIPS is relatively inexpensive, as abovementioned advantage of NGS applies also for this approach.

6.4.1 Detection of Chromosomal Abnormalities with NGS

NIPS leverages the capability of quantifying DNA through NGS. This method stands out as the sole approach capable of detecting minute differences in DNA quantities. Within NIPS, both the gain and loss of copy numbers in cffDNA are determined, with maternal cfDNA serving as a reference. Given that the "fetal fraction," representing the amount of cffDNA compared to cfDNA, can be exceedingly small, an unconventional parameter is utilized to express the obtained results: the Z-score. A Z-score exceeding three standard deviations from the expected value for DNA fragments originating from a specific chromosome is indicative of a high-risk result for trisomy [1–3].

6.4.2 Challenges of NGS in NIPS

So far, there is the classical testing by NIPS for common viable aneuploidies like trisomy 13 (Patau syndrome, ORPHA 3378), 18 (Edwards syndrome, ORPHA 3380), and 21 (Down syndrome). Besides, there is a wide range of additional tests offered as NIPS to pregnant women. This includes NIPS for sex chromosomes, DiGeorge syndrome (OMIM #188400, ORPHA 567), other selected microdeletions and microduplications [13, 20], all other rare autosomal trisomies, genome-wide copy number variations, and/or specific sequence changes for fetal NIPS in rhesus blood group genotyping and other blood group factors [21].

Challenges to be mentioned are the following:

- NIPS accuracy for trisomy 21 is given in Europe with detection rates of 99% and a false-positive rate of 0.2% or lower [17, 18]—still, up to 10% of trisomy 21 NIPS results were false positive, when summarizing 750,000 NIPS results worldwide. Why this accuracy is already lower for trisomies 13 and 18 is still unclear [1–3].
- High NIPS false-positive rates for gonosomal aneuploidies are partly explainable by the fact that NGS data obtained is presupposing that the pregnant woman has a karyotype 46,XX. As 10% or more of pregnant women have an undetected mosaic karyotype with unknown percentage of cells with a karyotype 45,X or 47,XXX, this leads to wrong NIPS result interpretation [1–3].
- Other challenges of NIPS, like feto-placental discordance, maternal undetected malignancy, and vanishing twins, were already mentioned.
- Besides, NIPS is not recommended for twin pregnancies, when sonographic (soft) markers are detected or when considering termination of pregnancy [1–3, 22, 23].

Some of these problems can partially by solved by "smart NGS" approaches, which try to separate placental cffDNA from maternal cfDNA by the length of the fragments.

6.5 NGS Techniques Used in NIPS

NGS techniques, while crucial for NIPS, are not the focus of this chapter. For detailed information, we recommend consulting specific references [19]. It's essential to understand that NGS techniques, serving as the foundation for NIPS, can vary significantly. The NGS utilized may be based on different approaches, also known as platforms, each with its own strengths and limitations.

NGS can be performed using different methods such as targeted massive parallel sequencing (t-MPS), shotgun massive parallel sequencing (s-MPS), or single nucleotide polymorphism (SNP)-based whole-genome sequencing (WGS). It's crucial to note that only SNP-based NGS approaches allow NIPS to detect triploidy [1–3]. Other approaches for NIPS, such as NGS amplification of selected regions and microarray, are detailed elsewhere [23].

6.6 Future Directions and Conclusion

Recent studies have indicated [24] that cffDNA presents distinct characteristics, potentially aiding its differentiation from maternal cfDNA in NGS. This could pave the way for future methods to distinguish cffDNA from cfDNA. Additionally, an increasingly discussed topic in the literature is the potential application of AI to enhance NIPS detection rates [25]. However, it's important to note that while AI may offer improvements, it cannot address the fundamental biological, medical, and ethical challenges associated with NIPS.

NIPS serves as a screening rather than a diagnostic tool and is not intended to replace existing approaches. It must find its place alongside other prenatal care methods. While scientifically intriguing, its primary value lies in providing accessible and relatively inexpensive screening for countries lacking adequate sonographic equipment. However, reliance solely on NIPS for decisions regarding abortion can lead to termination of pregnancies without sufficient justification, with termination rates varying widely depending on the tested condition. Consequently, the ethical implications of NIPS remain unresolved and are often overlooked by medical professionals, policymakers, and the public.

References

1. Liehr T, Lauten A, Schneider U, Schleussner E, Weise A. Noninvasive prenatal testing (NIPT)—when is it advantageous to apply? Biomed Hub. 2017;2:458432.
2. Liehr T. Non-invasive prenatal testing, what patients do not learn, may be due to lack of specialist genetic training by gynecologists and obstetricians? Front Genet. 2021;12:682980.
3. Liehr T, Harutyunyan T, Williams H, Weise A. Non-invasive prenatal testing in Germany. Diagnostics (Basel). 2022;12:2816.
4. Tijo JH, Levan A. The chromosome number of man. Hereditas. 1956;42:1–6.
5. Liehr T. Human genetics—edition 2020: a basic training package. Epubli; 2020.
6. Liehr T. Molecular cytogenetics in the era of chromosomics and cytogenomic approaches. Front Genet. 2021;12:720507.
7. Darouich AA, Liehr T, Weise A, Schlembach D, Schleußner E, Kiehntopf M, Schreyer I. Alpha-fetoprotein and its value for predicting pregnancy outcomes—a re-evaluation. J Prenat Med. 2015;9:18–23.
8. Liehr T, Mrasek K, Klein E, Weise A. Modern high throughput approaches are not meant to replace 'old fashioned' but robust techniques. J Genet Genomes. 2017;1:e101.
9. Weise A, Mrasek K, Pentzold C, Liehr T. Chromosomes in the DNA era: perspectives in diagnostics and research. Med Genet. 2019;31:8–19.
10. Caspersson T, Zech L, Johansson C, Modest EJ. Identification of human chromosomes by DNA-binding fluorescent agents. Chromosoma. 1970;30:215–27.
11. Liehr T. Fluorescence in situ hybridization (FISH)—application guide. In: Liehr T, editor. 2nd ed. Springer; 2017. isbn:978-3662529577.
12. Liehr T. Uniparental disomy is a chromosomic disorder in the first place. Mol Cytogenet. 2022;15:5.
13. Weise A, Mrasek K, Klein E, Mulatinho M, Llerena JC Jr, Hardekopf D, Pekova S, Bhatt S, Kosyakova N, Liehr T. Microdeletion and microduplication syndromes. J Histochem Cytochem. 2012;60:346–58.
14. Liehr T, Othman MA, Rittscher K, Alhourani E. The current state of molecular cytogenetics in cancer diagnosis. Expert Rev Mol Diagn. 2015;15:517–26.
15. Wulff CB, Gerds TA, Rode L, Ekelund CK, Petersen OB, Tabor A, Danish Fetal Medicine Study Group. Risk of fetal loss associated with invasive testing following combined first-trimester screening for Down syndrome: a national cohort of 147,987 singleton pregnancies. Ultrasound Obstet Gynecol. 2016;47:38–44.

16. Krishna I, Badell M, Loucks TL, Lindsay M, Samuel A. Adverse perinatal outcomes are more frequent in pregnancies with a low fetal fraction result on noninvasive prenatal testing. Prenat Diagn. 2016;36:210–5.
17. Evans MI, Andriole S, Curtis J. The epidemic of abnormal copy number variant cases missed because of reliance upon noninvasive prenatal screening. Prenat Diagn. 2018;38:730–4.
18. Hochstenbach R, van Binsbergen E, Schuring-Blom H. A survey of undetected, clinically relevant chromosome abnormalities when replacing postnatal karyotyping by whole genome sequencing. Eur J Med Genet. 2019;62:103543.
19. Ungelenk M. Sequencing approaches. In: Liehr T, editor. Cytogenomics. Academic Press; 2021. pp. 87–122.
20. Tian W, Yuan Y, Yuan E, Zhang L, Liu L, Li Y, et al. Evaluation of the clinical utility of extended non-invasive prenatal testing in the detection of chromosomal aneuploidy and microdeletion/microduplication. Eur J Med Res. 2023;28:304.
21. Alford B, Landry BP, Hou S, Bower X, Bueno AM, Chen D, et al. Validation of a non-invasive prenatal test for fetal RhD, C, c, E, K and Fy^a antigens. Sci Rep. 2023;13:12786.
22. Benn P, Cuckle H. Overview of noninvasive prenatal testing (NIPT) for the detection of fetal chromosome abnormalities; differences in laboratory methods and scope of testing. Clin Obstet Gynecol. 2023;66:536–56.
23. Hui L, Ellis K, Mayen D, Pertile MD, Reimers R, Sun L, et al. Position statement from the International Society for Prenatal Diagnosis on the use of non-invasive prenatal testing for the detection of fetal chromosomal conditions in singleton pregnancies. Prenat Diagn. 2023;43:814–28.
24. Chiu RWK, Lo YMD. Cell-free fetal DNA coming in all sizes and shapes. Prenat Diagn. 2021;41:1193–201.
25. Lee J, Lee SM, Ahn JM, Lee TR, Kim W, Cho EH, et al. Development and performance evaluation of an artificial intelligence algorithm using cell-free DNA fragment distance for non-invasive prenatal testing (aiD-NIPT). Front Genet. 2022;13:999587.

Proteomic Biomarkers of Maternal Plasma and Their Use in Noninvasive Prenatal Testing (NIPT)

7

Indu Sharma, Anuradha Sharma, Reena Tomer, and Subhas Chandra Saha

7.1 Introduction

The phrase "cell-free fetal DNA" (cffDNA) refers to the tiny DNA fragments that are believed to be produced by the regular death of placental cells [1]. It is interesting to note that maternal plasma levels of the cffDNA increase with gestational age and can be detected in the mother's blood as early as the first trimester of pregnancy [2]. During pregnancy, cffDNA makes up approximately 3–13% of the total amount of maternal cell-free DNA [3]. The amount of cffDNA in the plasma of women who had just undergone a caesarean section was measured by Lo et al. in 1999 using polymerase chain reaction (PCR) [4].

cffDNA is pregnancy-specific as it is eliminated from the mother's plasma few hours after the placenta is delivered [5, 6]. Fetal fraction (FF) is the measure of the proportion of fetal DNA to total DNA in maternal plasma, which fluctuates and rises with gestational age. According to Wang et al. [7], the FF has a significant influence on how trustworthy NIPT outcomes are. The median value of FF at 10 weeks of gestation is roughly 10% [7, 8].

Major research conducted in the 1960s explored the potential use of analyzing the cytology of amniotic fluid to identify the gender and karyotype of the fetus. In the 1980s, the first trimester became the new standard for prenatal diagnosis instead of second trimester, owing to the widespread use of chorionic villi sampling (CVS) for fetal karyotyping [5]. Since then, amniocentesis and CVS have become increasingly popular for the prenatal diagnosis of genetic disorders. However, due to their

I. Sharma (✉) · A. Sharma · R. Tomer
Department of Zoology, Panjab University, Chandigarh, India
e-mail: indu2702@pu.ac.in

S. C. Saha
Department of Obstetrics and Gynaecology, PGIMER, Chandigrah, India

R. A. Rather, S. C. Saha (eds.), *Non-invasive Prenatal Screening (NIPS) in Clinical Practice*, https://doi.org/10.1007/978-981-97-6402-0_7

invasive nature and the potential risk of miscarriage, these treatments faced a serious disadvantage [9].

Noninvasive prenatal testing (NIPT), sometimes referred to as cell-free DNA testing or noninvasive prenatal screening (NIPS), is a significant advancement in the field of screening tests for fetal chromosomal disorders and is usually carried out by utilizing a peripheral sample of the mother's blood during pregnancy (Fig. 7.1).

NIPT, being the most recent development in prenatal diagnostics, is meant to help medical professionals manage pregnancies and provide prenatal counselling, as well as help prospective parents make thoughtful and informed decisions about their unborn child. According to Taylor-Phillips et al. [10], it is currently marketed in 61 countries in Europe, Asia, Africa, North America, and South America. The majority of NIPT are performed from the ninth to tenth week of gestation onward; if the test is done too early in the gestation, there may not be enough cffDNA in the sample, and the test may fail [10].

Approximately 1 in 150 live infants have chromosomal abnormalities, such as aneuploidy, translocations, duplications, and deletions [11]. The noninvasive diagnosis of fetal genetic disorders using cffDNA is now clinically feasible, thanks to the recent development of next-generation sequencing analytical techniques. cffDNA testing is done as a first-line screening test for maternal preference, advanced maternal age (AMA; 35 years), or in cases where fetal aneuploidy occurred in a previous pregnancy, if there was a high-risk first or second trimester screening result, or if there were markers of aneuploidy or fetal abnormalities detected by ultrasound [12, 13].

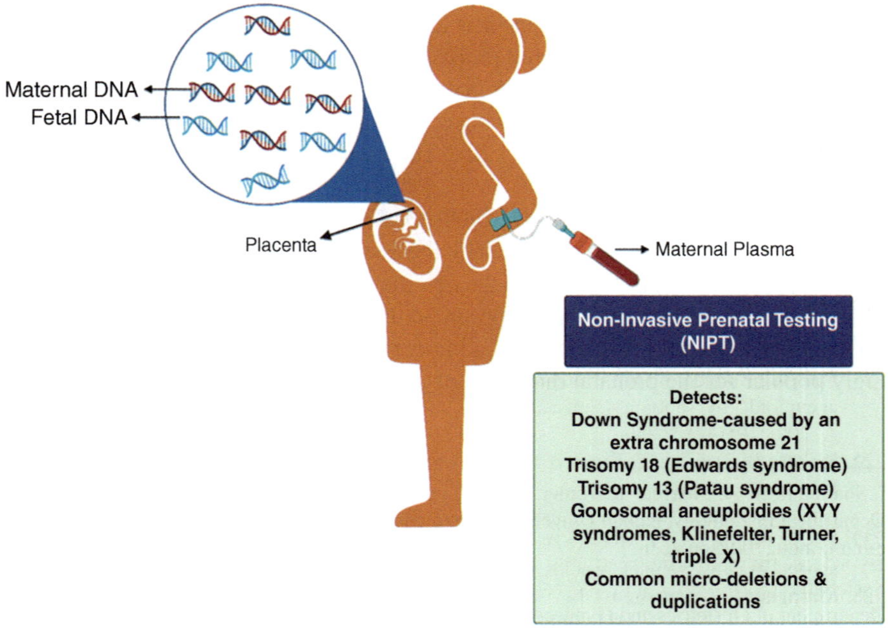

Fig. 7.1 Illustration showing the clinical benefits of noninvasive prenatal testing (NIPT)

The NIPT is now recognized as the most reliable screening test for trisomy 21, 18, and 13 and has recently been approved in twin pregnancies for the same. NIPT offers the highest reported detection rate from Down syndrome, followed by the lower detection rates for trisomy 13 and 18 [14]. Detection rates of 99%, 96%, and 91% have been reported for trisomy 21, 18, and 13, respectively, with false-positive rates of 0.09%, 0.13%, and 0.13%, respectively [15]. The recent clinical uses of NIPT include the diagnosis of structural chromosomal abnormalities and other widespread genetic diseases. There is mounting evidence that NIPT can be used in whole exome sequencing, in situations of hereditary monogenic illnesses with unidentified underlying causes [16].

The development of biomarkers based on multi-omics for prenatal diagnosis has led to substantial advancements in NIPT, which is revolutionizing the medical field. A few potential biomarkers for aneuploidy, preeclampsia, intrauterine growth retardation, and congenital anatomical defects have been identified by the researchers [17]. Proteomics-based biomarker identification for prenatal detection of fetal anomalies has advanced significantly during the past 10 years. Preterm delivery, preeclampsia, intra-amniotic infection, and fetal stress are some of the diseases and pregnancy-related disorders that have been studied using proteomics as a platform.

Therefore, the current chapter aims to present an updated review of recent research advancements in the area of proteomics-based fetal diagnosis. It also seeks to highlight the developments made in NIPT in the analysis of fetal proteins to identify illness states and contribute to our knowledge of prenatal health.

7.2 Proteomic Biomarkers of Maternal Plasma

Based on their physicochemical properties, molecular biomarkers can be classified as nucleic acids, peptides, proteins, lipids, metabolites, and other small molecules. According to the American Medical Association and the Office of Cancer Clinical Proteomics Research at the National Cancer Institute, the terms proteome and genome were combined to form the word proteome. The term *prote-* was derived from the word protein, while the suffix *-ome* was derived from genome. Consequently, proteomes are collections of proteins that are produced by many genomes and organisms [18]. Therefore, proteomics refers to the study of all the proteins that are expressed in a cell, tissue, or an individual [19]. It offers useful data on a variety of subjects, such as protein profiles, expression levels, sites of alteration, and protein interactions in pathophysiological events.

7.2.1 Proteomic Technologies Used in Biomarker Discovery

Due to the advancement of proteomic technology and its application to a variety of disorders, there has been substantial progress attained in identifying clinically useful biomarkers and novel therapeutic targets [20]. The discovery of proteins or protein patterns that can be used as markers of fetal genetic defects or other complications (Table 7.1) is the goal of the extensive research process that goes into the

Table 7.1 Proteomic biomarkers detected for various fetal chromosomal abnormalities and other pregnancy-related problems

Profile	Time of diagnosis	Proteomic biomarkers detected	References
Down syndrome	10–13 weeks	PAPP-A, AFP, β-hCG, ceruloplasmin, inter-alpha-trypsin inhibitor heavy chain H4, complement proteins C1s subcomponent, C4-A, C5, C9, and kininogen 1	Heywood et al. [21]; Kolialexi et al. [22]
Turner syndrome	15–20 weeks	Complement C1S (C1S), Complement C3 (CO3), Clusterin (CLUS), Afamin (AFAM), Hyaluronan Binding Protein 2 (HABP2), Ig alpha-1chain C region (IGHA1), CD5 antigen-like protein (CD5), Sex hormone binding globulin (SHBG), Haptoglobin (HPT), Kininogen (KNG1), Transthyretin (TTHY), and Immunoglobulin J chain (IGJ)	Kolialexi et al. [23]
Klinefelter syndrome	15–20 weeks	Androgen-binding protein (ABP), alpha-1- antitrypsin, ceruloplasmin, zinc-alpha-2-glycoprotein, apolipoprotein 1(ApoA1), plasma retinol binding protein (PRBP), gelsolin, and vitamin D binding protein	Anagnostopoulos et al. [24]
Edward syndrome	15–20 weeks	Vitamin D binding protein, alpha-antitrypsin protein (A1AT), insulin-like growth factor binding protein (IGFP-1), and transthyretin (TTR)	Vasani and Kumar [25]
Gestational diabetes mellitus	24–28 weeks	IGFBP-1, apo CIII, and apo AII	Hou et al. [26]; Singh et al. [27]; Kim et al. [28]
Congenital heart disease	18–22 weeks	LMNA, FLNA, TPM4, and ACTG1	Chen et al. [29]
Preeclampsia	After 20 weeks	APOE, apolipoprotein E; FN1, fibronectin; IGFBP4, insulin-like growth factor-binding protein 4; PDHX, pyruvate dehydrogenase protein X component; and SERPINA5, plasma serine protease inhibitor	Chen et al. [30]
Preterm PROM	Before 37 weeks	Mesothelin, IGFBP2, and placental protein 14 (PP14)	Wang et al. [31]

identification and characterization of proteomic biomarkers. To find a biomarker for a particular disease, proteomic analysis involves three key phases:

1. Protein extraction and separation
2. Protein identification
3. Protein verification [32]

The final processes are as follows:

1. Database searching
2. Evaluation of protein-protein interactions (PPIs)
3. Statistical analysis which will be discussed later in the chapter [33]

7.2.2 Proteomic Biomarkers for Chromosomal Aneuploidies

7.2.2.1 Turner Syndrome

In a study, Kolialexi and colleagues used a noninvasive prenatal diagnosis (NIPD) technique to look for potential biomarkers for Turner syndrome (TS). They examined the protein composition of samples of maternal plasma from expectant mothers carrying TS- affected fetuses with those bearing normal fetuses. They employed a proteomic approach that involves 2D-E separation of proteins and analysis of those proteins using mass spectrometry (MS) to find any proteins that displayed differential expression between the two groups.

According to the study, TS subjects had significant alterations in the levels of many proteins. It was discovered that certain proteins, including Complement C1S (C1S), Complement C3 (CO3), Clusterin (CLUS), Afamin (AFAM), Hyaluronan Binding Protein 2 (HABP2), Ig alpha-1chain C region (IGHA1), CD5 antigen-like protein (CD5), Sex hormone binding globulin (SHBG), and Haptoglobin (HPT), had significantly higher levels in TS fetuses. On the other hand, Kininogen (KNG1), Transthyretin (TTHY), and Immunoglobulin J chain (IGJ) were found to exhibit comparatively lower levels [23].

7.2.2.2 Down Syndrome

Currently, amniocentesis, which is conducted between 15 and 20 weeks of pregnancy, or CVS, which is usually done between 11 and 14 weeks of pregnancy, are the two procedures most frequently used to diagnose Down syndrome, a common chromosomal abnormality. However, it is crucial to remember that both of these procedures carry a 1% risk of miscarriage. Pregnant women are also assessed for Down syndrome using an array of markers, including plasma protein levels, ultrasound findings, and maternal age [34]. Pregnancy-associated plasma protein A (PAPP-A), alpha-fetoprotein (AFP), human chorionic gonadotropin (hCG) and its subunits, unconjugated estriol (uE3), and inhibin A are a few of the screening markers that are employed.

In a different investigation, Heywood and colleagues analyzed maternal plasma samples taken during the first and second trimesters of pregnancy from women bearing either normal or Down syndrome-affected fetuses [21]. They found that seven protein biomarkers—ceruloplasmin, inter-alpha-trypsin inhibitor heavy chain H4, complement proteins C1s subcomponent, C4-A, C5, C9, and kininogen 1—had elevated levels in the maternal plasma of Down syndrome fetuses in the second trimester. These protein biomarkers were discovered to be more highly expressed during these pregnancies when compared to control pregnancies, suggesting that they may serve as markers for Down syndrome [21].

7.2.2.3 Klinefelter's Syndrome

Klinefelter's syndrome is characterized by the presence of an extra chromosome, which results in a 47XXY pattern rather than the standard 46XY pattern in the male karyotype. In a study by Anagnostopoulos et al. [24], it was found that people with Klinefelter's syndrome have higher-than-normal levels of the follicle-stimulating hormone (FSH). Androgen-binding protein (ABP), alpha-1-antitrypsin, ceruloplasmin, and zinc-alpha-2-glycoprotein are all upregulated as a result of this increase in FSH. On the other hand, it was shown that people with Klinefelter's syndrome have downregulated levels of proteins like apolipoprotein 1 (ApoA1), plasma retinol binding protein (PRBP), gelsolin, and vitamin D binding protein. This issue can be identified between 15 and 20 weeks of pregnancy.

In their investigation, the researchers combined the techniques of matrix-assisted laser desorption/ionization time-of-flight mass spectrometry (MALDI-TOF-MS) and two-dimensional gel electrophoresis (2DE). According to the conclusion of the study, if these proteins are correctly identified, it might be a major discovery. Furthermore, a reliable technique could be developed in the future to find precursor biomarkers for Klinefelter's syndrome [24].

7.2.2.4 Edward Syndrome

Edward syndrome (sometimes referred to as trisomy 18) is a rare genetic disease which causes a variety of medical issues in a growing fetus. In a research study, experts analyzed trisomy 18-related biomarkers from the 15 to 20 weeks of pregnancy using cutting-edge methods such 2DE-DIGE combined with MALDI-TOF-MS and Western blotting [35]. According to their research, trisomy 18 was found to express a protein known as ApoA1, which has been linked to cognitive deficits. On the other hand, proteins like transthyretin (TTR), insulin-like growth factor binding protein (IGFP-1), alpha-antitrypsin protein (A1AT), and vitamin D binding protein (VDBP), were discovered to be downregulated. Nevertheless, it is crucial to highlight that this study has limitations because these proteins are also linked to cardiovascular illness or lipid metabolism, which means they may overlap with biomarkers for other diseases [25].

7.2.3 Proteomic Biomarkers for Preeclampsia

Blood samples from pregnant women with either early-onset or late-onset preeclampsia as well as samples from healthy women were studied by researchers in a study. They examined the samples using mass spectrometry and found that numerous proteins were expressed differently in preeclampsia-affected women compared to the control group. These proteins played a role in the metabolism of fat, blood coagulation, and inflammation. The study identified specific proteins as possible biomarkers for preeclampsia, including APOE, FN1, IGFBP4, PDHX, and SERPINA5 [30].

In a different study, Blankley et al. used the isobaric tagging technique to find possible biomarker proteins in blood samples from nulliparous pregnant women who later developed preeclampsia at 15 weeks of pregnancy. According to the findings, a protein known as pregnancy-specific beta-1-glycoprotein 9 (PSG9) may serve as a trustworthy biomarker for detecting preeclampsia in pregnant women [36].

Kolialexi and coworkers in their study found 12 proteins to be differentially expressed in the plasma of women who subsequently developed early-onset preeclampsia (EOPE) as compared to controls. Alpha-1-antitrypsin (A1AT), CD5 antigen-like molecule (CD5L) Keratin, type I cytoskeletal 9 (K1C9), Myeloid cell nuclear differentiation antigen (MNDA), Transferrin (TRFE), and Vitamin D-binding protein (VTDB) were upregulated, whereas Alpha-2-HS-glycoprotein (FETUA), Beta-2- glycoprotein 1 (APOH), Complement factor B (CFAB), Haptoglobin (HPT), Vitronectin (VTNC), and Zinc-alpha-2-glycoprotein (ZA2G) were downregulated [22].

7.2.4 Proteomic Biomarkers for Other Pregnancy Complications

7.2.4.1 Congenital Heart Disease

According to Llurba et al., abnormalities in heart development in congenital heart disease (CHD) are linked to an imbalance of angiogenic and antiangiogenic factors. In their initial research, they suggested to use placental growth factor (PIGF) as a first-trimester biomarker to test for valvular and conotruncal abnormalities [37]. In the second study, fetal heart, umbilical cord, and mother blood samples obtained in the second and third semesters were examined for angiogenic factors and hypoxia indicators [38]. In contrast to controls, damaged fetal cardiac samples had significantly higher levels of VEGF-A, s-Flt1, HIF-2α, HO-1, and SOD-1 mRNA expression. Surprisingly, PlGF and HIF-1α showed no difference.

Chen et al. used serum proteomic analysis to look for circulating biomarkers of fetal CHD in maternal blood. During the discovery phase, they carried out shotgun quantitative proteomics utilizing the isobaric tags for relative and absolute quantification (iTRAQ) technique to compare the protein profiles of pooled serum samples from the affected and control groups. The cytoskeleton pathway proteins LMNA, FLNA, TPM4, and ACTG1 were among a group of biomarkers they identified that were significantly lower in the maternal serum of CHD patients compared to healthy controls. Interesting results showed that the blood level of LMNA increased with gestational age in pregnant women while it decreased in nonpregnant women, suggesting a potential role in pregnancy [29].

7.2.4.2 Gestational Diabetes Mellitus

A growing number of women are now getting diagnosed with gestational diabetes mellitus (GDM), which may prove to be detrimental for both the mother and the unborn child. There is an eightfold increased risk of diabetes in offspring of mothers with GDM. Currently, an oral glucose tolerance test is used to identify GDM between the 24 and 28 weeks of pregnancy. Researchers have revealed the decreased amount of the protein IGFBP-1, which is present in the amniotic fluid of mothers with GDM. Alternately, a rise in insulin levels in the amniotic fluid has also been identified as a biomarker for GDM. According to several studies [26, 27, 39], this alteration in IGFBP-1 levels may increase IGF-1 levels and cause anomalies in metabolic processes.

In 2012, Kim and colleagues made a significant breakthrough by utilizing proteomic technology. They provided preliminary evidence in favor of the hypothesis that biomarkers detected in the maternal circulation during the first weeks of the second trimester may foretell the onset of GDM weeks before the disorder is formally diagnosed. According to the study's findings, the apo CIII and apo AII biomarkers were discovered to be linked to the emergence of GDM [28]. This finding creates opportunities for early detection and treatment of GDM, empowering medical personnel to control the condition and reduce any potential repercussions.

7.2.4.3 Cervical Insufficiency

In a case of cervical insufficiency during the second trimester of pregnancy, an emergency cerclage is the only approach to continue pregnancy and save the fetus. The most reliable method for determining whether rescue cerclage will be successful in women has long been the analysis of amniotic fluid (AF) samples obtained during amniocentesis. This examination evaluates the AF's inflammatory or infectious status and searches for indications of decidual hemorrhage [40–42]. The clinical application of second trimester amniocentesis is constrained by the fact that it is an invasive procedure that raises the possibility of membrane rupture.

Dan and coworkers carried out a groundbreaking study that focused on using plasma sample analysis as an alternative to address this constraint. They found that IGFBP-2, PSG4, and LXN plasma concentrations could function as reliable biomarkers to forecast unfavorable pregnancy outcomes after rescue cerclage. These biomarkers revealed excellent accuracy when employed as a combined multi-biomarker panel [43].

7.2.4.4 Premature Rupture of Membranes

One common pregnancy problem that frequently results in maternal and perinatal morbidity is premature rupture of the membranes (PROM). Injecting an intra-amniotic dye, such as indigo carmine, is the most precise way to diagnose PROM. However, this approach is intrusive and could result in an infection or an abortion. Mesothelin, IGFBP2, and placental protein 14 (PP14) were shown in a study to have potential value for detecting PROM [31]. With a sensitivity and specificity of 100% and 87.5%, respectively, the study's findings showed that PP14 was the best biomarker for identifying preterm PROM (PPROM), with a cutoff value of 0.008 μg/mL; however, mesothelin and IGFBP2 were found to be insufficient to diagnose PPROM.

7.3 Proteomic Biomarkers of Maternal Plasma and Their Use in NIPT

7.3.1 Methods for Proteomic Biomarker Analysis

Recent advances in molecular biotechnology have enhanced the possibility of identifying and validating biomarkers or molecules that are related to different biological or medical outcomes. The process of finding new biomarkers often involves proteomics, which is the in-depth analysis of proteins in biological material [44]. These technologies provide complicated, high-dimensional functional and imaging data, which present a number of analytical challenges that must be successfully resolved for comparative proteomic studies that can offer potential biomarkers. The protein biomarker analysis is usually done following the steps below:

7.3.1.1 Collection and Preparation

First, samples of various kinds should be taken from a group of patients. Consequently, pleural effusion, blood, tissue, saliva, urine, and tissue interstitial fluid samples are taken from each patient. Second, the pretreatment and preparation of samples vary on their type—for example, pretreatment and preparation of tissue sample distinct compared to that of blood samples. Depending on the sample, it is necessary to use various lysis buffers, saline, and digesting procedures. The surroundings and circumstances needed to preserve the samples useful for the testing also varied [45].

7.3.1.2 Extraction and Separation of Proteins

Proteomic methods for extracting and separating proteins include 2DE, laser-capture microdissection (LCM), and 2D-DIGE.

First, the method most frequently used for separating proteins was 2DE [40, 41]. Furthermore, the 2DE methodology is experiencing a surge in popularity attributed to its capacity for integration with a variety of analytical techniques and bioinformatics technologies [46]. Proteins are concentrated in the first dimension based on their isoelectric point or inherent charge. Using sodium dodecyl sulfate-polyacrylamide gel electrophoresis (SDS-PAGE), proteins are subsequently separated in the second dimension according to their relative molecular masses. This method is used to assess the composition of protein components, validate protein sample homogeneity, and purify proteins for various applications. The speed at which proteins migrate in SDS-PAGE depends on factors such as the size, shape, charge of the protein, and the pore size of the gel matrix [47]. 2DE is not suitable for effectively separating integral membrane proteins. Its utility lies in the detection of membrane-associated proteins and membrane proteins with low hydrophobicity, rather than achieving effective separation for integral membrane proteins [48–50].

Second, LCM is a technique employed in proteomic research to precisely isolate specific cells from a sample for further analysis. LCM has proven to be a dependable method of obtaining tissue cellular populations, delivering excellent sensitivity and precision for multiplex genomic and proteomic analysis. LCM has been utilized to collect data for a wide range of research concerns, including entomology, agricultural studies, embryology, metabolic disease, cardiac disease, neuroscience, infectious disease, and cancer [51]. Proteomic researchers employ a robust approach by combining LCM with 2DE to explore novel targets and biomarkers.

The technique known as two-dimensional fluorescence difference gel electrophoresis (2D-DIGE) relies on the labeling of individual protein samples with specific fluorophores (such as Cy3, Cy5, or Cy2) that covalently bind to the epsilon amino group of lysine residues. The ability to separate all three protein samples (Cy3-, Cy5-, and Cy2-labeled) on a single 2D gel is a definite advantage of the 2D-DIGE technique [52, 53].

Both one-dimensional gel electrophoresis (1DE) and 2DE are considered valuable proteomic techniques for protein extraction and separation. In 1DE, proteins are separated primarily based on their charge, while in 2DE, proteins are separated based on their molecular weight. However, these approaches do have certain limitations in terms of protein separation [54]. This work describes 2D-DIGE, which uses fluorescent labeling to segregate proteins based on their molecular weight or charge. Proteins can be seen using an MS-compatible staining process such as glutaraldehyde-free silver staining protocols, classic Coomassie blue staining, and colloidal. Coomassie staining is commonly utilized during sample preparation for subsequent mass MS analysis. Traditional silver staining methods that use glutaraldehyde to cross-link proteins to acrylamide gels are incompatible with MS [55, 56].

7.3.1.3 Identification of Proteins

Label-Free Proteomics

The signal intensity produced by the peptides in the mass spectrometer is used for label-free quantification. The accuracy and precision of proteomic data have been steadily improving, while throughput has improved, resulting in an ever-increasing influx of data. This is owing to advancements in liquid chromatography-mass spectrometry (LC-MS) speed, sensitivity, and resolution, as well as developments in data processing that benefited from increased computational power and better algorithms [57, 58]. Proteomics is widely used in many fields of study, including the investigation of biological mechanisms, the finding of biomarkers, and drug creation. Bottom-up proteomics is the most generally employed technique in LC-MS-based proteomics, in which proteins are first proteolytically degraded into peptides, which are then separated using reversed-phase high-performance liquid chromatography [59]. After fragmentation in a collision cell, the masses of the peptides are determined at the MS1 level, while the masses of the matching fragment ions are obtained at the MS2 level. The peptides in the sample can be identified by matching the precursor ions measured in MS1 and the fragment ions measured in MS2 with the theoretical masses generated from in silico digestion of the corresponding protein sequence [60]. Label-free quantitative proteomics reveals abnormal expression levels of LRG, C9, FN, A1AT, and AGP1 in the plasma of colorectal cancer patients [61].

Label-Based Proteomics

It is critical for label-free techniques to generate repeatable patterns across all samples and various time points, as well as to design software tools that can reliably align and match the associated patterns across samples. An MS/MS-based method using isobaric tags for relative and absolute quantification (iTRAQ) or tandem mass tag (TMT) is one of the most extensively utilized for labeling free amine groups at the peptide level for global comparative proteomics [62]. The main advantage of the TMT labeling method is that it allows for a high degree of multiplexing because it can analyze up to 16 samples in a single evaluation. Each mass tag's intensity reflects each sample's relative concentration of the original labeled peptide [63]. TMTs have been widely used in shotgun proteomics for cancer research on a variety of materials including cells, tissues, and bodily fluids [64]. Isobaric labeling TMT proteomics has recently been applied to investigate hepatitis B virus-related hepatocellular carcinoma (HCC) using patient samples with liver tumors and adjacent healthy tissue. Phosphoproteomic methods revealed that PYCR2 and ADH1A are linked to metabolic reprogramming in HCC and that Aldolase A phosphorylation increases glycolysis and proliferation in CTNNB1-mutated HCC cells. This research has provided mechanistic insight into the development of successful treatments for the clinical treatment of HCC [65].

Selected reaction monitoring is a targeted mass spectrometry technique that is emerging as a supplement to untargeted shotgun methods in the field of proteomics. It is especially beneficial when preset sets of proteins, such as those found in cellular networks or sets of putative biomarkers, must be evaluated consistently, reproducibly, and quantitatively.

As a quantitative proteomic approach, stable isotope labeling by amino acids in cell culture (SILAC) has grown in popularity. It is a metabolic labeling approach that involves the incorporation of isotope-labeled amino acids into proteins in vivo during translation [66]. Differentially labeled samples are mixed immediately after cell lysis and before any further processing, minimizing quantitative mistakes caused by handling distinct samples in parallel. Because it is a metabolic labeling method, SILAC has the lowest experimental variability. The isotope reagent is employed in the initial step of sample preparation (cell culture), and the labeled proteins are created by metabolic processes [49, 50].

Recently, a new approach known as neutron encoding (NeuCode) has been developed to further improve multiplexing capacity in LC-MS-based quantitative proteomics by using the modest mass differences (low milliDalton range) of distinct isotopologs for quantification. In theory, up to 39 distinct lysine isotopologs can be introduced into cell culture and multiplexed for examination [67].

Mass Spectrometry

MS is a label-free, untargeted method for identifying peptides, proteins, lipids, and metabolites that may be able to identify these molecules as coming from a particular cell. The benefits of high sensitivity and specificity, the ability to recognize and quantify a variety of compounds, and accurate structural identification have made MS an important analytic tool for proteomics [68].

A mass spectrometer has three parts: an ion generator, a mass analyzer, and an ion detection system. It utilizes matrix-assisted laser desorption/ionization (MALDI) and electrospray ionization (ESI) as ionization sources and time-of-flight (TOF), ion trap, quadrupole, and Fourier transform ion cyclotron (FTIC) as mass analyzers for protein identification [69]. MALDI involves crystallizing the sample with a matrix, exciting it with a laser, and releasing the analyte ions into the gas phase [70]. ESI ionizes the analyte from a solution using a high voltage needle spray [71].

Tandem mass spectrometry, sometimes known as MS/MS, is the combination of two separate MS separation principles. Following liquid phase separation, individual peptides that have been digested by trypsin are fragmented in tandem MS analysis [72]. Tandem MS instruments such as triple quadrupole, quadrupole ion trap, Fourier transform ion-cyclotron resonance, or quadrupole time-of-flight are employed to generate spectra of peptide fragment ions in LC-MS/MS or nanospray investigations using electrospray ionization (ESI) [73].

MALDI-TOF-MS is widely considered as a highly potent technology for protein identification in proteomic research. It particularly excels in studying high-molecular-weight proteins and offers distinct advantages and disadvantages compared to other methods. One notable advantage of MALDI-TOF-MS is its high-throughput capability, allowing for the simultaneous identification of multiple proteins [74]. This approach is also simple to implement and has good selectivity and sensitivity. The disadvantages of this approach include development restrictions, a narrow mass window range, inconsistent sample preparation, and distinct preference peaks for identical samples [75].

Surface-enhanced laser desorption/ionization time-of-flight mass spectrometry (SELDI-TOF-MS) is a high selectivity approach that can evaluate complex biological molecules even in small amounts, making it particularly useful for identifying serum-derived biomarkers. This approach utilizes ProteinChip arrays that feature surfaces with anion exchange, cation exchange, hydrophilic, or hydrophobic properties [76]. However, SELDI-TOF-MS has numerous drawbacks, including poor repeatability of results due to the usage of many chips and noise.

In addition, it is worth noting that this technique may not provide the same level of accuracy as other MS techniques commonly used in proteomics. Moreover, it is not compatible with proteins of high molecular weight [77].

Protein Microarray

Protein microarrays (or protein chips) track protein interactions and activities on a large scale. They utilize a support surface (e.g., glass slide, nitrocellulose membrane, bead, or microtiter plate) with an array of capture proteins [78]. Fluorescently labeled probe molecules are added, and their interaction with immobilized proteins produces a detectable signal using a laser scanner [79]. Protein microarrays are fast, automated, cost-effective, and sensitive, requiring minimal sample and reagent quantities [80]. While fluorescence labeling is commonly used, label-free detection technologies such as surface plasmon resonance, carbon nanotubes, carbon nanowire sensors, and microelectromechanical system cantilevers are also available [81].

Protein chips are poised to make a substantial contribution to the area of proteomics by providing an alternate method for analyzing low-abundance proteins and enabling high-throughput applications for biomarker detection. Protein chips, on the other hand, differ from conventional approaches such as 2DE or LC-MS/MS screening in that they can only provide data on a limited selection of proteins chosen by the investigator [82].

7.3.1.4 Data Analysis and Interpretation

There are various services that search for proteins using MS/MS data to uncover previously discovered or identified biomarkers. These servers include Mascot, Sequest MS-Tag, PepProb, Phenyx, Andromeda, Sonar, Uniprot, and Poptiam [83, 84].

To explore the overlap among the identifications obtained by each algorithm, peptide and protein identifications were compared using different search engines—Mascot, SEQUEST, and MaxQuant [85]. Andromeda is a new peptide search engine that employs a probabilistic scoring model. It outperforms Mascot, a popular commercial search engine, on proteome data, according to sensitivity and specificity study based on target decoy searches [86].

In gel-free approaches such as ICAT and MudPIT, samples are analyzed directly by MS, whereas in gel-based proteomics (2DE and 2D-DIGE), the protein spots are extracted from the gel and then digested with trypsin. The resultant peptides are then separated by LC or examined directly analyzed by MS. Search engines like Mascot and Sequest are utilized to match empirically obtained peptide masses with the peptide fingerprints of known proteins stored in databases [87]. The objective of this stage is to establish the physical association between identified proteins acting as biomarkers and their interactions. STRING software can be used to perform protein-protein interactions. This technique, also known as peptide mass mapping or peptide mass fingerprinting, which relies on the detection of peptide masses obtained after enzymatic digestion of either a single protein or a combination of proteins, is the initial stage in protein identification [88, 89].

7.4 Clinical Application of Proteomic Biomarkers in NIPT

7.4.1 Comparison with Current NIPT Methods

Proteomic biomarkers offer unique advantages and complement other current NIPT methods in several ways (Table 7.2). The following points compare the proteomic biomarkers with the two widely used NIPT methods: cell-free DNA (cffDNA) analysis and ultrasound-based screening:

In summary, while proteomic biomarkers face technical challenges and require further validation, they hold promise for expanding the detection capabilities of NIPT and enabling early detection and management of a wider range of fetal abnormalities.

Table 7.2 Comparison of cffDNA and ultrasound-based screening with proteomic biomarker based NIPS

Features	Cell-free DNA (cffDNA) analysis/ ultrasound-based screening	Proteomic biomarkers
Sensitivity and specificity	Primarily focuses on identifying chromosomal abnormalities	Provide additional information on fetal health, including the detection of microdeletions, microduplications, and other genetic disorders
Scope of detection	cffDNA primarily detects chromosomal abnormalities, while ultrasound screening detects fetal anatomy, growth, and position and assesses the placenta	Identify a broader range of fetal abnormalities by analyzing specific proteins associated with various genetic disorders and fetal health conditions
Technical challenges	Relies on the DNA sequencing and imaging techniques	Require advanced proteomic technologies, such as mass spectrometry or immunoassays, which may be more complex and costly
Complementary information	Ultrasound-based screening can detect structural abnormalities and assess fetal growth	Provide molecular insights into the underlying causes of certain abnormalities or identify conditions that may not be detectable through imaging alone
Clinical application	The most widely used NIPT method due to its well-established clinical utility and reliability	Still in the research and development stage; require further validation studies

7.4.2 Clinical Use of Proteomic Biomarkers in NIPT

Over the past decade, prenatal screening using cffDNA sequencing has brought about a revolution in obstetric care, significantly decreasing the need for invasive diagnostic procedures like amniocentesis for genetic abnormalities [90]. NIPT was initially employed to detect fetal trisomies, specifically Down syndrome, Edwards syndrome, and Patau syndrome [91]. However, a lot of commercial companies now provide screening for a variety of additional disorders, such as sex chromosome aneuploidies, targeted microdeletions, and rare autosomal aneuploidies, which are autosomal conditions with anomalies in the genome's copy number. NIPT has varying degrees of success in spotting various disorders [92, 93].

Screening and diagnosis during pregnancy, for neonatal genetic illnesses such as Down syndrome and thalassemia, are two applications of PCR methods for genetic diseases. Using a higher-resolution pico-droplet digital PCR (dPCR), Chang et al. developed an innovative NIPT procedure capable of predicting the genotype of a second fetus based on the known genotype of the first baby. This approach is particularly applicable to autosomal recessive monogenic disorders [94].

7.4.3 Ethical and Social Considerations

Internationally, opinions on the ethical acceptability and clinical value of various professional and medical organizations' recommendations vary [95]. Sex chromosome aneuploidies also bring up challenging ethical problems for medical professionals, potential parents, and future kids. NIPT can provide information about fetal conditions or abnormalities that may prompt decisions regarding pregnancy continuation or selective termination. The availability of this information raises ethical questions regarding reproductive autonomy, the right to make decisions about one's body and pregnancy, and the potential societal implications of selective termination based on certain conditions. Moreover, the identification of certain genetic conditions through NIPT can potentially lead to stigmatization and discrimination of individuals or communities associated with these conditions. It is important to promote understanding, tolerance, and inclusive attitudes to ensure that the use of NIPT does not reinforce negative social attitudes or contribute to discrimination against individuals with certain genetic conditions. As it is currently not integrated into publicly sponsored programs, ethical considerations surrounding this issue primarily revolve around matters concerning consumer choice. These include discussions about the limits of reproductive autonomy, the importance of information accessibility for informed consent, and the facilitation of autonomous decision-making [96].

7.5 Future Directions

7.5.1 Potential Areas of Research for Proteomic Biomarkers in NIPT

Proteomic biomarkers may improve NIPT by revealing important details regarding fetal growth and probable anomalies. The following are prospective research areas for proteomic biomarkers in NIPT [97]:

- Identification of novel protein biomarkers
- Fetal organ-specific proteomic signatures
- Screening for chromosomal abnormalities
- Prediction of preterm birth risk
- Assessment of placental health and function
- Evaluation of fetal growth and development
- Personalized medicine in NIPT

Numerous commercial tests are now accessible, due to the ongoing development and enhancement of algorithms for data sequencing and analysis (Fig. 7.2).

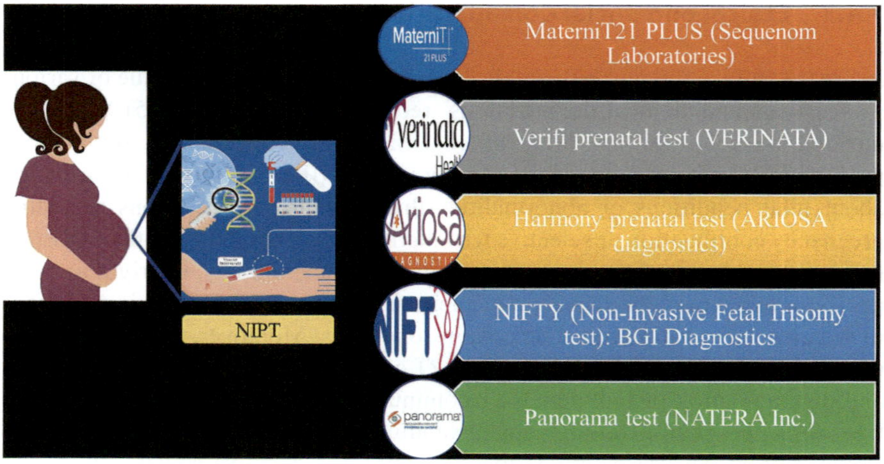

Fig. 7.2 Various commercial tests of NIPT available in the market

7.5.2 Novel Proteomic Technologies for Biomarker Discovery

Although DNA- and RNA-based technologies have enormous potential, they are unable to reliably anticipate protein expression patterns or the subsequent processing of these proteins on their own. Surprisingly, a number of studies have revealed a poor link between protein and gene expression. As a result, many research organizations and companies are working to create protein biochips to simplify proteomics using two different approaches. The first format is like DNA biochips, where immobilized recognition molecules like antibodies are arranged in high-density arrays. The binding of target proteins is observed, frequently using fluorescence. However, this method has drawbacks due to the lack of sufficient compatible binding molecules as the development of recognition molecules capable of interacting with every gene product is a difficult task.

In a different protein biochip format, proteins are immediately detected by mass MS after being captured by biochemical or intermolecular interactions. This technique is used in the Proteinchip Biomarker System, a technology created by Ciphergen Biosystems and made accessible for purchase. In order to properly compare samples from various treatment or illness groups when analyzing protein expression, it is crucial to constantly observe a certain set of proteins. Compared to other proteomic methods, the Proteinchip System offers a number of advantages. It selectively identifies a subset of proteins according to their distinct characteristics, such as their isoelectric point or hydrophobicity, using chromatographic ProteinChip arrays [98]. This novel method does not require particular recognition molecules and enables the identification of hundreds of proteins from a single location on the biochip.

Contemporary techniques like mass spectrometry and liquid chromatography have revolutionized the field of proteomics. These techniques are antibody-independent, guaranteeing great accuracy and making them suitable for in-depth studies of the human plasma proteome. Their ability to conduct dynamic testing using as little as 1 μL of plasma gives them a competitive advantage. MS-based proteomics, which allows for the detection and quantification of thousands of proteins inside a single sample of interest, has completely changed the way that research operates [99]. Geyer et al. also claimed that employing this technique, there is still an excellent opportunity to find more than 200 biomarkers among human plasma proteins [100].

In a recent study, Ngo et al. [101] developed a method for anticipating gestational age and preterm birth by examining fetal cell-free RNA transcripts isolated from maternal plasma [101]. In a different study, Liang et al. [102] showed that artificial intelligence can diagnose a pediatric disease with as much precision as the pediatricians [102]. It is intriguing to speculate that the future evaluation of fetal anomalies might be entirely transformed by the revolutionary biomarker discovery and cutting-edge machine learning.

7.5.3 Translation of Proteomic Biomarkers into Clinical Practice

Collaboration between researchers, physicians, regulatory bodies, and industry partners is essential for the dynamic and iterative process of translating proteomic biomarkers into clinical practice. It is a journey that goes through several phases of validation, standardization, and implementation with the goal of enhancing patient care and outcomes by putting proteomic knowledge to use in common clinical situations (Fig. 7.3).

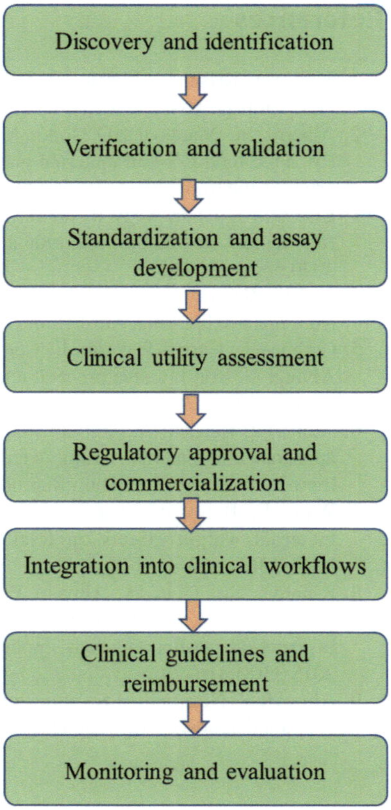

Fig. 7.3 Major steps involved in the translation of proteomic biomarkers into clinical practice

7.6 Conclusion

NIPT is typically recommended as a screening test rather than a diagnostic test. With the current advancements going on in the field, it is possible that soon NIPT may be applicable to detect a wider range of genetic disorders. Therefore, it will be crucial for healthcare professionals to offer written information to the patients regarding the test and have in-depth conversations that emphasize its drawbacks. These discussions should address a range of topics, including the test's sensitivity, false-positive rate, negative and positive predictive values, probable accidental discoveries, scheduling concerns, and the subsequent management procedures.

Proteomic research is advancing, but further research is needed to validate proteomic markers for enhancing or complementing current genetic-based techniques in NIPT. Improved sensitivity, especially with MS, enables the detection of previously undetectable analytes. Future biomarker tests will require automation and miniaturization for wide dissemination and acceptance. Discovering additional biomarkers will enhance diagnostic accuracy and enable specialized prenatal treatment approaches. However, challenges such as case heterogeneity, detecting reliable markers from mixed fetal and maternal analytes, disease overlap, and low prevalence of cases due to small sample sizes must be considered when developing new NIPT approaches.

References

1. Martin A, Krishna I, Martina B, Samuel A. Can the quantity of cell-free fetal DNA predict preeclampsia: a systematic review. Prenat Diagn. 2014;34(7):685–91.
2. Alberry M, Maddocks D, Jones M, Abdel Hadi M, Abdel-Fattah S, Avent N, Soothill PW. Free fetal DNA in maternal plasma in anembryonic pregnancies: confirmation that the origin is the trophoblast. Prenat Diagn. 2007;27(5):415–8.
3. Biró O, Rigó J Jr, Nagy B. Noninvasive prenatal testing for congenital heart disease–cell-free nucleic acid and protein biomarkers in maternal blood. J Matern Fetal Neonatal Med. 2020;33(6):1044–50.
4. Lo YD, Zhang J, Leung TN, Lau TK, Chang AM, Hjelm NM. Rapid clearance of fetal DNA from maternal plasma. Am J Hum Genet. 1999;64(1):218–24.
5. Carbone L, Cariati F, Sarno L, Conforti A, Bagnulo F, Strina I, Pastore L, Maruotti GM, Alviggi C. Non-invasive prenatal testing: current perspectives and future challenges. Genes (Basel). 2020;12(1):15.
6. Guibert J, Benachi A, Grebille AG, Ernault P, Zorn JR, Costa JM. Kinetics of SRY gene appearance in maternal serum: detection by real time PCR in early pregnancy after assisted reproductive technique. Hum Reprod. 2003;18(8):1733–6.
7. Wang E, Batey A, Struble C, Musci T, Song K, Oliphant A. Gestational age and maternal weight effects on fetal cell-free DNA in maternal plasma. Prenat Diagn. 2013;33(7):662–6.
8. Kinnings SL, Geis JA, Almasri E, Wang H, Guan X, McCullough RM, Bombard AT, Saldivar JS, Oeth P, Deciu C. Factors affecting levels of circulating cell-free fetal DNA in maternal plasma and their implications for noninvasive prenatal testing. Prenat Diagn. 2015;35(8):816–22.
9. Alfirevic Z, Navaratnam K, Mujezinovic F. Amniocentesis and chorionic villus sampling for prenatal diagnosis. Cochrane Database Syst Rev. 2017;9(9):CD003252.

10. Taylor-Phillips S, Freeman K, Geppert J, Agbebiyi A, Uthman OA, Madan J, Clarke A, Quenby S, Clarke A. Accuracy of non-invasive prenatal testing using cell-free DNA for detection of Down, Edwards and Patau syndromes: a systematic review and meta-analysis. BMJ Open. 2016;6(1):e010002.
11. Renga B. Non-invasive prenatal diagnosis of fetal aneuploidy using cell free fetal DNA. Eur J Obstet Gynecol Reprod Biol. 2018;225:5–8.
12. American College of Obstetricians and Gynecologists Committee on Genetics. Committee opinion no. 545: noninvasive prenatal testing for fetal aneuploidy. Obstet Gynecol. 2012;120(6):1532–4.
13. Kimelman D, Confino R, Confino E, Shulman LP, Zhang JX, Pavone ME. Do patients who achieve pregnancy using IVF-PGS do the recommended genetic diagnostic testing in pregnancy? J Assist Reprod Genet. 2018;35(10):1881–5.
14. Gil MM, Accurti V, Santacruz B, Plana MN, Nicolaides KH. Analysis of cell-free DNA in maternal blood in screening for aneuploidies: updated meta-analysis. Ultrasound Obstet Gynecol. 2017;50(3):302–14.
15. Santorum M, Wright D, Syngelaki A, Karagioti N, Nicolaides KH. Accuracy of first-trimester combined test in screening for trisomies 21, 18 and 13. Ultrasound Obstet Gynecol. 2017;49(6):714–20.
16. Alberry MS, Aziz E, Ahmed SR, Abdel-Fattah S. Non-invasive prenatal testing (NIPT) for common aneuploidies and beyond. Eur J Obstet Gynecol Reprod Biol. 2021;258:424–9.
17. Wang T, Xu J, Wang L, Cui X, Yan Y, Tang Q, Wu W. Prenatal diagnosis: the main advances in the application of identification of biomarkers based on multi-omics. In: Ectopic pregnancy and prenatal diagnosis; 2022.
18. Sabino F, Hermes O, Auf dem Keller U. Body fluid Degradomics and characterization of basic N-Terminome. Methods Enzymol. 2017;585:177–99.
19. Altelaar AF, Munoz J, Heck AJ. Next-generation proteomics: towards an integrative view of proteome dynamics. Nat Rev Genet. 2013;14(1):35–48.
20. Lin YH, Eguez RV, Torralba MG, Singh H, Golusinski P, Golusinski W, Masternak M, Nelson KE, Freire M, Yu Y. Self-assembled STrap for global proteomics and salivary biomarker discovery. J Proteome Res. 2019;18(4):1907–15.
21. Heywood WE, Madgett TE, Wang D, Wallington A, Hogg J, Mills K, Avent ND. 2D-DIGE analysis of maternal plasma for potential biomarkers of Down syndrome. Proteome Sci. 2011;9:56.
22. Kolialexi A, Tsangaris GT, Sifakis S, Gourgiotis D, Katsafadou A, Lykoudi A, Marmarinos A, Mavreli D, Pergialiotis V, Fexi D, Mavrou A, Papaioanou GK, Papantoniou N. Plasma biomarkers for the identification of women at risk for early-onset preeclampsia. Expert Rev Proteomics. 2017;14(3):269–76.
23. Kolialexi A, Anagnostopoulos AK, Papantoniou N, Vougas K, Antsaklis A, Fountoulakis M, Mavrou A, Tsangaris GT. Potential biomarkers for turner in maternal plasma: possibility for noninvasive prenatal diagnosis. J Proteome Res. 2010;9(10):5164–70.
24. Anagnostopoulos AK, Kolialexi A, Mavrou A, Vougas K, Papantoniou N, Antsaklis A, Kanavakis E, Fountoulakis M, Tsangaris GT. Proteomic analysis of amniotic fluid in pregnancies with Klinefelter syndrome foetuses. J Proteome. 2010;73(5):943–50.
25. Vasani A, Kumar MS. Advances in the proteomics of amniotic fluid to detect biomarkers for chromosomal abnormalities and fetomaternal complications during pregnancy. Expert Rev Proteomics. 2019;16(4):277–86.
26. Hou RL, Zhou HH, Chen XY, Wang XM, Shao J, Zhao ZY. Effect of maternal lipid profile, C-peptide, insulin, and HBA1c levels during late pregnancy on large-for-gestational age newborns. World J Pediatr. 2014;10(2):175–81.
27. Singh A, Subramani E, Datta Ray C, Rapole S, Chaudhury K. Proteomic-driven biomarker discovery in gestational diabetes mellitus: a review. J Proteome. 2015;127(Pt A):44–9.
28. Kim SM, Park JS, Norwitz ER, Lee SM, Kim BJ, Park CW, Jun JK, Kim CW, Syn HC. Identification of proteomic biomarkers in maternal plasma in the early second tri-

mester that predict the subsequent development of gestational diabetes. Reprod Sci. 2012;19(2):202–9.
29. Chen L, Gu H, Li J, Yang ZY, Sun X, Zhang L, Shan L, Wu L, Wei X, Zhao Y, Ma W, Zhang H, Cao S, Huang T, Miao J, Yuan Z. Comprehensive maternal serum proteomics identifies the cytoskeletal proteins as non-invasive biomarkers in prenatal diagnosis of congenital heart defects. Sci Rep. 2016;6:19248.
30. Chen H, Aneman I, Nikolic V, Karadzov Orlic N, Mikovic Z, Stefanovic M, Cakic Z, Jovanovic H, Town SEL, Padula MP, McClements L. Maternal plasma proteome profiling of biomarkers and pathogenic mechanisms of early-onset and late-onset preeclampsia. Sci Rep. 2022;12(1):19099.
31. Wang Y, Luo H, Che G, Li Y, Gao J, Yang Q, Zhou B, Gao L, Wang T, Liang Y, Zhang L. Placental protein 14 as a potential biomarker for diagnosis of preterm premature rupture of membranes. Mol Med Rep. 2018;18(1):113–22.
32. Liu W, Yang Q, Liu B, Zhu Z. Serum proteomics for gastric cancer. Clin Chim Acta. 2014;431:179–84.
33. Alharbi MG. Protein biomarkers in autistic children: a review. Asian Journal of Biochemistry, Genetics and Molecular Biology. 2022;12(1):1–17.
34. Bahtiyar MO, Copel JA, Mahoney MJ, Buhimschi IA, Buhimschi CS. Proteomics: a novel methodology to complement prenatal diagnosis of chromosomal abnormalities and inherited human diseases. Am J Perinatol. 2007;24(3):167–81.
35. Hsu TY, Lin H, Hung HN, Yang KD, Ou CY, Tsai CC, Cheng HH, Chung SH, Cheng BH, Wong YH, Chou AK, Hsiao CC. Two-dimensional differential gel electrophoresis to identify protein biomarkers in amniotic fluid of Edwards syndrome (trisomy 18) pregnancies. PLoS One. 2016;11(1):e0145908.
36. Blankley RT, Fisher C, Westwood M, North R, Baker PN, Walker MJ, Williamson A, Whetton AD, Lin W, McCowan L, Roberts CT, Cooper GJ, Unwin RD, Myers JE. A label-free selected reaction monitoring workflow identifies a subset of pregnancy specific glycoproteins as potential predictive markers of early-onset pre-eclampsia. Mol Cell Proteomics. 2013;12(11):3148–59.
37. Llurba E, Syngelaki A, Sánchez O, Carreras E, Cabero L, Nicolaides KH. Maternal serum placental growth factor at 11–13 weeks' gestation and fetal cardiac defects. Ultrasound Obstet Gynecol. 2013;42(2):169–74.
38. Llurba E, Sánchez O, Ferrer Q, Nicolaides KH, Ruíz A, Domínguez C, Sánchez-de-Toledo J, García-García B, Soro G, Arévalo S, Goya M, Suy A, Pérez-Hoyos S, Alijotas-Reig J, Carreras E, Cabero L. Maternal and foetal angiogenic imbalance in congenital heart defects. Eur Heart J. 2014;35(11):701–7.
39. Star J, Canick JA, Palomaki GE, Carpenter MW, Saller DN Jr, Sung CJ, Tumber MB, Coustan DR. The relationship between second-trimester amniotic fluid insulin and glucose levels and subsequent gestational diabetes. Prenat Diagn. 1997;17(2):149–54.
40. Lee J, Lee JE, Choi JW, Han MH, Seong SY, Park KH, Park JW. Proteomic analysis of amniotic fluid proteins for predicting the outcome of emergency cerclage in women with cervical insufficiency. Reprod Sci. 2020a;27(6):1318–29.
41. Lee PY, Saraygord-Afshari N, Low TY. The evolution of two-dimensional gel electrophoresis-from proteomics to emerging alternative applications. J Chromatogr A. 2020b;1615:460763.
42. Park JC, Kim DJ, Kwak-Kim J. Upregulated amniotic fluid cytokines and chemokines in emergency cerclage with protruding membranes. Am J Reprod Immunol. 2011;66(4):310–9.
43. Dan K, Lee JE, Han D, Kim SM, Hong S, Kim HJ, Park KH. Proteomic identification of biomarkers in maternal plasma that predict the outcome of rescue cerclage for cervical insufficiency. PLoS One. 2021;16(4):e0250031.
44. Amiri-Dashatan N, Koushki M, Abbaszadeh HA, Rostami-Nejad M, Rezaei-Tavirani M. Proteomics applications in health: biomarker and drug discovery and food industry. Iran J Pharm Res. 2018;17(4):1523–36.

45. Belczacka I, Latosinska A, Metzger J, Marx D, Vlahou A, Mischak H, Frantzi M. Proteomics biomarkers for solid tumors: current status and future prospects. Mass Spectrom Rev. 2019;38(1):49–78.
46. Pomastowski P, Buszewski B. Two-dimensional gel electrophoresis in the light of new developments. TrAC Trends Anal Chem. 2014;53:167–77.
47. Chang L, Wang D, Peng C, Wang Q, Xu B, Tong Z. A method for high-concentration agarose gel preparation and its application in high-resolution separation of low-molecular-weight nucleic acids and proteins. Int J Biol Macromol. 2023;231:123358.
48. Paulo JA. Sample preparation for proteomic analysis using a GeLC-MS/MS strategy. J Biol Methods. 2016;3(3):e45.
49. Wang H, Yang L, Liu M, Luo J. Protein post-translational modifications in the regulation of cancer hallmarks. Cancer Gene Ther. 2023a;30(4):529–47.
50. Wang M, Cai J, Chen J, Liu J, Geng X, Yu X, Yang J. PCR techniques and their clinical applications. 2023b.
51. Liotta LA, Pappalardo PA, Carpino A, Haymond A, Howard M, Espina V, et al. Laser capture proteomics: spatial tissue molecular profiling from the bench to personalized medicine. Expert Rev Proteomics. 2021;18(10):845–61.
52. Shevchenko G, Konzer A, Musunuri S, Bergquist J. Neuroproteomics tools in clinical practice. Biochim Biophys Acta. 2015;1854(7):705–17.
53. Meleady P. Two-dimensional gel electrophoresis and 2D-DIGE. Difference gel electrophoresis: methods and protocols. 2018;3–14.
54. Abdallah C, Dumas-Gaudot E, Renaut J, Sergeant K. Gel-based and gel-free quantitative proteomics approaches at a glance. Int J Plant Genomics. 2012;2012:494572.
55. Echan LA, Speicher DW. Protein detection in gels using fixation. Curr Protoc Protein Sci. 2002;Chapter 10:10.5.1–10.5.18.
56. Gundry RL, White MY, Murray CI, Kane LA, Fu Q, Stanley BA, Van Eyk JE. Preparation of proteins and peptides for mass spectrometry analysis in a bottom-up proteomics workflow. Curr Protoc Mol Biol. 2009;Chapter 10:Unit10.25. https://doi.org/10.1002/0471142727.mb1025s88.
57. Afzal M, Sielaff M, Curella V, Neerukonda M, El Hassouni K, Schuppan D, et al. Characterization of 150 wheat cultivars by LC-MS-based label-free quantitative proteomic. Plants (Basel). 2021;10(3):424.
58. Templeton EM, Pilbrow AP, Kleffmann T, Pickering JW, Rademaker MT, Scott NJ, et al. Comparison of SPEED, S-trap, and in-solution-based sample preparation methods for mass spectrometry in kidney tissue and plasma. Int J Mol Sci. 2023;24(7):6290.
59. Fang W, Du Z, Kong L, Fu B, Wang G, Zhang Y, Qin W. A rapid and sensitive single-cell proteomic method based on fast liquid-chromatography separation, retention time prediction and MS1-only acquisition. Anal Chim Acta. 2023;1251:341038.
60. Macklin A, Khan S, Kislinger T. Recent advances in mass spectrometry based clinical proteomics: applications to cancer research. Clin Proteomics. 2020;17:17. https://doi.org/10.1186/s12014-020-09283-w.
61. Verathamjamras C, Chantaraamporn J, Sornprachum T, Mutapat P, Chokchaichamnankit D, Mingkwan K, et al. Label-free quantitative proteomics reveals aberrant expression levels of LRG, C9, FN, A1AT and AGP1 in the plasma of patients with colorectal cancer. Clin Proteomics. 2023;20(1):1–16.
62. Dayon L, Affolter M. Progress and pitfalls of using isobaric mass tags for proteome profiling. Expert Rev Proteomics. 2020;17(2):149–61.
63. Chen X, Sun Y, Zhang T, Shu L, Roepstorff P, Yang F. Quantitative proteomics using isobaric labeling: a practical guide. Genomics Proteomics Bioinformatics. 2021;19(5):689–706.
64. Sivanich MK, Gu T-J, Tabang DN, Li L. Recent advances in isobaric labeling and applications in quantitative proteomics. Proteomics. 2022;22(19–20):2100256.
65. Kwon YW, Jo HS, Bae S, Seo Y, Song P, Song M, Yoon JH. Application of proteomics in cancer: recent trends and approaches for biomarkers discovery. Front Med. 2021;8:747333.

66. Lee JM, Hammarén HM, Savitski MM, Baek SH. Control of protein stability by post-translational modifications. Nat Commun. 2023;14(1):201.
67. Armbruster M, Grady S, Agongo J, Arnatt CK, Edwards JL. Neutron encoded derivatization of endothelial cell lysates for quantitation of aldehyde metabolites using nESI-LC-HRMS. Anal Chim Acta. 2022;1190:339260.
68. Zhang L, Vertes A. Single-cell mass spectrometry approaches to explore cellular heterogeneity. Angew Chem Int Ed. 2018;57(17):4466–77.
69. Danzi F, Pacchiana R, Mafficini A, Scupoli MT, Scarpa A, Donadelli M, Fiore A. To metabolomics and beyond: a technological portfolio to investigate cancer metabolism. Signal Transduct Target Ther. 2023;8(1):137.
70. Gopal J, Muthu M. Matrix-assisted laser desorption ionization-time of flight mass spectrometry (MALDI-TOF MS) for the analysis of mushroom polysaccharides of clinical, nutritional, and medicinal significance-contemplating the achievements and eliciting future trends. TrAC Trends Anal Chem. 2023;164:117095.
71. Nakayama K, Li X, Shimizu K, Akamatsu S, Inoue T, Kobayashi T, et al. qShot MALDI analysis: a rapid, simple, convenient, and reliable quantitative phospholipidomics approach using MALDI-TOF/MS. Talanta. 2023;254:124099.
72. Larsen MR, Trelle MB, Thingholm TE, Jensen ON. Analysis of posttranslational modifications of proteins by tandem mass spectrometry: mass spectrometry for proteomics analysis. Biotechniques. 2006;40(6):790–8.
73. Deutsch EW, Lam H, Aebersold R. Data analysis and bioinformatics tools for tandem mass spectrometry in proteomics. Physiol Genomics. 2008;33(1):18–25.
74. Albrethsen J. Reproducibility in protein profiling by MALDI-TOF mass spectrometry. Clin Chem. 2007;53(5):852–8.
75. Lv P, Liu Z, Xu B, Tang C, Li X, Qin H, et al. Exploratory study on application of MALDI-TOF-MS to detect serum and urine peptides related to small cell lung carcinoma. Mol Med Rep. 2020;21(1):51–60.
76. Liu G, Yu J, Li C, Zhou X, Nie L, Wei Y, et al. Application of protein chip combined with SELDI-TOF-MS detection to investigate serum protein expression in patients with silicosis fibrosis. Exp Ther Med. 2019;17(3):2172–84.
77. Han Z, Wen L, Feng L. Screening of serum protein biomarkers in hemorrhagic cerebral infarction by surface-enhanced laser desorption/ionization time-of-flight mass spectrometry (SELDI-TOF-MS) technology. Ann Transl Med. 2020;8(18):1186.
78. Chen G, Yang L, Liu G, Zhu Y, Yang F, Dong X, et al. Research progress in protein microarrays: focussing on cancer research. Proteomics Clin Appl. 2023;17(1):2200036.
79. Hu S, Xie Z, Qian J, Blackshaw S, Zhu H. Functional protein microarray technology. Wiley Interdisciplinary Reviews: Systems Biology and Medicine. 2011;3(3):255–68.
80. Lee JR, Magee DM, Gaster RS, LaBaer J, Wang SX. Emerging protein array technologies for proteomics. Expert Rev Proteomics. 2013;10(1):65–75.
81. Avens HJ, Bowman CN. Development of fluorescent polymerization-based signal amplification for sensitive and non-enzymatic biodetection in antibody microarrays. Acta Biomater. 2010;6(1):83–9.
82. Chandramouli K, Qian PY. Proteomics: challenges, techniques and possibilities to overcome biological sample complexity. Hum Genomics Proteomics. 2009;2009:239204.
83. Calvano CD, Rigante E, Picca RA, Cataldi TRI, Sabbatini L. An easily transferable protocol for in-situ quasi-non-invasive analysis of protein binders in works of art. Talanta. 2020;215:120882.
84. Zhao Y, Lin YH. Whole-cell protein identification using the concept of unique peptides. Genomics Proteomics Bioinformatics. 2010;8(1):33–41.
85. Paulo JA, Kadiyala V, Brizard S, Banks PA, Conwell DL, Steen H. Short gel, long gradient liquid chromatography tandem mass spectrometry to discover urinary biomarkers of chronic pancreatitis. The Open Proteomics Journal. 2013;6:1.

86. Cox J, Neuhauser N, Michalski A, Scheltema RA, Olsen JV, Mann M. Andromeda: a peptide search engine integrated into the MaxQuant environment. J Proteome Res. 2011;10(4):1794–805. https://doi.org/10.1021/pr101065j.
87. Mandal N, Heegaard S, Prause JU, Honoré B, Vorum H. Ocular proteomics with emphasis on two-dimensional gel electrophoresis and mass spectrometry. Biol Proced Online. 2010;12:56–88.
88. Szklarczyk D, Franceschini A, Wyder S, Forslund K, Heller D, Huerta-Cepas J, et al. STRING v10: protein–protein interaction networks, integrated over the tree of life. Nucleic Acids Res. 2015;43(D1):D447–52.
89. Yuen HY, Jansson J. Normalized L3-based link prediction in protein–protein interaction networks. BMC Bioinformatics. 2023;24(1):59.
90. Reddy T, Esmail A, Chang JC, Ghobrial RM, Abdelrahim M. Utility of cell-free DNA detection in transplant oncology. Cancers (Basel). 2022;14(3):743.
91. Yin L, Tang Y, Lu Q, Pan A, Shi M. Application value of NIPT for uncommon fetal chromosomal abnormalities. Mol Cytogenet. 2020;13:1–7.
92. Haidong W, Zhijie Y, Picchiassi E, Tarquini F, Coata G, You W, … Di Renzo GC. Non-invasive prenatal testing of fetal aneuploidies using a new method based on digital droplet PCR and cell free fetal DNA. medRxiv. 2020–12.
93. Tian M, Feng L, Li J, Zhang R. Focus on the frontier issue: progress in noninvasive prenatal screening for fetal trisomy from clinical perspectives. Crit Rev Clin Lab Sci. 2023;60:1–22.
94. Chang MY, Ahn S, Kim MY, Han JH, Park HR, Seo HK, et al. One-step noninvasive prenatal testing (NIPT) for autosomal recessive homozygous point mutations using digital PCR. Sci Rep. 2018;8(1):2877.
95. Kater-Kuipers A. Principles, practices and promises of prenatal screening: ethical and social aspects of non-invasive prenatal testing (NIPT) and the expansion of the scope of prenatal screening. 2020.
96. Johnson T. Considering the collective in ethical decision-making concerning non-medical uses of noninvasive prenatal testing. Am J Bioeth. 2023;23(3):23–5.
97. Kotsopoulou I, Tsoplou P, Mavrommatis K, Kroupis C. Non-invasive prenatal testing (NIPT): limitations on the way to become diagnosis. Diagnosis (Berl). 2015;2(3):141–58.
98. Chapman K. The proteinchip® biomarker system from ciphergen biosystems: a novel proteomics platform for rapid biomarker discovery and validation. 2002.
99. Aebersold R, Mann M. Mass-spectrometric exploration of proteome structure and function. Nature. 2016;537(7620):347–55.
100. Geyer PE, Kulak NA, Pichler G, Holdt LM, Teupser D, Mann M. Plasma proteome profiling to assess human health and disease. Cell Syst. 2016;2(3):185–95.
101. Ngo TTM, Moufarrej MN, Rasmussen MH, Camunas-Soler J, Pan W, Okamoto J, Neff NF, Liu K, Wong RJ, Downes K, Tibshirani R, Shaw GM, Skotte L, Stevenson DK, Biggio JR, Elovitz MA, Melbye M, Quake SR. Noninvasive blood tests for fetal development predict gestational age and preterm delivery. Science. 2018;360(6393):1133–6.
102. Liang H, Tsui BY, Ni H, Valentim CCS, Baxter SL, Liu G, Cai W, Kermany DS, Sun X, Chen J, He L, Zhu J, Tian P, Shao H, Zheng L, Hou R, Hewett S, Li G, Liang P, Zang X, et al. Evaluation and accurate diagnoses of pediatric diseases using artificial intelligence. Nat Med. 2019;25(3):433–8.

Noninvasive Prenatal Screening (NIPS) of Chromosomal Abnormalities

8

Latika Chawla and Surekha Thakur

8.1 Introduction

In 1997, Lo et al. published a groundbreaking study demonstrating the existence of fetal DNA in maternal plasma [1]. This discovery laid the foundation for further exploration of its potential applications in prenatal diagnostics. Noninvasive prenatal screening (NIPS) for common fetal aneuploidies (e.g., trisomy 21, 18, and 13 and 46XO) involves analysis of the short fragments of cell-free fetal DNA (cffDNA) that circulate in maternal bloodstream. This screening modality has significantly reduced the need for invasive tests including chorionic villous sampling (CVS) and amniocentesis. cffDNA is also utilized for fetal sex determination, Rh genotyping, and detection of monogenic disorders. This chapter will discuss in detail the origin of cffDNA, genetic platforms used for analysis, clinical applications, and implications of NIPS in detection of chromosomal aneuploidies.

8.2 Source

Apoptosis of the syncytiotrophoblasts gives rise to cell-free DNA (cfDNA) that leaks into the maternal circulation during pregnancy [2]. Since fetus and placenta arise from the same embryo, thus they are considered to be genetically identical, and this DNA is considered to be cell-free "fetal" DNA. An exception to this may be rare cases of confined placental mosaicism where the placenta and fetus may have different genetic makeup. To a lesser extent, cffDNA also arises as a result of apoptosis of fetal erythroblasts in fetal circulation that enters maternal circulation [2]. cffDNA appears in maternal circulation as early as 5 weeks' period of gestation

L. Chawla (✉) · S. Thakur
Department of Obstetrics and Gynaecology, AIIMS, Rishikesh, Uttarakhand, India

R. A. Rather, S. C. Saha (eds.), *Non-invasive Prenatal Screening (NIPS) in Clinical Practice*, https://doi.org/10.1007/978-981-97-6402-0_8

and almost always by 9 weeks [3]. As the pregnancy advances, placental mass increases; thus, the concentration of cffDNA in maternal circulation also increases [4].

The size of cffDNA fragments can vary, but they are generally short, typically ranging from around 100–200 base pairs (bp) in length. Studies indicate that nearly 99% of fetal DNA is under 313 bp in length, while cell-free maternal DNA typically exhibits a medium length ranging from approximately 400–500 bp [5, 6].

8.3 Characteristics of cffDNA

The extensive study of cfDNA in plasma has found significant applications in oncology and general medicine, with the most notable success observed in prenatal screening for fetal aneuploidy [7]. Maternal plasma cfDNA encompasses both maternal and fetal cfDNA sources. Noninvasive prenatal testing (NIPT) for fetal aneuploidy utilizes high-throughput sequencing methods, often referred to as "next-generation sequencing" (NGS), to assess the proportional representation of each chromosome in plasma cfDNA. In a pregnant woman with a normal chromosomal complement (euploid), any deviation from the expected chromosome profile in plasma cfDNA, characterized by an excess or deficiency of cfDNA fragments from a specific chromosome, indicates the potential presence of fetal trisomy or monosomy, respectively [8].

An essential determinant influencing the reporting accuracy of NIPT is the fetal fraction (FF), representing the percentage of plasma cfDNA originating from the fetoplacental unit. Typically, the average FF ranges from 10 to 15% between 10 and 20 weeks of gestation [8]. The measurement of FF is critical in NIPT screening, ensuring an adequate presence of placental cffDNA in maternal plasma to yield meaningful results. Low FF may lead to test failure or a no-call result, with reported no-call rates varying from 1 to 8%. Among these instances, 50–60% of patients achieve success in second sampling. Notably, a concern associated with no-call reports is the elevated risk of aneuploidy, ranging from 2.7 to 23.3% [9].

There is large variation in the FF from individual to individual. Causes of low FF are early gestation age, suboptimal sample collection, maternal obesity, abnormal fetal karyotype, in vitro fertilization (IVF), and twin pregnancy. The difference in demographic parameters like age, BMI, and placental mass contributes to the low FF in IVF.

Conditions that augment the maternal contribution and diminish the placental contribution have the potential to decrease FF. The recommended minimum FF threshold for reliable NIPT is 4%. Recent attention has focused on investigating whether elevated levels of cffDNA might serve as an early indicator for placental pathologies, including spontaneous preterm birth and preeclampsia [10].

Maternal weight stands out as the most acknowledged and clinically impactful factor affecting FF. The decrease in FF correlated with maternal weight is attributed to an escalation in maternal adipocytes, accompanied by inflammation and apoptosis. Additionally, inflammatory conditions such as systemic lupus erythematosus, vitamin B12 deficiency, and the usage of low molecular weight heparin (LMWH) contribute to heightened levels of total cfDNA [8]. T21 is linked to a higher proportion of FF in maternal plasma, whereas T18 and T13 exhibit lower FF. Consequently, the detection rate is comparatively lower for T18 and T13 compared to T21 due to their association with lower FF. Notably, aneuploidies detected after a failed report typically include T18, T13, T16, T21, and monosomy X [8]. Suboptimal sample collection may happen as a result of a half-filled vial, use of wrong vial for blood collection, or inappropriate storage of sample.

8.4 Methods of Testing

8.4.1 Whole Genome Sequencing (WGS)/Massively Parallel Sequencing (MPS)

NIPT using this method provides the most informative results and is the most commonly used method [11]. It has lower failure rates as compared to targeted methodology. This method sequences the entire genome over cffDNA fragments. The fragments of chromosome of origin are identified through alignment with human genome build. The proportion of the fragments of specific chromosome will depend on the FF in maternal blood. If FF is higher, the cffDNA fragments will also be higher, and if FF is lower, the cffDNA fragments will also decrease. This method may use shotgun sequencing in which the whole genome is broken into small fragments and then sequenced.

8.4.2 Targeted Sequencing

In this method, only the chromosomes of interest are targeted, so it needs less sequencing to be aligned and hence uses less resource. This targeted approach requires more amplification rounds making it more complicated as compared to WGS. The approach for targeted sequencing includes the following:

8.4.2.1 Single Nucleotide Polymorphism (SNP) Analysis

This method reads thousands of highly polymorphic SNPs located on the chromosome of interest which will make specific patterns based on the maternal and fetal genotypes. In case of extra chromosome or triploidy, additional pattern from the SNP will be formed. It identifies the aneuploidy in twin pregnancies, twin zygosity, and a vanished dizygotic twin [12]. However, this cannot be used in settings like pregnancy after egg donation, pregnancy in the recipient of bone marrow, or any organ transplant due to the presence of additional confounding chromosome in maternal blood.

8.4.2.2 Array Based

It is based on microarray platform in which multiple unique loci on the targeted chromosome are quantified and sequenced. In case of trisomy in fetus, the sequence counts of that specific chromosome will increase.

8.4.2.3 Rolling Circle Amplification

This method targets selected fragments of cffDNA [13]. The probe specifically designed for these fragments bind with each targeted chromosome. Using this method, these products are amplified into fluorescent products and can be viewed on an automated microscope plate and then counted. Like other methods, counts of the fragments will increase with the trisomy in fetus.

8.5 Clinical Utility of NIPT

8.5.1 Screening Common Aneuploidies

The most common trisomies comprise of T21, T18, and T13. These constitute 71% of all prenatally detected chromosomal aneuploidies [14]. Various studies have promoted the application of NIPT for detecting these common trisomies. It has been found that the detection rate of NIPT for T21 is 99.2% for singleton pregnancy and 93.7% for twin pregnancies. However, the detection rates for T18 and T13 are 96.3% and 91%, respectively [15].

NIPT is the most sensitive screening method for detecting these aneuploidies, and it has effectively reduced invasive testing to detect these trisomies by 92.8% [15, 16]. The detection rate (DR) and false-positive rate (FPR) depend on the type of aneuploidy. A recent meta-analysis reported that, for common aneuploidies detected using cell-free fetal DNA (cffDNA), the detection rates (DR) and false positive rates (FPR) were as follows: 99.5% DR and 0.05% FPR for trisomy 21 (T21), 97.7% DR and 0.04% FPR for trisomy 18 (T18), and 96.1% DR and 0.06% FPR for trisomy 13 (T13) [17, 18].

8.5.2 Sex Chromosome Aneuploidy

Sex chromosome aneuploidy (SCA) includes disorders arising from atypical numbers of X and Y chromosomes, such as Turner syndrome (45,X), triple X syndrome (47,XXX), Klinefelter syndrome (47,XXY), and Jacob's syndrome (47,XYY) [19]. These conditions occur in approximately 1 in every 400 live births [20]. Early screening and prenatal detection of SCA enable timely interventions that can significantly enhance the affected child's quality of life. Various studies indicate that NIPT is a dependable method for detecting SCA, though instances of false negatives and positives have been documented. A comprehensive review and meta-analysis assessed the efficacy of cffDNA in the prenatal identification of 45,X,

Table 8.1 Showing sensitivity, specificity, and positive predictive value for prenatal detection of 45,X, 47,XXY, 47,XXX, and 47,XYY

Sex chromosome aneuploidy	Sensitivity	Specificity	Positive predictive value
45 X	98.8% (95% CI 94.6–100%)	99.4% (95% CI 98.7–99.9%)	14.5% (95% CI 7.0–43.8%)
47 XXY	100% (95% CI 99.6–100%)	100% (95% CI 99.9–100%)	97.7% (95% CI 78.6–100%)
47 XXX	100% (95% CI 96.9–100%)	99.9% (95% CI 99.7–100%)	61.6% (95% CI 37.6–95.4%)
47 XYY	100% (95% CI 91.3–100%)	100% (95% CI 100–100%)	100% (95% CI 76.5–100%)

47,XXY, 47,XXX, and 47,XYY, highlighting differences in sensitivity, specificity, and positive predictive values (PPV) [21] (Table 8.1).

The negative predictive values for all four SCAs were found to be over 99.99%, despite the occurrence of false negatives [21]. Given that SCAs are rarer compared to other chromosomal abnormalities, their PPV are also correspondingly lower [22]. Several factors contribute to the high rate of false-positive results, including the high homology between X and Y chromosomes which complicates detection, as well as the presence of Y chromosome segments that are similar to other chromosomes, reducing sequencing accuracy. Additionally, biases in guanine and cytosine content on the X chromosome, random X chromosome inactivation, confined placental mosaicism, and maternally derived SCAs can also decrease the specificity and detection accuracy of NIPT [23]. Despite these advancements, some challenges remain. Due to the FPR associated with NIPT for SCAs, the number of unnecessary invasive prenatal diagnostic procedures may increase, particularly for detecting 45,X [24].

8.5.3 Other Chromosome Aneuploidy

The efficacy of NIPT in detecting other chromosomal aneuploidies is limited. The overall PPV for these aneuploidies is reported to be 12.50%, which is significantly lower compared to the PPV for common autosomal aneuploidies and SCAs. In 2016, the American College of Medical Genetics and Genomics (ACMG) recommended that NIPT should not be used for detecting autosomal aneuploidies other than T21, T18, and T13 [25].

8.5.4 Copy Number Variants

NIPT has the capability to detect copy number variations (CNVs) such as cri du chat syndrome (5p deletion) and DiGeorge syndrome (22q11.2 microdeletion), but it is prone to false-positive results. Therefore, confirmatory validation through karyotype analysis or chromosome microarray analysis is still necessary [26]. The PPV for CNV detection has been reported at 32.14%, with a FPR of 0.15% [27].

8.6 Implementation of NIPT

8.6.1 Primary Screening

It is characterized by offering cffDNA testing as the initial screening test for patients. The impressive sensitivity and specificity of NIPT establish it as a highly effective method for screening common chromosomal aneuploidies. Substantial clinical evidence endorses the use of NIPT in the broader population, and several recently published guidelines advocate for its utilization as the primary screening test for all pregnant women [28, 29].

Some practitioners have restricted the adoption of cffDNA as the initial screening test due to various factors. Key considerations in cffDNA screening involve cost implications, availability of suitable pretest counseling, and concerns that individuals might opt for pregnancy termination following a positive screen without pursuing diagnostic testing, despite the lower PPV of the test in younger patients.

8.6.2 Secondary Screening

It is characterized by offering a noninvasive screening test as a follow-up to patients identified as high or intermediate risk in a preceding screening test. Factors indicating the need for secondary screening are maternal age of 35 years or older, previous baby with aneuploidy, abnormal serum screening test results like high or intermediate risk on first trimester combined screen or high-risk on quadruple screen, a positive family history of aneuploidy or bad obstetric history [30]. In the presence of certain soft markers, e.g., thick nuchal fold, aberrant right subclavian artery, and hyperechoic bowel, NIPS testing is offered after appropriate counseling. The advantage of cffDNA as a secondary screening method is significant decline in the utilization of invasive diagnostic tests. Since 2012, cffDNA screening has led to a reduction ranging from 40 to 76% in the number of invasive procedures for prenatal genetic diagnosis [31]. The disadvantage is that it has to rely on the sensitivity and specificity of the previous screening test. This means it will not offer testing to those patients who screen negative in initial screening.

8.6.3 Models Utilizing NIPT as Secondary Screening

There are two screening models, contingent and reflexive, that combine serum and cffDNA screening for the general pregnant population. While both models initiate the process with serum screening, they differ in risk cutoffs and follow-up testing approaches. The common goals of these models are to enhance the DR beyond the usual outcomes of serum screening and decrease the FPRs associated with it. These testing strategies aim to enhance the screening performance of cfDNA testing and reduce the costs below those caused by universally offering cfDNA testing.

8.6.3.1 Contingent Model

In the contingent model, all pregnant individuals undergo serum screening testing during the first trimester [32]. Based on the screening results, patients are categorized as either high-risk, low-risk, or intermediate-risk group determined by a predefined cutoff. The high-risk group is subsequently presented with options for further evaluation, including invasive testing, noninvasive cffDNA testing, or opting for no additional testing. Conversely, the low-risk population is considered routine antenatal patients without further testing. A distinct intermediate-risk group is identified. The newly established intermediate-risk group constitutes approximately 10–15% of the screened population. These individuals are notified of their intermediate risk and are offered cffDNA screening following counseling. Diagnostic testing is recommended for those in the high-risk group or intermediate-risk group if cffDNA testing returns positive. This strategic approach significantly boosts the detection rate by 10–15% and reduces the FPR (0.1–0.2%) for patients undergoing secondary screening. However, an inherent disadvantage is the dependency on patient compliance, as individuals with high risk or intermediate risk need to return for counseling, additional sampling, and cfDNA testing.

8.6.3.2 Reflex Model

In the reflexive model, patient provides two blood samples during the first trimester: one serum sample for the first trimester combined serum screening and one for possible cffDNA testing. Following the serum screening analysis, the high-risk cohort is presented with the option of invasive or cffDNA testing. Conversely, the low-risk group undergoes routine antenatal care. For the intermediate-risk population, cffDNA testing is automatically performed without necessitating a return visit for sample collection. Results are subsequently communicated to patients, accompanied by appropriate counseling. This process not only conserves time and resources but also ensures a predictable enhancement in DRs [33]. However, the disadvantage of this model is the additional expense caused by collecting samples from all patients, even though only 10–12% will undergo reflexive testing. The implementation of noninvasive prenatal aneuploidy screening through cffDNA in both high-risk and general population settings requires careful consideration of several factors.

8.7 Implementation Issues

Several factors must be considered when offering noninvasive prenatal aneuploidy screening in either high-risk or general population settings. Essential to this process is the necessity for effective patient education and counseling. Here are some of the key points to be discussed:

- **Genetic counseling:** Ideally, genetic counseling would precede testing; however, resource constraints make this impractical for the general pregnancy population. Key pretest discussions with patients considering cfDNA screening should cover screening's optional nature, the distinction between screening and diagnostic tests, basic principles of cfDNA technology, conditions detectable by screening and their clinical features, reporting formats, test performance metrics (e.g., DR, FPR, test failure rate), test limitations, incidental findings, timing, and the need to confirm abnormal screening results before contemplating termination.
- Patients must grasp that cffDNA screening is a choice, and their decision depends on how they weigh the benefits of obtaining aneuploidy information against the emotional and physical risks associated with the screening and subsequent testing process.
- It is crucial for patients to comprehend the fundamental difference between screening and diagnostic testing. While cffDNA sequencing serves as a screening test with inherent false-positive and false-negative results, targeting specific aneuploidies, diagnostic procedures like amniocentesis can provide a comprehensive diagnosis, including the detection of mosaicism and microdeletions/microduplications.
- Patients should also be informed about cost considerations, the slightly longer turnaround time compared to other screening methods, and the potential for incidental findings.
- Addressing the patient's desire for fetal sex results is essential, emphasizing that cffDNA screening primarily focuses on major fetal chromosomal anomalies, with fetal sex information being an additional aspect. Testing for the sex of the fetus is however not possible in the Indian scenario as per Indian PNDT act.
- Consideration of laboratory issues is equally important in implementation. Patients should be made aware of commercial marketing influences; be given the option to choose a laboratory based on various factors such as ease of use, turnaround time, and customer service; and be informed about the content and format of laboratory reports.

8.8 Test Failures

Test failures in cffDNA screening exhibit varying rates influenced by population characteristics, test methodology, and the routine testing of a second sample post-initial failure. Typically, no-result rates fall within the range of 1–3% [34]. Common causes for test failure encompass inadequate fetal DNA, FF below 4%, and insufficient sequenced and/or aligned fragments. Low FF accounts for up to 50% of failures. Notably, IVF pregnancies may experience double the failure rate compared to naturally conceived pregnancies [35]. Another reason for a test failure is the laboratory method used. Laboratory performance standards, prioritizing accuracy over failure rates vary, impacting reported failure rates.

8.9 False Positive

A false positive in cffDNA testing occurs when the test indicates a chromosomal abnormality while the fetus remains unaffected. Studies suggest a 90% PPV for common aneuploidies (trisomy), implying that 10% of patients with positive test results will not have a chromosomal abnormality [35]. Factors of false-positive reports are as follows:

- Confined placental mosaicism: This may occur in up to 1–2% of pregnancies [36] and is more likely with monosomy X and T13 than for T21/T18 [37].
- Maternal mosaicism: Most cffDNA testing methods assume that the mother has a normal karyotype, but this is not always true. With the advancing age, there can be small proportion of cells that have lost one X chromosome, and this may be detected in the cffDNA testing and misdiagnosed as a SCA.
- Maternal cancer: Pregnant individuals with a malignancy may release detectable amounts of cell-free tumor DNA into their bloodstream [38].
- Maternal copy number variant: In such cases, cffDNA sequencing may create a positive result if the maternal duplication is notably large and occurs on a relevant chromosome.
- Transplant recipient: In cases of transplant recipients receiving tissue (either bone marrow or organ) from a male donor, cffDNA testing may erroneously label a female fetus as male [39].
- Maternal blood transfusion: If a pregnant woman undergoes a blood transfusion from a male donor within 4 weeks before the cffDNA blood draw, there is a risk of incorrectly identifying a female fetus as male.
- Technical issues: Technical challenges, such as infrequent sample mix-ups or other errors, have the potential to result in false-positive or false-negative outcomes.

8.10 False Negative

- **Confined placental mosaicism**: This is mostly observed in trisomy 13 and 18 and rarely for trisomy 21.
- **Borderline low fetal fraction**: A FF that is low but still within adequacy (e.g., between 3 and 5%) may result in a minimal difference between the expected and observed percentage of chromosome fragments. If an insufficient number of fragments are sequenced, this difference may go unnoticed, leading to incorrect reporting of results as screen negative.
- **Technical issues**: Assay-related technical challenges can introduce difficulties in identifying certain aneuploidies, impacting the accuracy of the test result.

8.11 Pretest Counseling

Pretest counseling for NIPT is crucial, aiming to empower patients with comprehensive information for well-informed decisions [40]. The counseling points include the following:

1. **Optional Testing**: Patients are informed that screening and diagnostic testing for fetal aneuploidy are optional. The decision to proceed depends on individual perceptions of benefits weighed against emotional and physical risks.
2. **Define Screening**: Clarify that NIPT is a screening test, categorizing women into higher or lower risk categories. It cannot diagnose or eliminate the chance of a specific chromosome condition, and it poses no increased risk for miscarriage.
3. **Review Clinical Features**: Describe the clinical features of conditions included in the screening with sensitive, neutral language. Trisomy 21, 18, and 13 should be discussed, highlighting the increased risks for miscarriage, congenital anomalies, and severe disabilities. Sex chromosome aneuploidies should also be addressed.
4. **Describe Technology**: Detail the testing methods, focusing on MPS. Explain the analysis of cfDNA fragments in maternal blood to determine chromosome abnormalities.
5. **Reporting Format**: Inform patients about when and how results will be provided, considering variations among NIPT laboratories. Reports include qualitative results and, in some cases, numeric residual risk. Non-reportable results may necessitate a repeat test or an invasive testing if the laboratory already performed a rerun on a "no-call result" before sharing the final report.
6. **Sensitivity**: Patients are educated about the sensitivity of NIPT, which varies depending on the laboratory used and according to the aneuploidy.
7. **FPR and Confirmation**: Emphasize that NIPT is not diagnostic but a screening test. Patients are encouraged to consider confirmatory diagnostic testing, and termination based solely on abnormal NIPT results should not be considered.
8. **Predictive Values**: Patients are made aware of the variability in positive predictive values, influenced by factors like specificity, sensitivity, maternal age, and gestational age.
9. **Limitations**: NIPT does not test for all genetic syndromes or aneuploidies. Diagnostic testing through CVS or amniocentesis is highlighted for comprehensive assessment.
10. **Incidental Findings**: Rarely, NIPT may raise suspicion for maternal or fetal conditions beyond targeted aneuploidies, including maternal chromosome differences or malignancies.
11. **Timing**: NIPT can be performed after 10 weeks of gestation, and discussions include laboratory-specific turnaround times and legal limits for pregnancy termination.

8.12 Posttest Follow-Up

1. **Screen Positive**: It denotes that there is a higher chance for the baby to have specific chromosomal abnormality. Upon receiving a positive cffDNA screening, confirming fetal aneuploidy becomes imperative. The choice between invasive methods for confirmation, such as CVS or amniocentesis, remains a point of controversy. Notably, amniocyte analysis is considered more definitive due to its accurate representation of the fetal genotype [41]. Crucially, obtaining a conventional G-banded karyotype is emphasized in cffDNA screen-positive cases. A notable aspect is that 5–10% of pregnancies featuring common aneuploidy reveal a parent with a balanced translocation, significantly impacting reproductive counseling. Consequently, if FISH or microarray yields a positive result, it is advised to complement it with a conventional karyotype.
2. **Screen Negative**: Conversely, a screen-negative result suggests a diminished risk of aneuploidies within the test panel, although it does not eliminate the prospect of other chromosomal abnormalities. Generally, invasive diagnostic testing is not offered routinely to screen-negative patients unless specific indications emerge, such as the detection of a fetal anomaly during ultrasound examination. In such instances, the recommendation shifts toward invasive testing. Importantly, this guideline does not extend to fetuses identified solely with isolated soft markers.
3. **No-Call or No Result**: In cases where the cffDNA test yields no result, constituting 1–5% of cases, patients are presented with three viable options. Firstly, they can opt to repeat the cfDNA test after 7 days, provided it is permissible to do so. Alternatively, patients may consider invasive procedures such as amniocentesis or CVS coupled with diagnostic testing. The American College of Obstetricians and Gynecologists recommends comprehensive counseling in such situations. This counseling should highlight the heightened risk associated with test failure and offer extensive genetic evaluation through ultrasound and diagnostic testing [42].

8.13 NIPS in Twins

NIPS has demonstrated increased efficacy in detecting fetal aneuploidy in twins compared to traditional screening methods [43]. While there is a notable rise of up to 35% in cffDNA levels in maternal circulation for twin pregnancies in comparison with singletons, studies suggest that each twin may have a lower cffDNA quantity than a singleton pregnancy. This discrepancy is attributed to unequal FFs from each twin pair [44–46]. Recent research indicates that NIPS yields detection rates for T21 in twins are almost equivalent to those in singleton pregnancies. However, there is an observed increase in the number of test results categorized as "no-call." While limited, available data on NIPT experience for other trisomies in twin pregnancies align with expectations for singleton pregnancies. In a recent statement, the International Society for Prenatal Diagnosis reported overall detection rates of 98.8%, 93.1%, and 75% for trisomy 21, trisomy 18, and trisomy 13, respectively [47]. It is important to note that distinguishing which twin, if any, is affected is impossible based solely on cffDNA analysis. Therefore, results are reported for the entire pregnancy, necessitating invasive testing to determine the affected twin, if applicable. cffDNA can serve as a reliable method for determining zygosity in twin pregnancies. In a recent study [12] involving 95 twin pregnancies, the reported sensitivity and specificity for establishing zygosity were both 100%. This proves particularly valuable in situations where chronicity assessment could not be done due to delayed commencement of prenatal care or any other reason.

8.14 Advantages and Limitations

NIPS has several advantages, making it a popular choice for many antenatal women. Here are some of the key advantages: non invasive, high sensitivity and specificity for detecting common aneuploidies, early detection of risk of chromosomal abnormalities, reduced false positive rates as complared to other screening methods and comprehensive screening of multiple chromosomal abnormalities.

NIPS, although a powerful tool for detecting fetal aneuploidies, does have some limitations. Notably, it is a screening test rather than a diagnostic one, and while it provides a high level of accuracy, it can produce false-positive or false-negative results. Confirmatory diagnostic testing, such as CVS or amniocentesis, is required to confirm NIPS findings, and medical termination of pregnancy in high-risk reports should be advised only after confirmatory testing. Furthermore, NIPS primarily screens for common chromosomal abnormalities like trisomies, potentially missing other genetic conditions or structural abnormalities. Additionally, NIPS does not accurately detect mosaicism, which occurs when an individual has a mixture of cells with different genetic makeup, leading to potential false-negative results. Moreover, while NIPS is valuable for screening, it may not provide detailed insights into structural abnormalities in fetuses, necessitating further tests such as microarray analysis for a comprehensive evaluation. Lastly, the cost of NIPS is higher compared to traditional screening tests, which may limit its accessibility for some individuals, especially in developing nations like ours.

References

1. Lo YM, Corbetta N, Chamberlain PF, et al. Presence of fetal DNA in maternal plasma and serum. Lancet. 1997;350(9076):485–7.
2. Sekizawa A, Samura O, Zhen DK, et al. Apoptosis in fetal nucleated erythrocytes circulating in maternal blood. Prenat Diagn. 2000;20:886.

3. Guibert J, Benachi A, Grebille AG, et al. Kinetics of SRY gene appearance in maternal serum: detection by real time PCR in early pregnancy after assisted reproductive technique. Hum Reprod. 2003;18:1733.
4. Lee SY, Kim SJ, Han SH, et al. A new approach of digital PCR system for non-invasive prenatal screening of trisomy 21. Clin Chim Acta. 2018;476:75–80.
5. Chan KC, Zhang J, Hui AB, et al. Size distributions of maternal and fetal DNA in maternal plasma. Clin Chem. 2004;50(1):88–92.
6. Li Y, Zimmermann B, Rusterholz C, Kang A, Holzgreve W, Hahn S. Size separation of circulatory DNA in maternal plasma permits ready detection of fetal DNA polymorphisms. Clin Chem. 2004;50(6):1002–11.
7. Jamshidi A, Liu MC, Klein EA, et al. Evaluation of cell-free DNA approaches for multi-cancer early detection. Cancer Cell. 2022;40(12):1537–1549.e12.
8. Hui L, Bianchi DW. Fetal fraction and noninvasive prenatal testing: what clinicians need to know. Prenat Diagn. 2019;40(2):155–63.
9. Palomaki GE, Kloza EM, Lambert-Messerlian GM, et al. Circulating cell free DNA testing: are some test failures informative? Prenat Diagn. 2015;35(3):289–93.
10. Rafaeli-Yehudai T, Imterat M, Douvdevani A, et al. Maternal total cell-free DNA in preeclampsia and fetal growth restriction: evidence of differences in maternal response to abnormal implantation. PLoS One. 2018;13(7):e0200360.
11. Jones KJ, Wang E, Bogard P, et al. Targeted cell-free DNA analysis with microarray quantitation for assessment of fetal sex and sex chromosome aneuploidy risk. Ultrasound Obstet Gynecol. 2018;51(2):275–6.
12. Norwitz ER, McNeill G, Kalyan A, et al. Validation of a single-nucleotide polymorphism-based non-invasive prenatal test in twin gestations: determination of zygosity, individual fetal sex, and fetal aneuploidy. J Clin Med. 2019;8(7):937.
13. Ericsson O, Ahola T, Dahl F, et al. Clinical validation of a novel automated cell-free DNA screening assay for trisomies 21, 13, and 18 in maternal plasma. Prenat Diagn. 2019;39(11):1011–5.
14. Ravitsky V, Roy M-C, Haidar H, et al. The emergence and global spread of noninvasive prenatal testing. Annu Rev Genomics Hum Genet. 2021;22(1):309–38.
15. Gil MM, Quezada MS, Revello R, Akolekar R, Nicolaides KH. Analysis of cell-free DNA in maternal blood in screening for fetal aneuploidies: updated meta-analysis. Ultrasound Obstet Gynecol. 2015;45(3):249–66.
16. Kostenko E, Chantraine F, Vandeweyer K, et al. Clinical and economic impact of adopting noninvasive prenatal testing as a primary screening method for fetal aneuploidies in the general pregnancy population. Fetal Diagn Ther. 2019;45(6):413–23.
17. Taylor-Phillips S, Freeman K, Geppert J, et al. Accuracy of non-invasive prenatal testing using cell-free DNA for detection of Down, Edwards and Patau syndromes: a systematic review and meta-analysis. BMJ Open. 2016;6(1):e010002.
18. Mackie FL, Hemming K, Allen S, Morris RK, Kilby MD. The accuracy of cell-free fetal DNA-based non-invasive prenatal testing in singleton pregnancies: a systematic review and bivariate meta-analysis. BJOG. 2017;124(1):32–46.
19. Zhang B, Lu BY, Yu B, et al. Noninvasive prenatal screening for fetal common sex chromosome aneuploidies from maternal blood. J Int Med Res. 2017;45(2):621–30.
20. Berglund A, Stochholm K, Gravholt CH. The epidemiology of sex chromosome abnormalities. Am J Med Genet C Semin Med Genet. 2020;184:202.
21. Shear MA, Swanson K, Garg R, et al. A systematic review and meta-analysis of cell-free DNA testing for detection of fetal sex chromosome aneuploidy. Prenat Diagn. 2023;43(2):133–43.
22. Hu H, Wang L, Wu J, et al. Noninvasive prenatal testing for chromosome aneuploidies and subchromosomal microdeletions/microduplications in a cohort of 8141 single pregnancies. Hum Genomics. 2019;13(1):14.

23. McNamara CJ, Limone LA, Westover T, Miller RC. Maternal source of false-positive fetal sex chromosome aneuploidy in noninvasive prenatal testing. Obstet Gynecol. 2015;125(2):390–2.
24. Xu Y, Chen L, Liu Y, et al. Screening, prenatal diagnosis, and prenatal decision for sex chromosome aneuploidy. Expert Rev Mol Diagn. 2019;19(6):537–42.
25. Gregg AR, Skotko BG, Benkendorf JL, et al. Noninvasive prenatal screening for fetal aneuploidy, 2016 update: a position statement of the American College of Medical Genetics and Genomics. Genet Med. 2016;18(10):1056–65.
26. Chen Y, Yu Q, Mao X, et al. Noninvasive prenatal testing for chromosome aneuploidies and subchromosomal microdeletions/microduplications in a cohort of 42,910 single pregnancies with different clinical features. Hum Genomics. 2019;13:60.
27. Pang Y, Wang C, Tang J, Zhu J. Clinical application of noninvasive prenatal testing in the detection of fetal chromosomal diseases. Mol Cytogenet. 2021;14(1):31.
28. Committee opinion no. 640: cell-free DNA screening for fetal aneuploidy. Obstet Gynecol. 2015;126(3):e31-e37.
29. Cherry AM, Akkari YM, Barr KM, et al. Diagnostic cytogenetic testing following positive noninvasive prenatal screening results: a clinical laboratory practice resource of the American College of Medical Genetics and Genomics (ACMG). Genet Med. 2017;19(8):845–50.
30. American College of Obstetricians and Gynecologists Committee on Genetics. Committee opinion no. 545: noninvasive prenatal testing for fetal aneuploidy. Obstet Gynecol. 2012;120(6):1532–4.
31. Hui L, Hutchinson B, Poulton A, Halliday J. Population-based impact of noninvasive prenatal screening on screening and diagnostic testing for fetal aneuploidy. Genet Med. 2017;19(12):1338–45.
32. Chitty LS, Wright D, Hill M, et al. Uptake, outcomes, and costs of implementing non-invasive prenatal testing for Down's syndrome into NHS maternity care: prospective cohort study in eight diverse maternity units. BMJ. 2016;354:i3426.
33. Wald NJ, Bestwick JP. Prenatal reflex DNA screening for Down syndrome: enhancing the screening performance of the initial first trimester test. Prenat Diagn. 2016;36(4):328–31.
34. Palomaki GE, Kloza EM. Prenatal cell-free DNA screening test failures: a systematic review of failure rates, risks of Down syndrome, and impact of repeat testing. Genet Med. 2018;20(11):1312–23.
35. Lee TJ, Rolnik DL, Menezes MA, et al. Cell-free fetal DNA testing in singleton IVF conceptions. Hum Reprod. 2018;33(4):572–8.
36. Kalousek DK, Howard-Peebles PN, Olson SB, et al. Confirmation of CVS mosaicism in term placentae and high frequency of intrauterine growth retardation association with confined placental mosaicism. Prenat Diagn. 1991;11(10):743–50.
37. Grati FR, Bajaj K, Malvestiti F, et al. The type of feto-placental aneuploidy detected by cfDNA testing may influence the choice of confirmatory diagnostic procedure. Prenat Diagn. 2015;35(10):994–8.
38. Snyder HL, Curnow KJ, Bhatt S, Bianchi DW. Follow-up of multiple aneuploidies and single monosomies detected by noninvasive prenatal testing: implications for management and counseling. Prenat Diagn. 2016;36(3):203–9.
39. Bianchi DW, Parsa S, Bhatt S, et al. Fetal sex chromosome testing by maternal plasma DNA sequencing: clinical laboratory experience and biology. Obstet Gynecol. 2015;125(2):375–82.
40. Sachs A, Blanchard L, Buchanan A, Norwitz E, Bianchi DW. Recommended pre-test counseling points for noninvasive prenatal testing using cell-free DNA: a 2015 perspective. Prenat Diagn. 2015;35(10):968–71.
41. Benn P, Malvestiti F, Grimi B, Maggi F, Simoni G, Grati FR. Rare autosomal trisomies: comparison of detection through cell-free DNA analysis and direct chromosome preparation of chorionic villus samples. Ultrasound Obstet Gynecol. 2019;54(4):458–67.
42. Rose NC, Kaimal AJ, Dugoff L, Norton ME, American College of Obstetricians and Gynecologists. Screening for fetal chromosomal abnormalities: ACOG practice bulletin, number 226. Obstet Gynecol. 2020;136(4):e48–69.

43. Bergstrand E, Borregaard Miltoft C, Tabor A. Performance of first trimester screening for trisomy 21 in twin pregnancies. Prenat Diagn. 2021;41(2):210–7.
44. del Mar GM, Quezada MS, Bregant B, Syngelaki A, Nicolaides KH. Cell-free DNA analysis for trisomy risk assessment in first-trimester twin pregnancies. Fetal Diagn Ther. 2014;35:204–11.
45. Bevilacqua E, Gil MM, Nicolaides KH, et al. Performance of screening for aneuploidies by cell-free DNA analysis of maternal blood in twin pregnancies. Ultrasound Obstet Gynecol. 2015;45:61–6.
46. Canick JA, Kloza EM, Lambert-Messerlian GM, et al. DNA sequencing of maternal plasma to identify Down syndrome and other trisomies in multiple gestations. Prenat Diagn. 2012;32:730–4.
47. Palomaki GE, Chiu RWK, Pertile MD, et al. International Society for Prenatal Diagnosis position statement: cell free (cf)DNA screening for Down syndrome in multiple pregnancies. Prenat Diagn. 2021;41(10):1222–32.

9 Noninvasive Prenatal Detection of Fetal Sex and RhD Blood Type

Akash Patel, S. R. Santosh Kumar,
and Chandrashekar Srinivasa

9.1 Introduction

Noninvasive Prenatal Screening (NIPS) offers a method to detect certain genetic disorders or abnormalities in a developing fetus without the need for invasive procedures like amniocentesis or chorionic villus sampling (CVS). Traditional prenatal diagnostic methods, such as amniocentesis and CVS, involve extracting fetal cells or genetic material from the amniotic fluid or placenta, respectively, which carry a small risk of complications, including miscarriage. In contrast, NIPS provides a safer alternative, utilizing maternal blood samples or other noninvasive methods to analyze fetal genetic material [1].

Over the years, the discovery and development of cell-free fetal DNA (cffDNA) in maternal blood have transformed NIPS. Throughout pregnancy, small fetal DNA fragments circulate in the mother's bloodstream, offering valuable genetic insights into the fetus. NIPS primarily screens for chromosomal abnormalities like Down syndrome (trisomy 21), Edwards syndrome (trisomy 18), and Patau syndrome (trisomy 13). Additionally, it can determine fetal sex and identify specific single-gene disorders when the genetic mutation is known [2].

Different methods are employed in NIPS. For instance, cffDNA involves analyzing DNA fragments from the fetus in maternal blood, screening for aneuploidies, and determining fetal sex. For instance, next-generation sequencing (NGS) enables

A. Patel
Department of Biotechnology, G.M. Institute of Technology, Davangere, Karnataka, India

S. R. Santosh Kumar
Department of Studies in Food Technology, Davangere University,
Davangere, Karnataka, India

C. Srinivasa (✉)
Department of Studies in Biotechnology, Davangere University, Davangere, Karnataka, India
e-mail: chandru.s@davangereuniversity.ac.in

R. A. Rather, S. C. Saha (eds.), *Non-invasive Prenatal Screening (NIPS) in Clinical Practice*, https://doi.org/10.1007/978-981-97-6402-0_9

sequencing of the fetal DNA in maternal blood, detecting chromosomal abnormalities and genetic mutations associated with inherited disorders [3]. Microarray analysis uses a DNA microarray to analyze maternal blood genetic material, identifying chromosomal abnormalities [4, 5]. These are the most widely used genomic methods in NIPS.

9.2 Fetal Sex and RhD Blood Type

Fetal sex and RhD blood type are two key elements in prenatal care and can be identified using a variety of procedures, including both invasive and noninvasive ways. Determining the sex of the fetus during pregnancy is important for a variety of reasons, like fetal sex identification might help detect potential dangers or genetic diseases that are more common in a specific sex. This knowledge enables healthcare providers to provide appropriate genetic counseling, conduct additional diagnostic testing as needed, and create tailored care plans [6].

The RhD blood type denotes the presence or absence of the RhD antigen in red blood cells. Complications can arise when the mother's RhD status and that of the fetus are incompatible. If a RhD-negative mother carries a RhD-positive pregnancy, the baby is at risk of developing RhD sensitization. During pregnancy or childbirth, fetal RhD-positive red blood cells may enter the mother's bloodstream and cause an immunological reaction. RhD sensitization can cause hemolytic disease of the newborn (HDN) in subsequent pregnancies if the fetus is RhD positive [7]. Determining the fetus's RhD blood type helps healthcare providers take preventive steps, such as providing RhD immune globulin (RhoGAM) to the mother, to lower the chance of sensitization and manage the pregnancy appropriately [7].

9.2.1 Biological Basis of Fetal Sex and RhD Blood Type Determination

The genetic information inherited from parents serves as the basic basis for fetal sex and RhD blood type determination (Fig. 9.1).

9.2.1.1 Molecular Basis of Fetal Sex Determination

Fetal sex is determined by the presence of sex chromosomes. Human cells usually have 23 pairs of chromosomes, including one set of sex chromosomes. Females have two X chromosomes (XX), and men have one X and one Y chromosome (XY). During fertilization, the father's sperm contains either an X or a Y chromosome, which determines the sex of the kid. Specific genes on the sex chromosomes are responsible for fetal sex determination. The SRY gene (sex-determining Region Y) is found on the Y chromosome and plays an important role in the development of male traits. If the fetus obtains an X chromosome from the father's sperm, it develops as a female due to the absence of the SRY gene [8]. If the fetus gets a Y chromosome containing the SRY gene, it develops into a male.

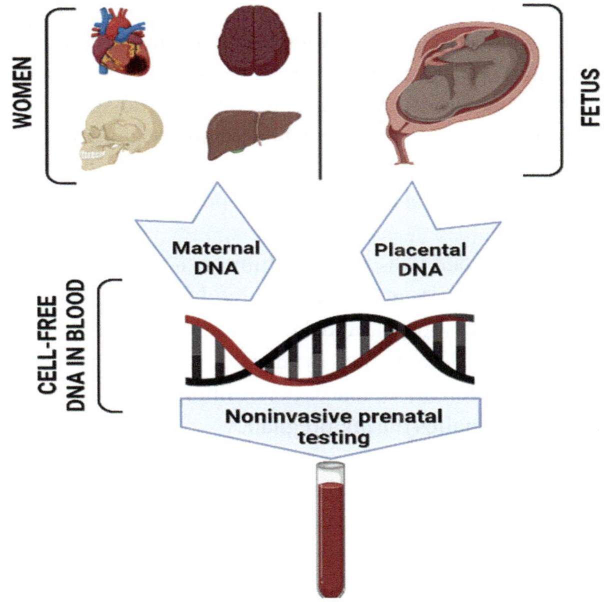

Fig. 9.1 Multiple sources of cell-free DNA in maternal plasma

9.2.1.2 Molecular Basis of RhD Blood Type Determination

The RhD blood type is defined as the presence or absence of the RhD antigen on the surface of red blood cells. The RhD antigen is encoded by the RHD gene, which is on chromosome 1. Individuals who inherit at least one functioning copy of the RHD gene from either parent are RhD positive (RhD+); those who receive two nonfunctional copies are RhD negative (RhD−) [9]. The genetic basis for determining fetal RhD blood type is the analysis of fetal DNA found in maternal blood. If the fetus has at least one functional RHD gene, the RhD antigen will be expressed on its red blood cells, indicating that it is RhD-positive. In contrast, if both copies of the RHD gene are nonfunctional, the fetus will not express the RhD antigen and will be labeled RhD−.

9.2.2 Clinical Significance of Fetal Sex and RhD Blood Type

The fetal sex and RhD blood type are clinically significant in prenatal care because they guide decision-making and influence pregnancy management. Here are the clinical implications of fetal sex and the RhD blood type.

9.2.2.1 Fetal Sex

- **Family planning:** Knowing the fetal sex enables parents to prepare emotionally and practically for the birth of a boy or a girl. It assists in making selections about names, clothing, and nursery setup.

- **Genetic counseling:** Fetal sex determination can aid in identifying potential dangers or genetic problems that are more common in a specific sex. It allows healthcare clinicians to provide appropriate genetic counseling, conduct additional diagnostic testing as needed, and design tailored care plans [10].
- **Psychological impact:** Understanding fetal sex can have psychological repercussions for parents, affecting their bond with the unborn child and allowing for early emotional connection and bonding.

Many genetic disorders exhibit sex-specific patterns of inheritance, meaning they disproportionately affect individuals of a particular sex. For instance, conditions like hemophilia and Duchenne muscular dystrophy primarily affect males, while others like Turner syndrome and congenital adrenal hyperplasia are more common in females. Hence, determining fetal sex is crucial to diagnose certain sex-related anomalies during pregnancy.

9.2.2.2 RhD Blood Type

- **RhD sensitization prevention:** Pregnancies in which the mother is RhD-negative and the fetus is RhD-positive increase the chance of RhD sensitization. Knowing the fetus' RhD blood type allows healthcare personnel to take preventive steps, such as giving RhD immune globulin (RhoGAM) to the mother [9]. This medication stops the mother from producing antibodies against the RhD antigen, lowering the likelihood of sensitization and problems in future pregnancies.
- **Management of subsequent pregnancies:** RhD blood type testing is critical for managing subsequent pregnancies. If the mother is sensitized to the RhD antigen, careful monitoring is essential in subsequent pregnancies to assess the risk of HDN. Knowing the fetal RhD blood type enables healthcare providers to build suitable treatment strategies to guarantee the fetus's well-being and provide required interventions if needed [11].
- **Blood transfusion planning:** RhD blood type information is critical in blood transfusion cases. RhD-negative individuals can get RhD-negative or RhD-compatible blood, whereas RhD-positive individuals can only receive RhD-positive blood. Knowing the fetal RhD blood type aids in the planning of transfusions when the fetus requires an intrauterine or neonatal blood transfusion.

9.3 Noninvasive Prenatal Screening (NIPS)

NIPS is the use of noninvasive techniques to identify and evaluate fetal genetic information while pregnant. Unlike invasive procedures, NIPS allows for fetal genetic testing without directly contacting the fetus, reducing hazards to both the mother and the unborn child. NIPS usually entails testing cffDNA in the maternal circulation. During pregnancy, tiny fetal DNA fragments are released into the mother's bloodstream [12]. These fragments can be separated and studied to learn about the fetus' genetic makeup and potential health issues. The cffDNA contains genetic material from the placenta and represents the fetal genotype [13].

9.3.1 Principles and Techniques of NIPS for Fetal Sex and RhD Blood Type

NIPS for fetal sex and RhD blood type diagnosis analyses cffDNA in the mother's bloodstream. The principles and techniques used in NIPD for fetal sex and RhD blood type are as follows:

9.3.2 Fetal Sex Determination

(a) **Quantitative PCR (qPCR):** This approach assesses the relative abundance of X and Y chromosome-specific DNA sequences in cffDNA. It uses primers specific to X and Y chromosome regions and real-time PCR to quantify the amplification products. Fetal sex can be identified by comparing the X and Y chromosomal sequence ratios.
(b) **NGS:** The method NGS can be used to determine fetal sex by sequencing millions of DNA fragments in cffDNA. By evaluating the read counts that correlate to the X and Y chromosomes, the relative abundance of each can be established, allowing for fetal sex identification [14].

9.3.2.1 Fetal RhD-Type Determination

(a) **PCR:** Several exons of the RHD gene exhibit high levels of polymorphism, characterized by the presence of multiple allelic variants within the population. Exons 4, 5, 7, and 10 are among the most polymorphic regions of the RHD gene, with numerous alleles identified that influence RhD antigen expression. Polymorphisms within these exons can result in different antigenic profiles, ranging from strong RhD positivity to complete RhD negativity. PCR assays target these exons. Exons 4, 5, 7, and 10 are among the most frequently targeted regions due to their polymorphic nature and relevance to RhD antigen expression.

It is crucial to highlight that while NIPD procedures for fetal sex and RhD blood type diagnosis are highly accurate, false-positive or false-negative results may occur in rare circumstances. Confirmatory testing may be required in some cases to confirm accurate results and inform proper medical management throughout pregnancy.

9.4 Conclusion

NIPS has evolved over time, reforming the process of prenatal screening and providing newfound ease. NIPS has been widely employed for the detection of chromosomal aneuploidies, monogenic disorders, and rare genetic conditions, showcasing significant potential in determining fetal sex and RhD status, critical aspects for expecting mothers from various perspectives. However, as NIPS continues to advance, it brings forth challenges that must be addressed to ensure its efficacy and reliability as a screening tool.

References

1. Alfirevic Z, Navaratnam K, Mujezinovic F. Amniocentesis and chorionic villus sampling for prenatal diagnosis. Cochrane Database Syst Rev. 2017;9(9):CD003252.
2. Allyse M, Sayres LC, Goodspeed TA, Cho MK. Attitudes towards non-invasive prenatal testing for aneuploidy among US adults of reproductive age. J Perinatol. 2014;34(6):429–34.
3. Avent ND, Chitty LS. Non-invasive diagnosis of fetal sex; utilisation of free fetal DNA in maternal plasma and ultrasound. Prenat Diagn. 2006;26(7):598–603.
4. Baird PA, Anderson TW, Newcombe HB, Lowry RB. Genetic disorders in children and young adults: a population study. Am J Hum Genet. 1988;42(5):677–93.
5. Benn P, Borrell A, Cuckle H, Dugoff L, Gross S, Johnson JA, et al. Prenatal Detection of Down Syndrome using Massively Parallel Sequencing (MPS): a rapid response statement from a committee on behalf of the Board of the International Society for Prenatal Diagnosis, 24 October 2011. Prenat Diagn. 2012;32(1):1–2.
6. Devaney SA, Palomaki GE, Scott JA, Bianchi DW. Noninvasive fetal sex determination using cell-free fetal DNA: a systematic review and meta-analysis. JAMA. 2011;306(6):627–36.
7. Devers PL, Cronister A, Ormond KE, Facio F, Brasington CK, Flodman P. Noninvasive prenatal testing/noninvasive prenatal diagnosis: the position of the National Society of Genetic Counselors. J Genet Couns. 2013;22(3):291–5.
8. Farrell RM, Agatisa PK, Nutter B. What women want: lead considerations for current and future applications of noninvasive prenatal testing in prenatal care. Birth. 2014;41(3):276–82.
9. Galbiati S, Smid M, Gambini D, Ferrari A, Restagno G, Viora E, et al. Fetal DNA detection in maternal plasma throughout gestation. Hum Genet. 2005;117(2–3):243–8.
10. Hill M, Lewis C, Jenkins L, Allen S, Elles RG, Chitty LS. Implementing noninvasive prenatal fetal sex determination using cell-free fetal DNA in the United Kingdom. Expert Opin Biol Ther. 2012;12(Suppl. 1):S119–26.
11. Chiu RW, Chan KC, Gao Y, Lau VY, Zheng W, Leung TY, et al. Noninvasive prenatal diagnosis of fetal chromosomal aneuploidy by massively parallel genomic sequencing of DNA in maternal plasma. Proc Natl Acad Sci U S A. 2008;105(51):20458–63.
12. Petersen OB, Vogel I, Ekelund C, Hyett J, Tabor A, Danish Fetal Medicine Study Group, Danish Clinical Genetics Study Group. Potential diagnostic consequences of applying non-invasive prenatal testing: population-based study from a country with existing first-trimester screening. Ultrasound Obstet Gynecol. 2014;43(3):265–71.
13. Lo KK, Karampetsou E, Boustred C, McKay F, Mason S, Hill M, Plagnol V, Chitty LS. Limited clinical utility of non-invasive prenatal testing for subchromosomal abnormalities. Am J Hum Genet. 2016;98(1):34–44.
14. Navaratnam K, Alfirevic Z, Royal College of Obstetricians and Gynaecologists. Amniocentesis and chorionic villus sampling: Green-top Guideline No. 8 July 2021: Green-top Guideline No. 8. BJOG. 2022;129(1):e1–e15.

Detection of Monogenic Disorders Using Noninvasive Prenatal Screening

10

Mohammad Taheri, Atefeh Harsij, Arian Askari, and Ashkan Pourtavakoli

10.1 Introduction

10.1.1 Monogenic Disorders

A Monogenic disorder is characterized by a genetic anomaly where a single gene is accountable for disease pathogenesis [1]. Such disorders typically result from mutations in a specific gene or de novo mutations occurring in parental germline cells. Common mutations observed in monogenic disorders include single nucleotide substitutions or microdeletions, although more complex mutations have also been linked to these conditions [2]. Currently, over 10,000 monogenic disorders or phenotypes stemming from mutations in a single gene have been documented. While these disorders are individually rare, affecting roughly 10 in every 1000 individuals, accurate diagnosis and effective management present significant challenges in the realm of medical genetics.

10.1.2 Noninvasive Prenatal Screening Applications for Monogenic Disorders

Early detection of cell-free fetal DNA (cffDNA) in the mother's bloodstream in the first trimester caused an incremental utilization of (noninvasive prenatal screening)

M. Taheri (✉)
Institute of Human Genetics, Jena University Hospital, Jena, Germany

Urology and Nephrology Research Center, Shahid Beheshti University of Medical Sciences, Tehran, Iran

A. Harsij · A. Askari · A. Pourtavakoli
Department of Medical Genetics, Shahid Beheshti University of Medical Sciences, Tehran, Iran

R. A. Rather, S. C. Saha (eds.), *Non-invasive Prenatal Screening (NIPS) in Clinical Practice*, https://doi.org/10.1007/978-981-97-6402-0_10

Table 10.1 Screening or diagnostic approaches for a number of monogenic disorders

Inheritance	Disorders	Gene	Mutation type	Screening or diagnostic test
Autosomal dominant	Achondroplasia	*FGFR3*	Point mutation	PCR-RED
	Thanatophoric dysplasia	*FGFR3*	Point mutation	PCR-RED
	Huntington disease	*HTT*	Triple repeat expansion (CAG)	QF-PCR
	Myotonic dystrophy – 1	*DMPK*	Triple repeat expansion (CTG)	PCR
Autosomal recessive	Cystic fibrosis	*CF*	Deletions and Point mutation	MEMO-PCR COLD-PCR AS-PCR
	Congenital adrenal hyperplasia	*CYP21*	Deletions and conversions	PCR
	Spinal muscular atrophy	*SMN1*	Deletions	NGS- RHDO
	Methylmalonic academia	*MUT* *MMADHC*	Deletions Point mutation	ddPCR
	Maple syrup urine disease	*BCKDH (A, B), DBT*	Point mutation	Targeted MPS NGS- RHDO
	Wilson's disease	*ATP7B*	Point mutation	cSMART-PCR
	β thalassemia	*HBB*	Point mutation	Nested RT-ASPCR AS-PCR dPCR NGS- RHDO
X-linked	Duchenne muscular dystrophy (DMD)	*DYS*	Point mutation	NGS- RHDO
	Hemophilia A	*F8*	Point mutation	ddPCR
	Hemophilia B	*F9*	Point mutation	ddPCR

PCR polymerase chain reaction, *PCR-RED* PCR-restriction enzyme digestion, *QF-PCR* quantitative fluorescence PCR, *COLD-PCR* coamplification at lower denaturation temperature PCR, *MEMO-PCR* mutant enrichment with 3′-modified oligonucleotides-PCR, *cSMART-PCR* circulating single-molecule amplification and resequencing technology-PCR, *AS-PCR* allele specific-PCR, *RT-ASPCR* real-time-allele-specific PCR, *NGS-RHDO* next-generation sequencing-relative haplotype dosage, *ddPCR* droplet digital PCR

NIPS to screen many monogenic disorders [3] (Table 10.1). Unlike the obtained results from cffDNA that require further confirmation using invasive methods, genetic tests performed based on positive family history or an abnormal finding in sonography are targeted and specific; NIPS results do not require validation and are used for screening directly. This neutralizes possible risks that are associated with invasive techniques [4]. Despite the application of cffDNA for aneuploidy screening, implementing NIPS to diagnose monogenic disorders has now gained official confirmation under specific terms and conditions from the National Health Service [5, 6].

10.1.2.1 Diverse Approaches in NIPS

Based on different factors, such as type of mutation, pattern of inheritance, and facilities, different kinds of NIPS methods are utilized. For instance, real-time PCR, targeted next-generation sequencing (NGS), amplicon-based NGS, digital PCR (dPCR), coamplification at lower denaturation temperature PCR (COLD-PCR), circulating single-molecule amplification and resequencing technology (cSMART), and mini sequencing are standard methodologies. The primary approach of NIPS is based on detecting mutations unrelated to maternal cell-free DNA (cfDNA), so it could be used for prenatal detection of mutations inherited from the paternal side and also the de novo ones [4]. When NIPS is performed to exclude paternal mutations, invasive testing is required for definitive diagnosis if the paternal mutation is detected to confirm whether or not the maternal mutation has also been inherited. This chapter provides a thorough examination of the use of NIPS for diagnosing monogenic disorders, including a comparison of various invasive approaches to NIPS and an analysis of their different aspects.

10.2 Technologies Utilized NIPS for Paternally Inherited and De Novo Mutations, as well as Fetal Sex Determination in X-Linked Recessive Disorders

Before investigating studies that have been done concerning NIPS of monogenic disorders, it is necessary to discuss different methods and technologies that are implemented in NIPS of monogenic disorders more straightforwardly. As mentioned earlier, one of the simplest and earliest approaches that are being used in NIPS regarding recognition of paternally inherited and de novo mutations are PCR-based techniques, such as real-time quantitative PCR (RT-qPCR), dPCR, and PCR-restriction enzyme digestion (PCR-RED) [7].

10.2.1 Digital PCR and Real-Time Quantitative PCR

In disorders with X-linked recessive inheritance, NIPS is widely used for sex determination in families at risk to rule out unnecessary interventions if the fetus is female. The most common method used for this purpose is qPCR, and in this technique, the target genes usually are sex-determining region Y (SRY) in the Y chromosome, DYS14, and copies of DAZ [8]. However, since the cffDNA level is low in the early months of pregnancy, using qPCR is not the first method of choice in these months. Therefore, a more advanced and precise method, droplet digital PCR (ddPCR), was recently developed [9]. The same materials and methods are used in ddPCR, a type of qPCR, but the reaction occurs in thousands of droplets on the nanometer scale instead of a PCR strip mixed with different analytes [10]. In ddPCR, the DNA molecule is extensively diluted so that there is only one or no DNA molecule in each droplet; therefore, it is evident that ddPCR has a higher sensitivity and specificity than the conventional qPCR. ddPCR is mainly used in situations like the early months of pregnancy when the amount of DNA is at its lowest level. Hence, ddPCR is also applicable for the sex determination of fetuses when there is a risk of inheriting an X-linked recessive condition. Duchene muscular dystrophy (DMD) and its milder form, Becker muscular dystrophy, in addition to hemophilia, are three good candidates for sex determination using NIPS. Congenital adrenal hyperplasia is also suitable for sex determination since pharmaceutical interventions are required solely in female fetuses [7, 11]. In addition, PCR-based techniques are also applicable for detecting paternally inherited mutations to exclude the inheritance of a specific mutation in the fetus. This approach is called paternal mutation exclusion [12].

10.2.2 PCR and Restriction Enzyme Digestion

Another example of a PCR-based technique that is utilized in NIPS is PCR-RED. In this method, after the PCR reaction, the products are digested with a restriction enzyme with a recognition site in the target gene. This method is widely used in NIPS of skeletal dysplasias like achondroplasia and thanatophoric dysplasia (TD), both caused by fibroblast growth factor receptor 3 (FGFR3) mutations [13, 14], which are discussed in detail in the following sections.

10.2.3 PCR-NGS

A more precise and accurate method extensively used in NIPS is a combination of PCR and NGS [7]. In this method, first, the gene of interest is amplified and then sequenced by NGS. Additionally, PCR-NGS is used in paternal mutation exclusion of autosomal recessive disorders such as cystic fibrosis and beta thalassemia [7, 15].

10.2.4 Applied Technologies in NIPS of Maternally Inherited Mutations

Previous sections were mainly focused on NIPS of paternal mutations or the de novo ones. Nevertheless, the main challenge lies within maternally inherited mutations. cfDNA is contaminated with maternal DNA and poses several challenges. PCR-NGS is also inefficient in these scenarios since it has a low reading depth. Therefore, further, more precise and accurate methods are required to elucidate the ratio of wild-type and mutant alleles and pinpoint the subtlest disturbances in this ratio. Effective techniques such as relative mutation dosage (RMD) and relative haplotype dosage analysis (RHDO) serve as good examples [7], which are discussed later in NIPS of autosomal recessive disorders.

10.3 NIPS for Autosomal Dominant Disorders with a Single Gene Etiology

The aim of NIPS in autosomal dominant disorders is to diagnose and identify mutant alleles inherited from either the father or mother in the cfDNA of maternal plasma to determine the status of fetal affliction. In NIPS, two different approaches exist for this category of disease. The paternal mutation detection/exclusion approach, using various methods and techniques developed in numerous studies, is performed in at-risk pregnancies to determine the status of fetal affliction in cases where the mutation is inherited paternally [16]. However, identifying maternally inherited mutant alleles in maternal plasma is more complex than detecting paternally inherited mutations due to the large amount of maternal mutant alleles present, requiring more precise and sensitive techniques. Two approaches used in such cases are RMD and RHDO [17]. However, it should be noted that these methods are not applicable in all autosomal dominant diseases. For instance, in diseases caused by expansions of trinucleotide repeats, the diagnosis of mutations is impossible using these methods due to the fragmented nature of DNA in maternal plasma [17]. Among the methods used in these conditions for paternal mutation detection/exclusion, PCR plus fragment analysis [18] and quantitative fluorescent PCR [19] can be mentioned.

10.3.1 NIPS for Mutation in Fibroblast Growth Factor Receptor 3 (FGFR3)

Mutations in FGFR3, located on the short arm of chromosome 4, lead to disruptions in bone growth, notably long bones, ultimately leading to dwarfism or condition, skeletal dysplasia. Achondroplasia and TD are two known conditions that result from FGFR3 mutations.

10.3.1.1 Achondroplasia

Achondroplasia is a monogenic disorder and is inherited in an autosomal dominant pattern. As mentioned earlier, mutations in the FGFR3 gene are responsible for this condition. Achondroplasia is the most common form of non-lethal skeletal dysplasia and occurs in about 5–15 in 100,000 newborns. Most cases of achondroplasia are sporadic and without a known familial history, mainly due to de novo mutation in paternal germline cells. c.1138G>A or p. Gly380Arg is responsible for approximately 98% of mutations, and 1% of them are due to c.1138G>C in FGFR3 [20, 21].

Since most of the ultrasonographic indicators of de novo achondroplasia, such as short limbs, trident hands, and frontal bossing, are presented in the third trimester of pregnancy, most of these conditions are not recognized in routine screenings for fetus anomalies. However, misdiagnosis is possible even in the third-trimester screenings [22, 23]. Despite extensive studies on deciphering the best indicators for achondroplasia in ultrasound, definitive diagnosis is only possible through mutation detection in FGFR3 or post-birth radiological features. In the past, definitive prenatal diagnosis was only possible through invasive techniques to obtain fetal DNA followed by sequencing. The advent of innovative and noninvasive techniques, such as analyzing cffDNA, which circulates in mothers' blood, has paved the way for a more flexible and cost-effective way of diagnosis. As mentioned earlier, since most of the circulating cffDNA in maternal blood originated from the mother, NIPS is mainly applicable for paternally transmitted mutations, and inherited mutations from the mother usually are not taken into account when considering NIPS. However, the pregnancy period positively correlates with the amount of cfDNA present in maternal plasma [24–26]. Since most achondroplasia indicators are reported in the third trimester of pregnancy when the concentration of cfDNA is at its peak, provided that they are detected through sonography, NIPS is also a suitable choice for diagnosis [20].

10.3.1.2 Thanatophoric Dysplasia (TD)

TD is another form of skeletal dysplasia with an autosomal dominant inheritance pattern, also due to FGFR3 mutations. This condition is the most common form of lethal dysplasia and occurs in about 2–3 in 100,000 newborns. Multiple mutations in FGFR3 are found to be the cause of disease, with the Arg248Cys, Tyr373Cys, and Lys650Glu being the most common ones. TD is categorized into type-1 and type-2, although they are overlapped in different aspects when considering molecular, radiological, and clinical features. Type-1 TD mainly occurs as substitutions like Arg248Cys or Tyr373Cys, and it affects the femur bone in a way that looks inclined to bend, and craniosynostosis barely occurs. Type-2 TD is the milder form of the disease which manifests itself with a straight femur and craniosynostosis, caused mainly by Lys650Glu substitution.

The accuracy of TD diagnosis via ultrasound findings is estimated to be around 40–88% [27]. TD is diagnosed in routine screenings performed in the second trimester of pregnancy, with clinical features like abnormal bones, macrocephaly, small chest, short fingers, frontal bossing, cloverleaf skull, and telephone receiver-shaped femora, diagnosed via sonography. However, extensive implementation of ultrasonographic techniques has also led to first-trimester diagnosis [28]. After the detection of the condition by ultrasound, molecular techniques are conducted to detect causative mutations. Until recently, obtaining genetic material was solely possible through invasive methods like amniocentesis or chorionic villus sampling (CVS) [29]. With the discovery of cffDNA in mother's blood, implementing NIPS as a definitive and noninvasive method with a lower risk of miscarriage is now possible.

10.3.1.3 PCR-Based or NGS Methods

In NIPS, different PCR-based methods like PCR and PCR-RED can be employed to identify the underlying mutation in conditions caused by a single specific mutation, such as achondroplasia. Conversely, PCR-based techniques in monogenic disorders like TD, which is caused by different kinds of mutations, can only identify a known mutation to further hamper its transmission in the family. In a monogenic condition like TD, in case of a de novo mutation, PCR-based techniques are not the method of choice since each PCR-RED reaction can identify only one mutation at a time (because of the unique nature of restriction enzyme recognition sites). That is why PCR-based techniques are helpful in identifying single and specific mutations and utilizing them is fruitless in conditions with a responsible de novo mutation [21].

Nevertheless, since most of the cfDNA in mother's circulation is of maternal origin, these simple molecular techniques are less precise than the more advanced and high-throughput methods like NGS. By utilizing NGS, the accuracy of NIPS has elevated, and its application in screening monogenic disorders like skeletal dysplasia is on the rise. NGS is able to detect any known or unknown variant in DNA, and by doing so, is the best method of choice when NIPS is considered [21].

10.3.2 NIPS Applications in Monogenic Disorders with Triple Repeat Expansion: Myotonic Dystrophy-1 (DM1) and Huntington Disease (HD)

Triple repeat expansion disorders are caused by expanding a triple nucleotide sequence in a specific gene. Most of these conditions affect the nervous system. In this section, we will be discussing NIPS's application in HD and DM1.

10.3.2.1 Huntington's Disease

HD is a progressive neurological disorder with symptoms like motor impairments and sensory and psychotic difficulties. Abnormal expansion of CAG trinucleotide in exon 1 on the HTT gene, located on chromosome 4p, is the leading cause of HD. Alleles with <27 CAG repeats are considered normal, whereas alleles with CAG repeats between 27 and 35 are pre-mutable, and pathogenic alleles usually have more than 35 repeats of CAG in the HTT gene. Pre-mutable alleles are unstable during the transmission to offspring (especially in a paternal way of inheritance), and there is a potential risk of HD development in the next generations. In pregnancies in which there is an affected person with HD in a family tree, prenatal diagnosis via PCR analysis of CAG repeats in HTT's exon is possible. This analysis is usually performed on the chorionic villus extracted DNA at 11–14 weeks of pregnancy [30]. As mentioned earlier, CVS is associated with an increased risk of miscarriage. By implementing maternal plasma for NIPS purposes, there is a promising future for using NIPS in diagnosing conditions caused by paternal repeat expansion.

One of the main studies related to this matter is a study conducted by Gonzalez et al. [30]. They investigated the possible application of NIPS in prenatal diagnosis of HD-affected fetuses with inherited paternal repeat expansions by methods such as semi-quantitative fluorescent PCR and the fetal DNA in maternal blood. This pioneering research paved the way for further applications of NIPS by using maternal plasma in HD diagnosis [30].

10.3.2.2 Myotonic Dystrophy-1

DM1 is the most frequent muscular dystrophy presented with clinical features like myotonia and progressive muscle weakness and degeneration. DM1 is caused by an increased number of CTG repeats in the 3′ UTR of the dystrophia myotonic protein kinase (DMPK) gene, located on 19q.13. Usually if the number of CTG repeats exceeds >50, it is considered a pathogenic state. Generally, molecular diagnosis of DM1 consists of analyzing CTG repeats in genomic DNA (gDNA), and the source of gDNA could be from the trophoblast cells [31].

Amicucci and colleagues were keen on identifying a way of prenatal diagnosis by implementing fetal DNA present in the maternal plasma [31]. Prior to this research, one study was also conducted to elucidate possible ways of noninvasive prenatal diagnosis of DM1. However, the study was restrained to trophoblast cells originating from the lower urinary tract sections [32]. The amount of available fetal DNA from trophoblast origin is lower than that obtainable from maternal plasma [25, 32]; therefore, it was concluded in the 2000s study to utilize maternal plasm as a noninvasive source of DM1 prenatal diagnosis, and it was stated that this noninvasive approach should substitute invasive ones in the first trimester of pregnancy. However, it is important to note that this method is only applicable in DM1 cases caused by paternally inherited alleles (alleles with an abnormal expansion of CTG repeats inherited from the father) [31].

10.3.2.3 NIPS Approaches Using Circulating Trophoblastic Fetal Cells (CFTCs) in Diseases Arising from Triple Repeat Expansions

As mentioned earlier, there has been an incremental trend in using cffDNA as a means of NIPS approach in monogenic disorders [33]; however, challenges still exist in cffDNA application for disorders that are caused by triplet repeat expansions, mainly due to fragmented nature of cffDNA and their trivial size [34, 35]. Therefore, theoretically, using filtered CFTCs from maternal blood for NIPS of these disorders has some advantages since they are not fragmented or contaminated by maternal DNA. However, it is worth mentioning that comprehensive and extensive application of CFTCs is troubled since they are very scarce in the maternal bloodstream and they undergo apoptosis most of the time; thus, in order to reach a broader application of CFTCs in triple repeat expansion disorders concerning NIPS, more advanced technologies are required for filtering and analyzing these cells [34].

10.4 NIPS for Monogenic Autosomal Recessive (AR) Disorders

As we mentioned in the introduction, early approaches in NIPS were mainly focused on mutations not present in maternal cfDNA, therefore, NIPS in autosomal dominant conditions was centered on paternally inherited mutations or de novo ones. Utilizing NIPS in autosomal recessive disorders is more challenging than autosomal dominant ones since most of the cfDNA in the mother's circulation originated from the mother herself; subsequently, high levels of maternal alleles exist. Conclusively, NIPS for AR disorders requires a more precise and accurate method for determining the ratio of wild-type and mutant alleles. In this regard, two different techniques, including RMD and RHDO, are helpful [36].

Generally, NIPS methods for determining AR conditions are categorized into two groups, mainly based on similar or different natures of inherited mutation from parents. The first approach, "paternal mutation detection/exclusion", is applied in situations where parents are carriers of different mutations, and the affected offspring is called a "compound heterozygote". As we discussed earlier, in this condition, if a paternally inherited mutation is detected in mother's circulation, additional invasive methods are needed in order to pinpoint whether the fetus has also inherited a maternal mutation or not. Conversely, if a paternally inherited variant is not found, the worst scenario would be the carrier status of the fetus for a maternal mutation, which does not require invasive tests [37, 38]. COLD-PCR and microarray techniques are amongst efficient methods that are utilized for paternal mutation detection/exclusion in NIPS of AR conditions. In COLD-PCR, the lowest amounts of mutated alleles can be amplified without altering the type and location of the variant. In paternal mutation detection/exclusion using COLD-PCR, paternal alleles are selectively amplified. Despite their effectiveness and high sensitivity, COLD-PCR and microarray techniques are limited to conditions when parental mutations differ in nature [39].

The second condition is when both mother and father carry the same mutation, and the affected fetus would be homozygous for that specific variant. In this situation, since maternally originated alleles are seen in cfDNA, fetus genotyping is challenging, and more precise methods capable of measuring wild type/mutated ratio are required. RMD and RHDO are useful for this purpose.

10.4.1 Relative Mutation Dosage

RMD is capable of measuring the ratio of wild-type and mutant alleles in cfDNA [26]. Consider a situation when parents are carriers of an AR condition (genotype: AA). In this scenario, if the fetus is affected, it would be a homozygous affected individual, demonstrated by an increased level of mutant allele observed in the RMD measurement. If the fetus is in a carrier status like the mother, the RMD measurement will show an even distribution of normal and mutant alleles. Moreover, lastly, if the fetus is in an unaffected homozygous status, RMD observation shows an elevated level of normal allele [4, 26]. Specific techniques such as RMD are developed in order to eliminate the involvement of invasive methods for diagnosing AR conditions. Therefore, approaches being used in RMD are required to fulfill adequate accuracy and specificity for screening the lowest amount of cfDNA in the mother's circulation [7]. Hence, methods such as dPCR and ddPCR add to RMD's specificity and accuracy [40]. ddPCR is one of the most sensitive tests when considering these situations and is capable of detecting the rare variant of fetal DNA even with the presence of maternal DNA. Although the implementation of ddPCR has gained promising results in NIPS of monogenic disorders, this technique is not widely used in clinical practice yet [7]. Some of these monogenic disorders include achondroplasia [41], cystic fibrosis [42], sickle cell anemia [26, 40], beta thalassemia [43–45], hemophilia [11, 46], and monogenic diabetes [47]. Recent advances have led to the advent of innovative generations of dPCR, such as ddPCR, and this approach is more reliable in detecting scarce variants than its former progenitors [7].

10.4.2 Relative Haplotype Dosage

RHDO is a combination of RMD and linkage analysis studies applicable for NIPS of mutations that are not solely inherited from the father [4]. On the other hand, in conditions where the causative mutation is complex in nature (large deletions or mutations in repetitive tandem repeats), RHDO might not work as efficiently. Therefore, in RHDO, on the one hand, different single nucleotide polymorphisms (SNPs) of a specific haplotype that are associated with the underlying gene are explored. On the other hand, allelic ratios that contain linked haplotypes of the mutant and normal genes are pinpointed. By doing so, the detection of parental haplotypes leads to the detection of haplotypes transferred to the fetus [4, 36, 48]. RHDO for NIPS of AR conditions requires sampling from both parents and proband, followed by an NGS on their genomic DNA and sequence analysis, and ultimately haplotype phasing for mutant and normal alleles. Nevertheless, in NIPS for X-linked disorders, NGS and haplotype phasing is exclusively performed on maternal and fetal DNA, and the father's sample is not required [49].

Disorders that are eligible for NIPS through RHDO include autosomal recessive conditions such as congenital adrenal hyperplasia [7], spinal muscular atrophy [50], cystic fibrosis [51], and X-linked Duchene muscular dystrophy [52]. Efforts are already underway for the optimization of RHDO for conditions like retinoblastoma [53], alpha and beta thalassemia [36], hemophilia [54], and mucopolysaccharidosis type II [54]. Since RHDO considers allelic ratio for haplotypes, the number of SNPs analyzed in RHDO is greater than RMD, and usually, it is a more dependable technique statistically [4]. Nevertheless, it should be noted that RHDO is also prone to errors and challenges. For instance, NIPS for autosomal recessive disorders is challenging since parents have similar haplotypes due to a shared ancestor [4]. Another challenging issue in RHDO is that, as we mentioned previously, this technique requires high-risk and low-risk alleles for haplotype phasing, which is achieved through sampling both parents and proband in AR conditions and mother/proband sampling in X-linked disorders. However, efforts have been made to reach a "proband-free RHDO by utilizing long-read sequencing" [4, 7]. Targeted locus amplification and microfluidics-based linked-read sequencing are among the methods being studied to reach a proband-free RHDO [7].

As mentioned above, one common difficulty faced in RHDO is the access to parents and proband samples for haplotype phasing so RMD has an advantage in this regard since it does not require DNA samples from the father and proband. However, RHDO is still advantageous in different aspects compared to RMD. For instance, in RMD, mutation analysis is limited to specific mutations at a specific locus, and this method is not applicable for complex mutations like significant structural changes and variants located in repetitive tandems. RHDO, on the other hand, is not limited to a specific area; because of haplotyping, it consists of a larger number of SNPs [7].

It is worth mentioning that other approaches, such as circulating cSMART and cfDNA barcode-enabled single-molecule test (cfBEST) for NIPS, are currently being investigated [55, 56]. cSMART has more sensitivity and specificity compared to RMD, but similarly, father and proband DNA samples are not required. One of the main studies done through cSMART is a proof-of-concept study on Wilson's disease, which is discussed in the next section [57].

In the next part, we will go through some of the NIPS-related studies conducted on monogenic disorders with autosomal recessive patterns of inheritance.

10.4.3 NIPS for Cystic Fibrosis

Cystic fibrosis (CF) is a severe autosomal recessive disease that typically presents in childhood. It affects several systems in the body, primarily the respiratory and digestive systems. CF is caused by mutations in the CFTR (cystic fibrosis transmembrane conductance regulator) gene located on chromosome 7q31.2. Over 2000 mutations causing CF have been identified, with the most common one being p.Phe508del mutation.

Invasive prenatal diagnosis methods with associated risks to the fetus are available for pregnancies in which both parents are known carriers of CF mutations [58]. Studies have investigated the possibility of NIPS for CF using cfDNA in maternal plasma, both in cases where the parental mutations are the same and where they are different. Some of these studies are discussed below.

Initially, NIPS for CF in populations with high disease prevalence, such as European populations, was performed using paternal mutation detection/exclusion, which was designed for about ten of the most common CF mutations in these populations [4, 59]. As a result, NIPS was only performed for couples who were carriers of different mutations, and the father carried one of these ten common mutations.

One of the studies conducted in the field of paternal mutation detection/exclusion in CF is the study by Guissart and colleagues in 2015 [58]. They conducted an investigation into a suitable approach for NIPS of CF in cases where the parents do not carry the same mutation in the CFTR gene and the father is a carrier of the common pPhe508del mutation. In such cases, if the fetus is affected, it will be considered "compound heterozygous." This study aimed to assess the effectiveness of the mutant enrichment with 3-modified oligonucleotides PCR coupled with fragment length analysis (MEMO-PCR-FLA) method in identifying the pPhe508del mutation inherited from the father. Additionally, multiplex fluorescent PCR was employed for quality control and to confirm the presence of cffDNA in the samples.

MEMO-PCR is a technique that uses an oligonucleotide modified at the 3′ end to prevent the wild-type allele from replicating, thus increasing the replication and enrichment of the mutant allele [60]. Using this test, if the pPhe508del mutation inherited from the father is not detected in the mother's plasma, the most pessimistic genotype of the fetus would be healthy heterozygote, and the mutation would be carried from the mother. Therefore, invasive tests are not necessary [58].

There are other methods that can be used in paternal mutation detection/exclusion, one of which we mentioned earlier is the COLD-PCR method, in which, like MEMO-PCR, enrichment of the mutant allele occurs. However, this approach is known to have a higher risk of reporting false negative results despite being cost-effective and efficient [39]. Both of these approaches are limited to cases where couples are carriers of different mutations.

In various other studies, other methods and techniques have also been proposed for the detection of paternally inherited fetal mutations in maternal plasma, some of which we will mention below:

1. Performing NIPS using NGS testing designed for the ten most common CF mutations on cfDNA in pregnancies where the father is a carrier of one of these ten mutations and the mother carries a different mutation [59].
2. Determining the fetal genotype for paternally inherited alleles using the SNaPshot method, which is a mini-sequencing-based method. This method can also be a suitable candidate for routine clinical use in NIPS of CF disease due to its high accuracy, sensitivity, ease of use, low cost, and fastness [61].

In addition, another study investigated the use of the allele-specific PCR (AS-PCR) method, which was designed specifically for detecting D1152H mutation in the CFTR gene, and also investigating approaches that increase the accuracy and sensitivity of this test, making it suitable for clinical application in NIPS of CF [62].

As the most prevalent mutation that causes CF is p.Phe508del, it has been estimated that in about 50% of carrier couples, both parents are carriers of the same mutation, and therefore, paternal mutation detection/exclusion is unsuitable for NIPS. In such cases, determining the fetal genotype with certainty is challenging, and techniques are needed in which the relative amounts of mutant alleles are determined and measured instead of just detecting the CF mutation [51].

One approach in this area was performing RHDO using whole genome NGS [9]. However, recently RHDO that uses data obtained from target-enriched NGS, which is more cost-effective than whole genome NGS, has been investigated [63].

In a study conducted in 2020, Chandler and colleagues [51] investigated the use of RHDO with NGS for its application in NIPS of CF in cases where parents are not eligible for paternal mutation detection/exclusion (regardless of the type of mutation that parents carry). As we have previously mentioned, this method has a high sensitivity, and since it is a method based on haplotyping and uses SNPs, it is statistically robust. Therefore, it can be a suitable approach in NIPS for the definitive diagnosis of CF deletions. However, RHDO still faces challenges such as the need for parental and proband samples, the possibility of nonrecombinant events, and twin or multiple pregnancies. The tested approach in this study is able to identify nonrecombinant events. If the mutation is located in a nonrecombinant region, invasive testing is recommended for a definitive diagnosis. Additionally, this method can still be used if the twins are monozygotic.

10.4.4 NIPS for Congenital Adrenal Hyperplasia

Congenital adrenal hyperplasia (CAH) is inherited in an autosomal recessive manner, and its severe form is known as "salt-wasting" since it causes imbalances in body electrolytes. CAH is primarily caused by mutations like deletions or conversions in the CYP21 gene, which is located on chromosome 6p21.3. CYP21 gene encodes 21-hydroxylase enzyme, and mutations in it lead to a deficiency in the synthesis of glucocorticoids and mineralocorticoids. In CAH, deficiency of the enzyme mentioned above leads to the accumulation of metabolic precursors and, consequently, increased production of androgens. This results in the masculinization of female fetuses [64].

One of NIPS's applications in managing CAH is utilizing it to determine the gender of the fetus in pregnancies where parents are carriers of the disease. Sex determination aims to administer dexamethasone treatment to control the synthesis of excess androgens and prevent virilization of the female fetus in utero, only if the fetus is a female. In cases where the fetus is a male, prenatal dexamethasone treatment is unnecessary [65]. In carrier couples, dexamethasone treatment is limited to female fetuses, regardless of whether or not they are affected. Since dexamethasone treatment may have side effects and is not completely safe, it is better to opt for an approach where treatment is provided only for affected female fetuses [64, 66].

One approach suggested in a study [66] is the utero or prenatal exclusion of CAH. This study was conducted on a pregnancy in which parents had a previous child with CAH diagnosis and the affected child had inherited a known mutation in CYP21 gene from the father and a de novo mutation from the mother. In this study, markers of polymorphisms associated with normal and mutant alleles from the father were used to determine the paternal inheritance of alleles in the cfDNA presented in the mother's plasma. Therefore, if it is determined that the fetus has inherited the normal paternal allele, CAH will not occur. Consequently, this approach has the potential to be implemented for NIPS of autosomal recessive diseases, such as CAH.

One of the challenges faced during NIPS using conventional approaches for CAH is the presence of pseudogene CYP21P, which is 98% homologous to CYP21. Therefore, a more accurate approach for NIPS could be a determination of parental haplotypes associated with the risk using a large number of heterozygous single-nucleotide polymorphic markers linked to the CYP21 gene, which leads to more accurate results [66, 67].

In a study conducted in 2014 [48], New et al. investigated the detection of mutations in CYP21A2 gene via haplotype analysis of parental DNA. They used massively parallel sequencing (MPS) of cfDNA in plasma from 14 at-risk pregnancies for CAH and successfully predicted the status of CAH in studied pregnancies.

Ma et al. also used a similar approach in the same year, using haplotypes of parents and sequencing of cfDNA in maternal plasma to predict fetal haplotypes and thus diagnose CAH in the fetus. This approach demonstrated high specificity and sensitivity subsequent to validation by invasive methods [68].

10.4.5 NIPS for Spinal Muscular Atrophy

Spinal muscular atrophy (SMA) is an autosomal recessive disease caused by mutations in the SMN1 gene located on chromosome 5q13.1, with various manifestations. The majority of disease-causing mutations include deletions in exons 7–8 of the SMN1 gene. Other mutations are also point mutations of SMN1. SMA, which is known as a neurodegenerative disease, leads to progressive muscle weakness and, ultimately, death [69].

Prenatal diagnosis of SMA in at-risk pregnancies is performed through invasive methods such as CVS and amniocentesis. These methods determine the copy number of the SMN1 gene containing exon 7 by techniques such as restriction fragment length polymorphism, multiplex ligation-dependent probe amplification (MLPA), and quantitative PCR [70]. However, in 4% of SMA carriers, two copies of the SMN1 gene are present in cis on one chromosome, which limits the applications of these methods [71].

A study by Parks et al. (2017) investigated a new method for NIPS of SMA disease [50]. In this study, targeted MPS was first performed on cfDNA from plasma samples obtained from 6 at-risk pregnancies for SMA and 10 normal pregnancies. RHDO analysis was then conducted on the data obtained from this sequencing to determine the fetal status. MLPA was performed on samples using invasive methods to confirm the results of NIPS with this approach. The results obtained from NIPS showed 100% sensitivity and specificity compared to those obtained from invasive methods.

As we mentioned before, determining the number of the SMN1 copies in PND and diagnosing carriers in cases where two copies of the SMN1 gene are located in one chromosome unit "in cis" pose a challenge. The method used in this study focuses on determining the status of the fetus regardless of parental carrier status. However, in cases where parental carrier status is not confirmed, if NIPS identifies the fetus as SMA-affected, further confirmatory tests using invasive methods are recommended to eliminate the possibility of false positive results.

10.4.6 NIPS for a Number of Metabolic Diseases

Prenatal diagnosis of metabolic disorders can greatly assist in prompt treatment and management of disease in newborns. Since many metabolic diseases are caused by defects in a single gene, prenatal diagnosis is possible using invasive tests, which carry a low but significant risk of miscarriage. The following will briefly discuss several studies regarding using NIPS in some metabolic diseases.

10.4.6.1 Methylmalonic Academia

Methylmalonic acidemia (MMA) is an autosomal recessive inherited disorder of organic acid metabolism, mainly affecting protein and fat metabolism. In most cases, deleterious mutations in the MUT gene, which is located on chromosome 6p12.3 and encodes methyl malonyl-CoA mutase enzyme, have been observed in

MMA patients. Since MMA can cause severe metabolic disturbances in affected newborns, early diagnosis is crucial to initiate timely disease management interventions and prevent severe irreversible damage. Therefore, screening programs for this disease are available before birth. However, noninvasive testing with low risk can also be helpful.

Gu et al. (2014) conducted a study to simultaneously measure the percentage of mutant alleles and fetal DNA in maternal cell-free DNA plasma [72]. Despite challenges in accurately measuring fetal DNA percentage, particularly in female fetuses without the Y chromosome, it is crucial for ensuring precision [46, 73]. To address this, the study proposed a method employing multiplex amplification with multiple polymorphic markers, independent of fetal gender, to determine fetal DNA content [72]. Moreover, the study utilized both direct and indirect approaches to assess the quantity of fetal mutant alleles. The direct method targeted specific mutations, applicable even in the absence of polymorphic markers, while the indirect method involved haplotyping using genetic markers linked to the mutant allele. Results indicated the feasibility and cost-effectiveness of this approach for diagnosing MMA through noninvasive testing [72].

One of the common causes of metabolic disorders in newborns is a combination of MMA with homocystinuria of the cblC type (MAHCD), which is inherited in an autosomal recessive pattern due to compound heterozygous or homozygous mutations in the MMADHC gene located on chromosome 2q23.

Han et al. investigated a haplotype-based method for NIPS in pregnancies at high risk of cblC deficiency [74]. They performed targeted sequencing to determine haplotypes linked to mutant alleles in parents and probands of 21 high-risk pregnancies. Then they used a hidden Markov model (HMM) to infer fetal genotypes based on parental haplotypes. This study confirmed the feasibility of using a haplotype-based approach for NIPS of this disease.

One of the challenges of NIPS using this approach is the occurrence of nonrecombinant events (NREs) that may complicate the interpretation of fetal genotypes. Although the approach used in this study can identify NREs, it cannot determine whether the event occurred in the fetus or proband. Therefore, if an NRE is detected using this method, invasive testing is recommended to determine the fetal genotype [74].

10.4.6.2 Maple Syrup Urine Disease

Clinical manifestations of Maple syrup urine disease (MSUD) with an autosomal recessive inheritance are caused by a branched-chain alpha-keto acid dehydrogenase (BCKDH) enzyme deficiency. Keto acids and branched-chain amino acids accumulate in the body due to the lack or complete loss of BCKDH enzyme, ultimately leading to toxicity. Mutations that cause BCKDH deficiency occur in the BCKDHA, BCKDHB, and DBT genes, which encode the E1α, E1β, and E2 subunits of the BCKDH enzyme. In 2014, You and colleagues [75] successfully performed NIPS using targeted MPS technologies on known genes and haplotyping methods with available parental and proband samples in a family at risk for

MSUD. However, further studies are needed to validate these methods in monogenic diseases fully.

10.4.6.3 Wilson's Disease

Wilson's disease (WD) is an autosomal recessive metabolic disorder of copper with a genetic mutation in the ATP7B gene, which encodes a copper-transporting ATPase involved in transferring copper from liver cells to the biliary system. One of the studies conducted on NIPS for monogenic diseases investigates the cSMART technique for determining fetal genotype in NIPS for pregnancies at risk for WD [57].

cSMART is an NGS-based diagnostic technique in which individual alleles are tagged with unique molecular identifiers, amplified in a circular fashion using inverse PCR, and then sequenced. Therefore, cSMART can determine the fetal genotype by measuring the level of mutant alleles in maternal plasma. Therefore, this approach can only be used for NIPS of monogenic disorders if the disease-causing mutations in the parents are known [57].

Ly et al.'s (2015) study investigated the cSMART approach in noninvasive prenatal test (NIPT) for determining fetal genotype in four pregnancies at risk for WD [57]. First, they used Sanger sequencing and whole-exome sequencing techniques to identify pathogenic parental mutations in the ATP7B gene in samples collected from the parents of these four families. They then used cSMART analysis in maternal plasma samples to determine the fetal genotypes. The results showed that NIPS results obtained using cSMART testing on cffDNA in maternal plasma were consistent with invasive testing results in these four pregnancies. Therefore, they concluded that this technique has the potential to be used for NIPS of other monogenic disorders with autosomal dominant, autosomal recessive, or X-linked inheritance. However, in cases where the disease-causing mutation involves large deletions or expansions of trinucleotide repeats, such as in DMD, where one or more exons of the relevant gene are often deleted, cSMART testing may face challenges. Nevertheless, cSMART can be helpful if markers of continuous polymorphism are attached to the mutant allele (continuous SNPs) or if the first or last nucleotide of deletion is known. In such cases, the practical application of cSMART in NIPS of these diseases requires further studies. Currently, another area where cSMART can be beneficial is detecting low levels of tumor DNA in cfDNA present in the plasma of individuals suspected of having cancer for early detection or for assessing tumor DNA levels to evaluate the individual's response to treatment.

10.4.7 NIPS in Different Kinds of Beta Thalassemia

10.4.7.1 Beta-Thalassemia and HBE Beta-Thalassemia

Beta-thalassemia is a common inherited blood disorder caused by autosomal recessive mutations in the HBB gene located on chromosome 11p15.4. These mutations can lead to reduced or absent beta-globin chain synthesis. Over 200 different and distinct mutations, which are rarely deletions, unlike the alpha-thalassemia-causing mutations, have been identified in the HBB gene [76]. Given the autosomal recessive inheritance of this disease, if both parents are carriers of one of the beta-thalassemia-causing mutations in the HBB gene, their offspring have a 25% chance of having beta-thalassemia major. The management of beta-thalassemia is difficult and expensive. Therefore, NIPS can be very helpful for this disease, especially in common regions such as Southeast Asia [77]. Like other autosomal recessive diseases, for beta-thalassemia, if the parents carry different mutations, paternal mutation detection/exclusion approaches are used to diagnose affected heterozygous fetuses [78, 79]. If the mutations are identical, techniques such as RMD or RHDO are suggested in various studies [26, 80, 81]. One of the common structural variants of hemoglobin caused by a point mutation c.79G>A in the HBB gene is Hb E. The Hb E can occur in combination with beta-thalassemia and lead to severe Hb E-beta-thalassemia disorder [82].

A study was conducted to investigate NIPS of Hb E-beta-thalassemia in pregnancies where the parents were carriers of different mutations. The approach used in the study was a combination of conventional PCR and nested real-time-allele-specific PCR (nested RT-ASPCR) designed for the Hb E mutation and two common beta-thalassemia mutations (CD 17 A>T β, CD 41/42 (-TTCT)). Results of cfDNA analysis in maternal plasma were matched entirely with the results obtained from invasive fetal DNA analysis [83].

In addition to AS-PCR, various other methods have been reported for paternal mutation detection/exclusion in different studies, including single-allele base extension reaction (SABER) [84], mass arrayed primer extension (APEX) [26, 85], dPCR, and MALDI-TOF mass spectrometry methods [86]. Since the fetal DNA content in maternal plasma cfDNA is low, these techniques have low sensitivity. However, the emergence of NGS-based technologies has increasingly led to an elevation in the sensitivity of noninvasive tests performed on maternal plasma cfDNA.

In a study, Lo and colleagues performed NIPS using massively parallel whole genome sequencing, an NGS-based method, in a pregnancy where parents were carriers of different mutations for beta-thalassemia [49]. They successfully diagnosed the healthy status of the fetus as a heterozygous carrier. Of course, targeted sequencing is much more cost-effective than whole genome sequencing. Other studies successfully performed NIPS for beta-thalassemia using targeted sequencing [36, 87, 88].

10.5 NIPS for Monogenic Disorders with X-linked Recessive Inheritance

The majority of individuals affected by X-linked disorders are male fetuses, while female fetuses are usually either carriers or non-manifesting. Therefore, if the fetus is female, invasive diagnostic methods are not required in pregnancies at risk. This reduces the need for such tests which have miscarriage risk. As determining fetal sex through ultrasound is not possible until around the 13th week of pregnancy, using NIPS for fetal sex determination, which is possible from weeks 7 to 12, is advised [89]. Therefore, identifying sequences specific to the Y chromosome in cfDNA pinpoints the fetus as male [90]. One of the widely used techniques for fetal sex determination in NIPS with high sensitivity in the eighth week of pregnancy is real-time PCR, which is designed for specific genes presented in the Y chromosome, such as SRY, DYS14, AMYLY, PAP, and DAZ [91]. However, if the fetus is identified as male, other molecular tests are needed to determine whether the fetus has inherited the mutant X chromosome from the mother or not. Since maternal plasma contains both fetal mutant alleles and high amounts of maternal-origin alleles, a vital requirement in these molecular tests is the ability to differentiate fetal alleles from the different types of maternal origin. Two common methods that have shown high sensitivity and specificity for NIPS of X-linked disorders are RMD and RHDO [17].

Another technique that has high sensitivity and can be beneficial in this area is ddPCR. As it is evident, in a pregnancy where the mother is a carrier of a mutation in her X chromosome (Xm Xn, m = mutant and n = normal), the mother's plasma will contain genotypes of XmXn and XmY, provided that the fetus is affected, and XmXn and XnY if the fetus is unaffected [17]. In this approach, the mother's plasma is diluted to the extent that each droplet contains either one or no DNA molecule. Each droplet contains all the necessary components for real-time PCR and two specific probes for mutant and normal alleles marked with different colors. Then, a PCR reaction is performed within thousands of nanoliter droplets. Each droplet containing DNA will be amplified and emit signals, then collected and analyzed to determine the mutant allele ratio (M) [17]. For a carrier female of an X-linked mutation in a normal state, *M* is considered to be 0.5 [92]. If a carried female becomes pregnant, an *M* value of greater than 0.5 indicates that the fetus is affected, while a value of *M* less than 0.5 indicates that the fetus is not affected [17].

Another approach that has the advantage of not requiring primers and specific probes for each specific mutation is RMD, which involves the simultaneous determination of mutant and normal alleles ratios [6]. Another alternative approach for detecting mutant alleles inherited from mothers in male fetuses is RHDO. This approach uses simultaneous analysis of genomic data obtained from maternal cfDNA presented in plasma and genomic data from parents and probands in the family, haplotype phasing, and determination of haplotypes related to mutant and normal alleles. Relative values of mutant and normal haplotypes are determined. The fetus is considered unaffected if the normal haplotype is more frequent than the mutant one. In contrast, if the mutant haplotype is more frequent than the normal one, the fetus is considered affected by a related X-linked disorder [17].

10.5.1 NIPS for Duchenne Muscular Dystrophy

DMD is an X-linked recessive inherited disorder. DMD is caused by detrimental mutations in the DYS gene located in Xp21 chromosomal region, which encodes dystrophin protein. Two of the most prominent clinical features of DMD are proximal muscle weakness and pseudohypertrophy of calf muscles. The onset of clinical symptoms in affected boys who have inherited a mutation in the DYS gene on their

X chromosome usually occurs in early childhood, typically between the ages of 2 and 4. The average lifespan of affected individuals is 20 years, especially if supportive measures are not taken, and they rarely live beyond 20 years old.

Due to difficulties in managing affected individuals and cost imposition to families and society, NIPT for DMD is necessary. Nevertheless, with the development of the first NIPT techniques for this disease, researchers were faced with the challenge that in cfDNA of maternal plasma, the number of mutant alleles that originated from the mother is higher than that of fetal-origin mutant alleles; thus, they faced challenges in detecting fetal-origin mutant alleles inherited from the mother. However, with the emergence of molecular techniques with higher sensitivity that focus on determining allele ratios, this problem has been somewhat resolved [50, 93]. NIPS was initially performed for pregnant women with a family history of DMD. Therefore, there was no need for confirmation using invasive tests in the case of a positive diagnosis for a male fetus [92].

Much progress has been made to the application of NIPT for DMD using cfDNA in maternal plasma [94]. Many challenges exist in these kinds of tests, from selecting eligible pregnant females to implementing the appropriate technique and method for diagnostic purposes and selecting complementary tests for further confirmation of the results [4]. Regarding DMD, research is ongoing to develop innovative and accurate diagnostic methods as well as treatment options to improve the quality of life of affected individuals.

There are two main approaches for NIPS of mutations that cause DMD: RMD and Haplotype-Based Approaches using RHDO. The RHDO method measures the relative values of associated haplotypes to the mutant and normal maternal alleles in cffDNA to determine fetal status for DMD [37]. Haplotype phasing of alleles is performed using both direct and indirect approaches in RHDO. However, the indirect method requires samples from both parents and the proband, which is challenging as these samples may not be accessible. While in some cases, indirect analysis has been performed without access to paternal samples; it has low accuracy in distinguishing carrier female fetuses from those who have inherited both normal alleles from their parents [95]. The direct method requires fewer capture probes and samples for analysis in each case. However, it increases the need for advanced and complex equipment, making it more expensive and limiting its wide application. Nevertheless, direct approaches have significant advantages over indirect ones, as they do not require paternal and proband sample collection. In haplotype-based NIPS, nonrecombinant events frequently occur, especially in the dystrophin gene. The RHDO method can determine these events, but the indirect method lacks this capability. The significant advantage of direct haplotype phasing over indirect is that it can determine whether the nonrecombinant event is present in the proband or fetus [96].

The RHDO method is used for identifying fetuses affected by DMD without considering the type of mutation. However, it has limitations when de novo mutation or mosaicism occurs in the maternal germline. In such cases, RHDO may be unable to detect affected fetuses [95]. On the other hand, the RMD approach for NIPS is not based on haplotypes, making it more cost-effective and unaffected by nonrecombinant events. RMD determines fetal status by measuring the ratio of mutant and normal alleles, regardless of parental origin, and can detect de novo mutations and ones caused by maternal germline mosaicism. However, RMD is an unsuitable method for detecting large deletions or insertions [94].

10.5.2 NIPS for Hemophilia

Hemophilia is one of the most prevalent monogenic disorders with X-linked recessive inheritance. There are two types of hemophilia, including type A and B. Clinical manifestations include spontaneous bleeding or bleeding due to injury or surgery in joints and muscles. Hemophilia A is caused by a mutation in the F8 gene located on Xq28 chromosomal region, which encodes coagulation factor VIII, resulting in a reduction of VIII levels in plasma. Hemophilia B is also caused by a mutation in the F9 gene located on Xq27.1, which also encodes coagulation factor IX [97]. More than 2000 different mutations in the F8 gene and approximately 1000 mutations in F9 have been reported for Hemophilia A and B, respectively. However, the frequency of these mutations differs from patient to patient. For example, nearly 50% of individuals with severe Hemophilia A have an inversion of intron 22 in the F8 gene, also known as flip inversion [98]. Prenatal testing is always available for pregnancies at risk of hemophilia, which has been performed mainly by invasive methods [99, 100].

NIPT for hemophilia has recently been studied using cfDNA analysis from maternal plasma to determine fetal genotype for variants and mutations in F8 and F9 genes, especially concerning common and severe mutations such as the inversion of intron 1 in F8 gene. This approach has been a research priority [46, 101]. Inversion of intron 1 in F8 gene occurs due to an intrachromosomal rearrangement between a small intragenic region called F8A and its homolog in the upstream telomeric region of the F8 gene [102, 103]. Direct detection of this inversion in using NIPT of at-risk pregnancies for hemophilia has not been feasible due to technical challenges in identifying breakpoint junctions. However, it is possible to identify through haplotyping-based approaches and analyses of allele imbalances [36, 48].

Tsui et al. have implemented an approach for NIPT of hemophilia based on measuring the mutant-to-normal allelic ratios in maternal plasma and analyzing allele imbalances to determine fetal status [46]. The amount of plasma's fetal DNA that affects these allele imbalances is important in these approaches, and therefore, it is essential to use highly accurate techniques for measuring allele ratios [26]. Two technologies with high sensitivity and accuracy are ddPCR and MPS [104].

In cases where maternal plasma is collected at earlier stages of pregnancy, the amount of fetal DNA in plasma is lower. In such cases, allele-counting-based strategies require more DNA molecules in plasma for accurate counting. ddPCR can even be performed when the percentage of fetal DNA is <10%; however, data analysis for determining fetal genotype has low statistical power, and therefore, MPS may be a better alternative [11]. Developing these advanced technologies for NIPT of hemophilia has improved the accuracy and reliability of prenatal diagnosis and managing pregnancies at risk.

RHDO is another approach used in the study conducted by Hudecova et al. for NIPS of hemophilia [11]. They first targeted the F8 gene region to multiply it. Then they determined the fetal hemophilia status by analyzing maternal plasma containing the flipped mutation using linked polymorphic markers and RHDO analysis.

Another sensitive and accurate approach in NIPS is targeted MPS, which is highly sensitive even when the percentage of fetal DNA content in maternal plasma is as low as 3% [97]. However, ddPCR technology is less expensive in NIPS for hemophilia. For noninvasive detection of fetal status for hemophilia, they first determine fetal sex by targeting Y chromosome-specific sequences or inherited paternal SNP polymorphic markers. Then, if the fetus is male, they utilize ddPCR technology to detect hemophilia mutations. This approach is applicable when fetal DNA percentage exceeds a threshold; otherwise, more digital analyses are required. In flipped mutations requiring haplotype-based analysis, ddPCR is impractical, and targeted MPS can be used instead. With reduced sequencing costs, this accurate and sensitive approach can potentially have greater applicability [11].

10.6 Genetic Counseling for NIPS of Monogenic Disorders

Knowledge about the fetus's condition is particularly important for families, especially the mother. All mothers are interested in knowing the health status of their unborn baby. Various methods have been developed to help health professionals with this, such as CT and ultrasound scans, prenatal screening, and other methods. However, the most important feature of these predictive methods is their safety for both the baby and the mother. NIPS, which uses cffDNA of the mother's blood to analyze fetal genotype, is a safe method that people will undoubtedly welcome if public awareness is raised about it.

The nature and scope of prenatal diagnosis are rapidly changing, and NIPS will gradually become available for various genetic disorders. In the upcoming future, it will be possible to establish a reliable and trustworthy protocol, but challenges arise. For some conditions, NIPS is expensive for families and arduous for technicians. Although the availability of NIPS tests is improving, transparent and reliable sources of information plus quick and fair access to them are required for patients and healthcare providers [4].

Targeted NIPS screening for FGFR3 mutation in the second trimester can distinguish TD from achondroplasia easily. The risk of recurrence of TD in subsequent pregnancies is low and may be due to germ-line mosaicism. However, parents often request definitive elimination of recurrence risk in subsequent pregnancies. Therefore, all relevant information regarding the risk of recurrent disease should be presented to them. The analysis of cffDNA in maternal plasma provides a definitive, accurate, and primary approach for eliminating recurrence risk. This situation is where NIPS can contribute to genetic counseling [14].

In families affected by HD, prenatal diagnosis may be needed. Prenatal diagnosis is performed by analyzing the number of CAG repeats in early pregnancy using PCR on DNA samples obtained from CVS. However, NIPS allows for a safer analysis and does not involve common risks associated with invasive tests [105].

Another issue to be considered is public acceptability. A survey of parents that were carriers of CF showed 56% of them were reluctant to invasive diagnostic tests, but 94.9% preferred NIPS for CF, including those who had previously refused invasive testing. It is evident that the elevation of public awareness leads to an increased acceptance rate of NIPS by couples [106].

NIPS can assist specialists in selecting an appropriate treatment approach for some metabolic disorders. Early diagnosis and intervention of them not only prevent complications but also minimize the risk for the mother. Conversely, prevention of drug usage can hamper side effects of it in the fetus and mother. For example, treatment with dexamethasone in CAH is only ideal for female fetuses, and male fetuses do not require this intervention [106].

Sometimes, NIPS alone may not be as informative as common tests such as amniotic fluid or CVS, but that does not mean it is entirely pointless. For example, in families with a history of thalassemia, if it is determined that the father has transmitted a defective allele, the mother will be encouraged to undergo an invasive test to confirm whether or not the fetus has also inherited the maternal mutation [107].

Another aspect of pregnancy is the mental health of the family, especially the pregnant mother. In a survey of families studied for thalassemia and sickle cell disease, it was evident that participants welcomed the psychological benefits of the safe NIPS test since there was less anxiety about decision making. Additionally, early testing provides more time for planning and preparing for the birth of an affected child. For women who intend to terminate their pregnancy, decisions become more complex as an attachment to pregnancy increases and moral and ethical concerns increase.

One notable point is that several participants raised whether such tests are accessible to everyone, particularly in low-income countries, where the cost and availability of such tests may pose a challenge.

10.7 Conclusion

In summary, the research conducted and the latest developments in the methodologies and techniques employed in NIPS of monogenic disorders offer a promising prospect for its extensive implementation in clinical practice. Despite the numerous challenges and ethical concerns associated with this field, it is anticipated that the comprehensive examination and resolution of these issues in forthcoming research endeavors will facilitate the incorporation of NIPS of monogenic disorders as a customary practice in prenatal care and management.

References

1. Gardner A, Stauffer S, Petley-Ragan L, Wismer P, Ungu DAK. Monogenic disorders. In: Labster virtual lab experiments: genetics of human diseases. Berlin, Heidelberg: Springer; 2019. p. 1–28.
2. Jensen PK. Monogenic hereditary diseases. Ugeskr Laeger. 2003;165(8):805–9.
3. Chitty LS, Bianchi DW. Noninvasive prenatal testing: the paradigm is shifting rapidly. Prenat Diagn. 2013;33(6):511–3.
4. Jenkins LA, Deans ZC, Lewis C, Allen S. Delivering an accredited non-invasive prenatal diagnosis service for monogenic disorders and recommendations for best practice. Prenat Diagn. 2018;38(1):44–51.
5. Hill M, Finning K, Martin P, Hogg J, Meaney C, Norbury G, et al. Non-invasive prenatal determination of fetal sex: translating research into clinical practice. Clin Genet. 2011;80(1):68–75.
6. Minear MA, Lewis C, Pradhan S, Chandrasekharan S. Global perspectives on clinical adoption of NIPT. Prenat Diagn. 2015;35(10):959–67.
7. Hanson B, Scotchman E, Chitty LS, Chandler NJ. Non-invasive prenatal diagnosis (NIPS): how analysis of cell-free DNA in maternal plasma has changed prenatal diagnosis for monogenic disorders. Clin Sci (Lond). 2022;136(22):1615–29.
8. D'Aversa E, Breveglieri G, Pellegatti P, Guerra G, Gambari R, Borgatti M. Non-invasive fetal sex diagnosis in plasma of early weeks pregnants using droplet digital PCR. Mol Med. 2018;24(1):14.
9. Pinheiro LB, Coleman VA, Hindson CM, Herrmann J, Hindson BJ, Bhat S, et al. Evaluation of a droplet digital polymerase chain reaction format for DNA copy number quantification. Anal Chem. 2012;84(2):1003–11.
10. Hindson BJ, Ness KD, Masquelier DA, Belgrader P, Heredia NJ, Makarewicz AJ, et al. High-throughput droplet digital PCR system for absolute quantitation of DNA copy number. Anal Chem. 2011;83(22):8604–10.
11. Hudecova I, Jiang P, Davies J, Lo YMD, Kadir RA, Chiu RWK. Noninvasive detection of F8 int22h-related inversions and sequence variants in maternal plasma of hemophilia carriers. Blood. 2017;130(3):340–7.
12. Pacault M, Verebi C, Lopez M, Vaucouleur N, Orhant L, Deburgrave N, et al. Non-invasive prenatal diagnosis of single gene disorders by paternal mutation exclusion: 3 years of clinical experience. BJOG: Int J Obstet Gynaecol. 2022;129(11):1879–86.
13. Lim JH, Kim MJ, Kim SY, Kim HO, Song MJ, Kim MH, Park SY, Yang JH, Ryu HM. Non-invasive prenatal detection of achondroplasia using circulating fetal DNA in maternal plasma. J Assist Reprod Genet. 2011;28(2):167–72.
14. Chitty LS, Khalil A, Barrett AN, Pajkrt E, Griffin DR, Cole TJ. Safe, accurate, prenatal diagnosis of thanatophoric dysplasia using ultrasound and free fetal DNA. Prenat Diagn. 2013;33(5):416–23.
15. Hill M, Twiss P, Verhoef TI, Drury S, McKay F, Mason S, et al. Non-invasive prenatal diagnosis for cystic fibrosis: detection of paternal mutations, exploration of patient preferences and cost analysis. Prenat Diagn. 2015;35(10):950–8.
16. Perlado S, Bustamante-Aragonés A, Donas M, Lorda-Sánchez I, Plaza J, Rodríguez de Alba M. Fetal genotyping in maternal blood by digital PCR: towards NIPS of monogenic disorders independently of parental origin. PLoS One. 2016;11(4):e0153258.
17. Mahdi Mortazavipour M, Mahdian R, Shahbazi S. The current applications of cell-free fetal DNA in prenatal diagnosis of single-gene diseases: a review. Int J Reprod Biomed. 2022;20(8):613–26.
18. van den Oever JME, van Minderhout IJHM, Harteveld CL, den Hollander NS, Bakker E, van der Stoep N, et al. A novel targeted approach for noninvasive detection of paternally inherited mutations in maternal plasma. J Mol Diagn. 2015;17(5):590–6.

19. González-González MC, Trujillo MJ, Rodríguez de Alba M, García-Hoyos M, Lorda-Sánchez I, Díaz-Recasens J, et al. Huntington disease–unaffected fetus diagnosed from maternal plasma using QF-PCR. Prenatal Diagn. 2003;23(3):232–4.
20. Chitty L, Griffin D, Meaney C, Barrett A, Khalil A, Pajkrt E, et al. New aids for the non-invasive prenatal diagnosis of achondroplasia: dysmorphic features, charts of fetal size and molecular confirmation using cell-free fetal DNA in maternal plasma. Ultrasound Obstet Gynecol. 2011;37(3):283–9.
21. Prior-de Castro C, Gómez-González C, Rodríguez-López R, Macher HC, Prenatal Diagnosis Commission and the Genetics Commission of the Spanish Society of Laboratory Medicine. Prenatal genetic diagnosis of monogenic diseases. Adv Lab Med. 2023;4(1):28–51.
22. Cordone M, Lituania M, Bocchino G, Passamonti U, Tomà P, Camera G. Ultrasonographic features in a case of heterozygous achondroplasia at 25 weeks' gestation. Prenat Diagn. 1993;13(5):395–401.
23. Modaff P, Horton VK, Pauli RM. Errors in the prenatal diagnosis of children with achondroplasia. Prenat Diagn. 1996;16(6):525–30.
24. Hahn S, Chitty LS. Noninvasive prenatal diagnosis: current practice and future perspectives. Curr Opin Obstet Gynecol. 2008;20(2):146–51.
25. Lo YM, Tein MS, Lau TK, Haines CJ, Leung TN, Poon PM, et al. Quantitative analysis of fetal DNA in maternal plasma and serum: implications for noninvasive prenatal diagnosis. Am J Hum Genet. 1998;62(4):768–75.
26. Lun FM, Tsui NB, Chan KC, Leung TY, Lau TK, Charoenkwan P, et al. Noninvasive prenatal diagnosis of monogenic diseases by digital size selection and relative mutation dosage on DNA in maternal plasma. Proc Natl Acad Sci U S A. 2008;105(50):19920–5.
27. Wang DC, Shannon P, Toi A, Chitayat D, Mohan U, Barkova E, Keating S, Tomlinson G, Glanc P. Temporal lobe dysplasia: a characteristic sonographic finding in thanatophoric dysplasia. Ultrasound Obstet Gynecol. 2014;44(5):588–94.
28. Khalil A, Pajkrt E, Chitty LS. Early prenatal diagnosis of skeletal anomalies. Prenat Diagn. 2011;31(1):115–24.
29. Delahaye S, Rosenblatt J, Costa JM, Bazin A, Bénifla JL, Jouannic JM. First-trimester molecular prenatal diagnosis of a thanatophoric dysplasia. Prenat Diagn. 2010;30(12–13):1222–3.
30. González-González MC, Trujillo MJ, Rodríguez de Alba M, Ramos C. Early Huntington disease prenatal diagnosis by maternal semiquantitative fluorescent-PCR. Neurology. 2003;60(7):1214–5.
31. Amicucci P, Gennarelli M, Novelli G, Dallapiccola B. Prenatal diagnosis of myotonic dystrophy using fetal DNA obtained from maternal plasma. Clin Chem. 2000;46(2):301–2.
32. Massari A, Novelli G, Colosimo A, Sangiuolo F, Palka G, Calabrese G, et al. Non-invasive early prenatal molecular diagnosis using retrieved transcervical trophoblast cells. Hum Genet. 1996;97(2):150–5.
33. Drury S, Hill M, Chitty LS. Cell-free fetal DNA testing for prenatal diagnosis. Adv Clin Chem. 2016;76:1–35.
34. Cayrefourcq L, Vincent M-C, Pierredon S, Moutou C, Imbert-Bouteille M, Haquet E, et al. Single circulating fetal trophoblastic cells eligible for non invasive prenatal diagnosis: the exception rather than the rule. Sci Rep. 2020;10(1):9861.
35. Hayward J, Chitty LS. Beyond screening for chromosomal abnormalities: advances in non-invasive diagnosis of single gene disorders and fetal exome sequencing. Semin Fetal Neonatal Med. 2018;23(2):94–101.
36. Lam KW, Jiang P, Liao GJ, Chan KC, Leung TY, Chiu RW, et al. Noninvasive prenatal diagnosis of monogenic diseases by targeted massively parallel sequencing of maternal plasma: application to β-thalassemia. Clin Chem. 2012;58(10):1467–75.
37. Verhoef TI, Hill M, Drury S, Mason S, Jenkins L, Morris S, et al. Non-invasive prenatal diagnosis (NIPS) for single gene disorders: cost analysis of NIPS and invasive testing pathways. Prenat Diagn. 2016;36(7):636–42.
38. Benn P. Non-invasive prenatal testing using cell free DNA in maternal plasma: recent developments and future prospects. J Clin Med. 2014;3(2):537–65.

39. Galbiati S, Monguzzi A, Damin F, Soriani N, Passiu M, Castellani C, et al. COLD-PCR and microarray: two independent highly sensitive approaches allowing the identification of fetal paternally inherited mutations in maternal plasma. J Med Genet. 2016;53(7):481–7.
40. Barrett AN, McDonnell TC, Chan KC, Chitty LS. Digital PCR analysis of maternal plasma for noninvasive detection of sickle cell anemia. Clin Chem. 2012;58(6):1026–32.
41. Orhant L, Anselem O, Fradin M, Becker PH, Beugnet C, Deburgrave N, et al. Droplet digital PCR combined with minisequencing, a new approach to analyze fetal DNA from maternal blood: application to the non-invasive prenatal diagnosis of achondroplasia. Prenat Diagn. 2016;36(5):397–406.
42. Debrand E, Lykoudi A, Bradshaw E, Allen SK. A non-invasive droplet digital PCR (ddPCR) assay to detect paternal CFTR mutations in the cell-free fetal DNA (cffDNA) of three pregnancies at risk of cystic fibrosis via compound heterozygosity. PLoS One. 2015;10(11):e0142729.
43. Constantinou CG, Karitzi E, Byrou S, Stephanou C, Michailidou K, Makariou C, et al. Optimized droplet digital PCR assay on cell-free DNA samples for non-invasive prenatal diagnosis: application to beta-thalassemia. Clin Chem. 2022;68(8):1053–63.
44. D'Aversa E, Breveglieri G, Boutou E, Balassopoulou A, Voskaridou E, Pellegatti P, et al. Droplet digital PCR for non-invasive prenatal detection of fetal single-gene point mutations in maternal plasma. Int J Mol Sci. 2022;23(5):2819.
45. Sawakwongpra K, Tangmansakulchai K, Ngonsawan W, Promwan S, Chanchamroen S, Quangkananurug W, et al. Droplet-based digital PCR for non-invasive prenatal genetic diagnosis of α and β-thalassemia. Biomed Rep. 2021;15(4):82.
46. Tsui NB, Kadir RA, Chan KC, Chi C, Mellars G, Tuddenham EG, et al. Noninvasive prenatal diagnosis of hemophilia by microfluidics digital PCR analysis of maternal plasma DNA. Blood. 2011;117(13):3684–91.
47. Caswell RC, Snowsill T, Houghton JAL, Chakera AJ, Shepherd MH, Laver TW, et al. Noninvasive fetal genotyping by droplet digital PCR to identify maternally inherited monogenic diabetes variants. Clin Chem. 2020;66(7):958–65.
48. New MI, Tong YK, Yuen T, Jiang P, Pina C, Chan KC, et al. Noninvasive prenatal diagnosis of congenital adrenal hyperplasia using cell-free fetal DNA in maternal plasma. J Clin Endocrinol Metab. 2014;99(6):E1022–30.
49. Lo YM, Chan KC, Sun H, Chen EZ, Jiang P, Lun FM, et al. Maternal plasma DNA sequencing reveals the genome-wide genetic and mutational profile of the fetus. Sci Transl Med. 2010;2(61):61ra91.
50. Parks M, Court S, Bowns B, Cleary S, Clokie S, Hewitt J, et al. Non-invasive prenatal diagnosis of spinal muscular atrophy by relative haplotype dosage. Eur J Hum Genet. 2017;25(4):416–22.
51. Chandler NJ, Ahlfors H, Drury S, Mellis R, Hill M, McKay FJ, et al. Noninvasive prenatal diagnosis for cystic fibrosis: implementation, uptake, outcome, and implications. Clin Chem. 2020;66(1):207–16.
52. Parks M, Court S, Cleary S, Clokie S, Hewitt J, Williams D, et al. Non-invasive prenatal diagnosis of Duchenne and Becker muscular dystrophies by relative haplotype dosage. Prenat Diagn. 2016;36(4):312–20.
53. Gerrish A, Bowns B, Mashayamombe-Wolfgarten C, Young E, Court S, Bott J, et al. Non-invasive prenatal diagnosis of retinoblastoma inheritance by combined targeted sequencing strategies. J Clin Med. 2020;9(11):3517.
54. Hui WW, Jiang P, Tong YK, Lee WS, Cheng YK, New MI, et al. Universal haplotype-based noninvasive prenatal testing for single gene diseases. Clin Chem. 2017;63(2):513–24.
55. Han M, Li Z, Wang W, Huang S, Lu Y, Gao Z, et al. A quantitative cSMART assay for noninvasive prenatal screening of autosomal recessive nonsyndromic hearing loss caused by GJB2 and SLC26A4 mutations. Genet Med. 2017;19(12):1309–16.
56. Yang X, Zhou Q, Zhou W, Zhong M, Guo X, Wang X, et al. A cell-free DNA barcode-enabled single-molecule test for noninvasive prenatal diagnosis of monogenic disorders: application to β-Thalassemia. Adv Sci (Weinh). 2019;6(11):1802332.

57. Lv W, Wei X, Guo R, Liu Q, Zheng Y, Chang J, et al. Noninvasive prenatal testing for Wilson disease by use of circulating single-molecule amplification and resequencing technology (cSMART). Clin Chem. 2015;61(1):172–81.
58. Guissart C, Dubucs C, Raynal C, Girardet A, Tran Mau Them F, Debant V, et al. Non-invasive prenatal diagnosis (NIPS) of cystic fibrosis: an optimized protocol using MEMO fluorescent PCR to detect the p.Phe508del mutation. J Cyst Fibros. 2017;16(2):198–206.
59. Modra LJ, Massie RJ, Delatycki MB. Ethical considerations in choosing a model for population-based cystic fibrosis carrier screening. Med J Aust. 2010;193(3):157–60.
60. Lee ST, Kim JY, Kown MJ, Kim SW, Chung JH, Ahn MJ, et al. Mutant enrichment with 3′-modified oligonucleotides a practical PCR method for detecting trace mutant DNAs. J Mol Diagn. 2011;13(6):657–68.
61. Bustamante-Aragones A, Gallego-Merlo J, Trujillo-Tiebas MJ, de Alba MR, Gonzalez-Gonzalez C, Glover G, et al. New strategy for the prenatal detection/exclusion of paternal cystic fibrosis mutations in maternal plasma. J Cyst Fibros. 2008;7(6):505–10.
62. Nasis O, Thompson S, Hong T, Sherwood M, Radcliffe S, Jackson L, et al. Improvement in sensitivity of allele-specific PCR facilitates reliable noninvasive prenatal detection of cystic fibrosis. Clin Chem. 2004;50(4):694–701.
63. Lo YM. Non-invasive prenatal diagnosis by massively parallel sequencing of maternal plasma DNA. Open Biol. 2012;2(6):120086.
64. White PC, Speiser PW. Congenital adrenal hyperplasia due to 21-hydroxylase deficiency. Endocr Rev. 2000;21(3):245–91.
65. Rijnders RJ, van der Schoot CE, Bossers B, de Vroede MA, Christiaens GC. Fetal sex determination from maternal plasma in pregnancies at risk for congenital adrenal hyperplasia. Obstet Gynecol. 2001;98(3):374–8.
66. Chiu RW, Lau TK, Cheung PT, Gong ZQ, Leung TN, Lo YM. Noninvasive prenatal exclusion of congenital adrenal hyperplasia by maternal plasma analysis: a feasibility study. Clin Chem. 2002;48(5):778–80.
67. Drury S, Mason S, McKay F, Lo K, Boustred C, Jenkins L, et al. Implementing Non-Invasive Prenatal Diagnosis (NIPS) in a National Health Service Laboratory; from dominant to recessive disorders. Adv Exp Med Biol. 2016;924:71–5.
68. Ma D, Ge H, Li X, Jiang T, Chen F, Zhang Y, et al. Haplotype-based approach for noninvasive prenatal diagnosis of congenital adrenal hyperplasia by maternal plasma DNA sequencing. Gene. 2014;544(2):252–8.
69. Lefebvre S, Bürglen L, Reboullet S, Clermont O, Burlet P, Viollet L, et al. Identification and characterization of a spinal muscular atrophy-determining gene. Cell. 1995;80(1):155–65.
70. Prior TW, Nagan N, Sugarman EA, Batish SD, Braastad C. Technical standards and guidelines for spinal muscular atrophy testing. Genet Med. 2011;13(7):686–94.
71. Ogino S, Wilson RB, Gold B. New insights on the evolution of the SMN1 and SMN2 region: simulation and meta-analysis for allele and haplotype frequency calculations. Eur J Hum Genet. 2004;12(12):1015–23.
72. Gu W, Koh W, Blumenfeld YJ, El-Sayed YY, Hudgins L, Hintz SR, et al. Noninvasive prenatal diagnosis in a fetus at risk for methylmalonic acidemia. Genet Med. 2014;16(7):564–7.
73. Fan HC, Blumenfeld YJ, Chitkara U, Hudgins L, Quake SR. Noninvasive diagnosis of fetal aneuploidy by shotgun sequencing DNA from maternal blood. Proc Natl Acad Sci U S A. 2008;105(42):16266–71.
74. Han L, Chen C, Guo F, Ye J, Peng Z, Qiu W, et al. Noninvasive prenatal diagnosis of cobalamin C (cblC) deficiency through target region sequencing of cell-free DNA in maternal plasma. Prenat Diagn. 2020;40(3):324–32.
75. You Y, Sun Y, Li X, Li Y, Wei X, Chen F, et al. Integration of targeted sequencing and NIPT into clinical practice in a Chinese family with maple syrup urine disease. Genet Med. 2014;16(8):594–600.
76. Giardine B, van Baal S, Kaimakis P, Riemer C, Miller W, Samara M, et al. HbVar database of human hemoglobin variants and thalassemia mutations: 2007 update. Hum Mutat. 2007;28(2):206.

77. Maryami F, Mahdian R, Jamali S, Karimi Arzanani M, Khatami S, Maryami F, et al. Comparisons between RT-PCR, real-time PCR, and in vitro globin chain synthesis by α/β ratio calculation for diagnosis of α- from β-thalassemia carriers. Arch Iran Med. 2013;16(4):217–20.
78. Yenilmez ED, Tuli A, Evrüke IC. Noninvasive prenatal diagnosis experience in the Çukurova Region of Southern Turkey: detecting paternal mutations of sickle cell anemia and β-thalassemia in cell-free fetal DNA using high-resolution melting analysis. Prenat Diagn. 2013;33(11):1054–62.
79. Saba L, Masala M, Capponi V, Marceddu G, Massidda M, Rosatelli MC. Non-invasive prenatal diagnosis of beta-thalassemia by semiconductor sequencing: a feasibility study in the sardinian population. Eur J Hum Genet. 2017;25(5):600–7.
80. Wang W, Yuan Y, Zheng H, Wang Y, Zeng D, Yang Y, et al. A pilot study of noninvasive prenatal diagnosis of alpha- and beta-thalassemia with target capture sequencing of cell-free fetal DNA in maternal blood. Genet Test Mol Biomarkers. 2017;21(7):433–9.
81. Camunas-Soler J, Lee H, Hudgins L, Hintz SR, Blumenfeld YJ, El-Sayed YY, et al. Noninvasive prenatal diagnosis of single-gene disorders by use of droplet digital PCR. Clin Chem. 2018;64(2):336–45.
82. Vichinsky E. Hemoglobin E syndromes. Hematol Am Soc Hematol Educ Program. 2007;(1):79–83. https://doi.org/10.1182/asheducation-2007.1.79.
83. Tungwiwat W, Fucharoen G, Fucharoen S, Ratanasiri T, Sanchaisuriya K, Sae-Ung N. Application of maternal plasma DNA analysis for noninvasive prenatal diagnosis of Hb E-beta-thalassemia. Transl Res. 2007;150(5):319–25.
84. Ding C, Chiu RW, Lau TK, Leung TN, Chan LC, Chan AY, et al. MS analysis of single-nucleotide differences in circulating nucleic acids: application to noninvasive prenatal diagnosis. Proc Natl Acad Sci U S A. 2004;101(29):10762–7.
85. Papasavva T, Kalikas I, Kyrri A, Kleanthous M. Arrayed primer extension for the noninvasive prenatal diagnosis of beta-thalassemia based on detection of single nucleotide polymorphisms. Ann N Y Acad Sci. 2008;1137:302–8.
86. Li Y, Di Naro E, Vitucci A, Grill S, Zhong XY, Holzgreve W, et al. Size fractionation of cell-free DNA in maternal plasma improves the detection of a paternally inherited beta-thalassemia point mutation by MALDI-TOF mass spectrometry. Fetal Diagn Ther. 2009;25(2):246–9.
87. Papasavva T, van Ijcken WF, Kockx CE, van den Hout MC, Kountouris P, Kythreotis L, et al. Next generation sequencing of SNPs for non-invasive prenatal diagnosis: challenges and feasibility as illustrated by an application to β-thalassaemia. Eur J Hum Genet. 2013;21(12):1403–10.
88. Xiong L, Barrett AN, Hua R, Tan TZ, Ho SS, Chan JK, et al. Non-invasive prenatal diagnostic testing for β-thalassaemia using cell-free fetal DNA and next generation sequencing. Prenat Diagn. 2015;35(3):258–65.
89. Vivanti AJ, Costa JM, Rosefort A, Kleinfinger P, Lohmann L, Cordier AG, et al. Optimal non-invasive diagnosis of fetal achondroplasia combining ultrasonography with circulating cell-free fetal DNA analysis. Ultrasound Obstet Gynecol. 2019;53(1):87–94.
90. Jang SS, Lim BC, Yoo SK, Shin JY, Seo JS, Hwang D, et al. Development of a common platform for the noninvasive prenatal diagnosis of X-linked diseases. Prenat Diagn. 2018;38(11):835–40.
91. Colmant C, Morin-Surroca M, Fuchs F, Fernandez H, Senat MV. Non-invasive prenatal testing for fetal sex determination: is ultrasound still relevant? Eur J Obstet Gynecol Reprod Biol. 2013;171(2):197–204.
92. Chitty LS, Mason S, Barrett AN, McKay F, Lench N, Daley R, et al. Non-invasive prenatal diagnosis of achondroplasia and thanatophoric dysplasia: next-generation sequencing allows for a safer, more accurate, and comprehensive approach. Prenat Diagn. 2015;35(7):656–62.
93. Gerson KD, O'Brien BM. Cell-free DNA: screening for single-gene disorders and determination of fetal rhesus D Genotype. Obstet Gynecol Clin N Am. 2018;45(1):27–39.

94. Xu Y, Li X, Ge HJ, Xiao B, Zhang YY, Ying XM, et al. Haplotype-based approach for noninvasive prenatal tests of Duchenne muscular dystrophy using cell-free fetal DNA in maternal plasma. Genet Med. 2015;17(11):889–96.
95. Kong L, Li S, Zhao Z, Feng J, Chen G, Liu L, et al. Haplotype-based noninvasive prenatal diagnosis of 21 families with Duchenne muscular dystrophy: real-world clinical data in China. Front Genet. 2021;12:791856.
96. Zaninović L, Bašković M, Ježek D, Katušić BA. Accuracy of non-invasive prenatal testing for Duchenne muscular dystrophy in families at risk: a systematic review. Diagnostics (Basel). 2023;13(2):183. https://doi.org/10.3390/diagnostics13020183.
97. Bowen DJ. Haemophilia A and haemophilia B: molecular insights. Mol Pathol. 2002;55(2):127–44.
98. Bagnall RD, Giannelli F, Green PM. Int22h-related inversions causing hemophilia A: a novel insight into their origin and a new more discriminant PCR test for their detection. J Thromb Haemost. 2006;4(3):591–8.
99. Chi C, Lee CA, Shiltagh N, Khan A, Pollard D, Kadir RA. Pregnancy in carriers of haemophilia. Haemophilia. 2008;14(1):56–64.
100. Huq FY, Kadir RA. Management of pregnancy, labour and delivery in women with inherited bleeding disorders. Haemophilia. 2011;17(Suppl. 1):20–30.
101. Engert A, Balduini C, Brand A, Coiffier B, Cordonnier C, Döhner H, et al. The European hematology association roadmap for European hematology research: a consensus document. Haematologica. 2016;101(2):115.
102. Naylor JA, Buck D, Green P, Williamson H, Bentley D, Giannelli F. Investigation of the factor VIII intron 22 repeated region (int22h) and the associated inversion junctions. Hum Mol Genet. 1995;4(7):1217–24.
103. Naylor J, Brinke A, Hassock S, Green PM, Giannelli F. Characteristic mRNA abnormality found in half the patients with severe haemophilia A is due to large DNA inversions. Hum Mol Genet. 1993;2(11):1773–8.
104. Chiu RW, Cantor CR, Lo YM. Non-invasive prenatal diagnosis by single molecule counting technologies. Trends Genet. 2009;25(7):324–31.
105. Bustamante-Aragones A, Trujillo-Tiebas MJ, Gallego-Merlo J, Rodriguez de Alba M, Gonzalez-Gonzalez C, Cantalapiedra D, et al. Prenatal diagnosis of Huntington disease in maternal plasma: direct and indirect study. Eur J Neurol. 2008;15(12):1338–44.
106. Haider S, Islam B, D'Atri V, et al. Structure-phenotype correlations of human CYP21A2 mutations in congenital adrenal hyperplasia. Proc Natl Acad Sci U S A. 2013;110:2605–10.
107. Xiong L, Barrett AN, Hua R, Tan TZ, Ho SSY, Chan JK, et al. Non-invasive prenatal diagnostic testing for β-thalassaemia using cell-free fetal DNA and next generation sequencing. Prenatal Diagn. 2015;35(3):258–65.

Screening of Placental Dysfunction Utilizing Cell-Free Nucleic Acids (cfNAs) of Maternal Plasma

11

Chakshu Chaudhry and Bharti Sharma

11.1 Introduction

Placental dysfunction is linked to various obstetric disorders, including hypertensive disorder of pregnancy, fetal growth restriction, and preterm labor. These conditions are primary contributors to stillbirths and neonatal mortality and morbidity. Early identification of at-risk pregnancies could facilitate timely interventions to improve placental function. Screening tests, such as history-based assessments, clinical parameters, and the evaluation of placental biomarkers like placental growth factor (PlGF) and soluble fms-like tyrosine kinase (sFLT), are employed to predict the likelihood of developing conditions like preeclampsia.

PlGF belongs to the vascular endothelial growth factor (VEGF) family and plays a crucial role in the development and maturation of the placental vascular system during normal pregnancy. However, the presence of the anti-angiogenic protein sFLT1 can counteract the activity of VEGF and PlGF, leading to placental dysfunction. In women with preeclampsia, the deficiency of PlGF may result from decreased expression, particularly toward the end of pregnancy. Additionally, the free level of PlGF decreases as it binds with sFLT1. Unfortunately, to date, there is no screening test available that can accurately predict most pregnancies at risk of placental dysfunction.

Noninvasive prenatal screening (NIPS) utilizing cell-free fetal DNA (cffDNA) represents a significant advancement in both research and clinical practice. It offers the capability to detect various fetal conditions, including aneuploidy, fetal gender, RhD genotype, and other genetic disorders. The aim of this chapter is to

C. Chaudhry
Department of Paediatrics, Maharishi Markandeshwar College of Medical Sciences, Sadopur, Ambala, Haryana, India

B. Sharma (✉)
Department of Obstetrics and Gynaecology, Postgraduate Institute of Medical Education and Research, Satellite Centre, Sangrur, Punjab, India

R. A. Rather, S. C. Saha (eds.), *Non-invasive Prenatal Screening (NIPS) in Clinical Practice*, https://doi.org/10.1007/978-981-97-6402-0_11

comprehensively review existing literature to ascertain the role of NIPS in screening and detecting pregnancies at risk of placental dysfunction. Furthermore, it seeks to identify the latest evidence available that can inform clinical care practices in this domain.

11.2 Basic Science of Placental Dysfunction

The placenta, often referred to as the "engine" of pregnancy, serves as a highly vascular organ facilitating the feto-maternal circulation and playing a vital role in transferring nutrients and oxygen to the fetus. Throughout placental development, two critical steps ensure optimal fetal growth: placental adherence and uterine remodeling. Initially, the placenta, primarily composed of trophoblast cells, adheres to the uterus through anchoring villi and adhesion molecules secreted by the inner layer of trophoblast known as cytotrophoblast [1]. Subsequently, the invasion of spiral arteries by trophoblast cells emerges as a key aspect of placental development, crucial for establishing uteroplacental blood flow. Endovascular trophoblasts penetrate the lumen of spiral arteries, while interstitial trophoblasts surround the arteries, shaping the placental bed. This invasion transforms the spiral arteries into dilated, low-resistance uteroplacental vessels. Any disruption in placental development can lead to placental dysfunction or insufficiency.

11.2.1 Causes of Placental Dysfunction

Placental dysfunction encompasses various conditions, with maternal medical disorders such as diabetes mellitus, chronic hypertension, autoimmune disorders (including anti-phospholipid antibody syndrome and systemic lupus erythematosus), maternal infections, anemia, and certain medications (such as anticoagulants) being among the most common [2]. While placental dysfunction is often observed in women with underlying chronic diseases, hypertensive disorders of pregnancy represent a significant cause of maternal and perinatal morbidity and mortality, affecting approximately 10% of pregnancies worldwide. Preeclampsia, a condition characterized by the maldevelopment of the placenta, can lead to placental dysfunction, resulting in preterm births and fetal growth restriction.

11.2.2 Pathophysiology of Placental Dysfunction

Placenta produces a range of steroid and protein hormones crucial for maternal physiological adaptation during pregnancy. Many of these placental biomarkers, historically used for screening chromosomal abnormalities, are now under study for their potential in predicting placental dysfunction. These biomarkers include human chorionic gonadotropin (hCG), progesterone, human placental lactogen, VEGF, VEGF-receptor (FLT1), PlGF, inhibin A, and pregnancy-associated plasma protein-A (PAPP-A).

Human Chorionic Gonadotropin (hCG): This glycoprotein is expressed by both syncytiotrophoblast and cytotrophoblast. The concentration of hCG typically peaks around the ninth week of gestation, with levels ranging between 50,000 and 100,000 mIU/mL. Elevated maternal serum levels are observed in multifetal gestation, fetal hemolytic anemia, gestational trophoblastic disease, and pregnancies with Down syndrome, while relatively lower levels are associated with ectopic pregnancy [3].

The production of PlGF, an angiogenic peptide is also found reduced in women who develop placental dysfunction in later gestation. Angiogenic and anti-angiogenic factors play critical roles in placental function and the development of complications like preeclampsia. In women predisposed to preeclampsia, the placenta tends to overproduce antiangiogenic peptides such as soluble sFLT1 and soluble endoglin. sFLT1 interferes with and reduces the levels of PlGF and vascular VEGF, leading to endothelial dysfunction. Consequently, elevated levels of sFLT1 are typically observed in women who later develop preeclampsia or fetal growth restriction. Soluble endoglin, another antiangiogenic peptide, inhibits the binding of transforming growth factor-beta to endoglin, an endothelial receptor crucial for nitric oxide-dependent vasodilation of uteroplacental vessels [4]. Additionally, the production of PlGF, an angiogenic peptide, is reduced in women who experience placental dysfunction later in gestation [5, 6].

Pregnancy-associated plasma protein-A (PAPP-A): PAPP-A is another crucial placental glycoprotein with metalloprotease activity. It functions by cleaving insulin-like growth factor binding proteins, thereby releasing insulin-like growth factor (IGF) [7]. This free IGF plays a pivotal role in placental development and fetal growth by regulating the uptake of glucose and amino acids [8]. PAPP-A primarily regulates IGF activity, and lower levels of PAPP-A have been observed in pregnancies associated with adverse outcomes such as spontaneous abortion, preterm delivery, fetal growth restriction, preeclampsia, and stillbirth, in addition to chromosomal aneuploidy [9].

Adverse pregnancy outcomes linked with placental dysfunction encompass fetal growth restriction, preeclampsia, and placental abruption. These complications induce hypoxia within the intervillous space, leading to oxidative stress and subsequent placental apoptosis. Prior to the manifestation of symptoms, elevated cffDNA may be detected in such conditions. This increased shedding of cffDNA is manifested by rising values in maternal circulation [10].

11.3 Overview of Cell-Free Nucleic Acids (cfNAs) in Maternal Plasma and Role of cfNAs in Placental Dysfunction

cfNAs in pregnancy refer to fragments of DNA and RNA that circulate freely in the bloodstream of pregnant individuals. These nucleic acids originate from both maternal and fetal compartments and are released into the maternal circulation through various physiological processes, including apoptosis, necrosis, and secretion. One of the most studied types of cfNAs in pregnancy is cffDNA, which originates primarily from the placenta and represents the genetic material of the developing fetus. In addition to cffDNA, other types of cfNAs, such as cell-free fetal RNA, also hold promise as biomarkers for various pregnancy-related conditions. Changes in the levels and profiles of these cfNAs have been associated with complications such as preeclampsia, fetal growth restriction, preterm birth, and gestational diabetes. The analysis of cfNAs in pregnancy offers a noninvasive approach to assess the genetic and molecular characteristics of the developing fetus and the maternal-placental unit.

11.3.1 DNA-Based Biomarker

cffDNA was discovered in 1997 and has since emerged as a screening modality for various maternal and fetal conditions. This technology has proven highly valuable in noninvasive prenatal screening for aneuploidy, fetal sex determination, RhD genotype, and detection of other genetic disorders [11]. cffDNA is essentially placental DNA, consisting of small fragments released into maternal circulation due to ongoing physiological apoptosis or necrosis of trophoblast cells throughout pregnancy. Its concentration in maternal plasma typically ranges between 3% and 13% of total cfDNA [12, 13]. The quantity of cffDNA in maternal circulation fluctuates with gestational age, aneuploidy, maternal factors such as obesity and race, as well as pregnancy complications like preeclampsia. In the context of aneuploidy detection, cffDNA screening demonstrates high sensitivity rates for Trisomy 21 (T21) (>99%), Trisomy 18 (T18) (96%), Trisomy 13 (T13) (92%), and sex chromosome aneuploidies (90%).

In cases of placental dysfunction, the heightened apoptosis and necrosis contribute to an elevation in the proportion of cffDNA present in maternal plasma. As such, cffDNA is increasingly recognized as a promising analyte for monitoring and screening various common conditions linked with placental dysfunction. These conditions encompass a range of issues such as preeclampsia, fetal growth restriction, preterm birth, ectopic pregnancy, gestational trophoblastic disease, and placenta accrete [14, 15].

With the advent of quantitative real-time polymerase chain reaction (qPCR), the quantification of cffDNA has become feasible with enhanced sensitivity and specificity. In cases of preeclampsia, cfDNA levels are observed to be nearly five times higher than those in normotensive women [16, 17]. This elevation in cffDNA levels is detected well before the onset of clinical symptoms, making it a potential biomarker for predicting preeclampsia [14, 16]. While some studies have noted an increase in cffDNA levels during the first trimester, others have observed this rise during the second trimester. However, according to systematic reviews and meta-analyses, the optimal timing for utilizing cffDNA as a biomarker is from the beginning of the second trimester, as detection rates before this period are not significantly significant [14, 17–19].

11.3.2 RNA-Based Biomarkers

Similar to cffDNA, fetal or placental-derived RNA can also be identified in maternal circulation. MicroRNAs (miRNAs) play crucial roles in normal placental development, and aberrant expression of miRNAs is linked to impaired placentation and compromised pregnancies. The emergence of high-throughput sequencing has enabled the classification of noncoding RNA (ncRNA) based on biogenesis pathways and modes of action, including small interfering RNAs (siRNAs), miRNAs, PIWI-interacting RNAs (piRNAs), endogenous small interfering RNAs (endo-siRNAs or esiRNAs), promoter-associated RNAs (pRNAs), and small nucleolar RNAs (snoRNAs) [20]. Unlike conventionally used Y chromosome fetal DNA markers, the analysis of circulating fetal RNA is unaffected by fetal sex. The levels of circulating fetal RNA are observed to be tenfold higher in pregnant women with preeclampsia compared to controls [10].

mRNA, while informative, faces challenges due to its rapid degradation and the need for meticulous handling procedures. In contrast, miRNA stands out as an attractive biomarker option owing to its stability and abundant presence. Studies employing microarray techniques have revealed changes in miRNA levels in pregnancies marked by spontaneous preterm births and preeclampsia, highlighting its potential as a diagnostic tool for identifying preeclampsia. It is crucial to recognize that alterations in miRNA expression vary depending on factors such as maternal gestational age, blood pressure levels, and uterine artery Doppler pulsatility index, emphasizing the complexity of its regulatory role in pregnancy-related conditions.

11.4 Methodology for Screening of Placental Dysfunction Utilizing cfNA

To predict placental dysfunction, quantifying cfNA is crucial. Fetal cfNA molecules are notably smaller than maternal nucleic acids, facilitating their distinction during quantification. Various tests are employed for cfDNA quantification, including qPCR), Nested PCR, Digital PCR, Sequencing, and Mass Spectrometry [21, 22]. Each of these techniques offers distinct advantages in detecting and measuring cfNA levels, contributing to the comprehensive assessment of placental health and potential dysfunction.

qPCR serves as a fundamental method for measuring DNA concentration, targeting specific pre-selected DNA sequences. Recent advancements have introduced spectrophotometry, fluorometry, and electrophoresis-based techniques, offering rapid and cost-effective alternatives [23]. Spectrophotometers, such as NanoDrop, gauge DNA concentration by assessing maximal absorbance at 260 nm. Fluorometers like Qubit and Quantus utilize fluorescent probes tailored to detect either double-stranded DNA or single-stranded DNA, enabling precise quantification even at low concentrations. Instruments like TapeStation and Bioanalyzer, based on electrophoresis, not only quantify cfNAs through fluorescence but also visualize their concentration relative to size [24]. Currently, cfNA quantification employs spectrophotometric and fluorimetric methods or PCR methods. Alternatively, innovative techniques such as next-generation sequencing or digital PCR offer enhanced capabilities. Numerous automated or semi-automated instruments and cfNA extraction kits are available, including tender, with four automated or semi-automated extraction methods: MagNA Pure 24 (Roche®), IDEAL (IDSolution®), LABTurbo 24 (Taigen®), and Chemagic 360 (Perkin Elmer®). These advancements collectively empower efficient and accurate quantification of cfNAs, facilitating comprehensive analysis in clinical and research settings.

11.4.1 Steps for cfNA Quantification

1. Blood Collection: A blood sample of 3 mL is required for cfNA quantification, drawn into EDTA tubes. It should be processed within 3 h of collection.
2. Plasma Preparation: After collection, plasma is obtained by centrifuging the sample at 1600 × *g* for 10 min at room temperature. The upper 1.5 mL of plasma is further centrifuged at 13,000 × *g* for 10 min to remove any residual blood components. The supernatant is then stored in 1.5 microtubes at −80 °C until further analysis.
3. DNA Isolation and Quantification: Robust cfDNA isolation is essential for reliable quantification. Several commercially available methods exist, utilizing different binding chemistries such as ethanol precipitation, anion-exchange resin coupled with ethanol precipitation, silica gel membrane binding, and magnetic silica particle binding technologies. The choice of method depends on factors like sample number and volume, required output, purity, and downstream applications. Most kits target DNA fragments ranging from 50 to 800 bp, covering the typical size range of cfDNA (130–170 bp). Notably, NanoDrop tends to report higher concentration estimates compared to Quantus and TapeStation methods.

11.5 Conclusion

cfNAs hold promise as a noninvasive biomarker for early prediction of placental dysfunction during pregnancy. However, extensive large-scale prospective studies are necessary to assess its validity, utility, and cost-effectiveness in clinical practice. Comparative studies with other biomarkers are essential to identify the most reliable and universally applicable marker for predicting adverse perinatal outcomes associated with placental dysfunction. Such research endeavors are critical for advancing our understanding and improving the management of pregnancies at risk of placental dysfunction.

References

1. Baschat AA. Fetal responses to placental insufficiency: an update. BJOG. 2004;111(10):1031–41.
2. Wardinger JE, Ambati S. Placental insufficiency. In: StatPearls. Treasure Island (FL): StatPearls Publishing; 2024. Available from: https://www.ncbi.nlm.nih.gov/books/NBK563171/. [Updated 2022 Oct 3].
3. Betz D, Fane K. Human chorionic gonadotropin. In: StatPearls. StatPearls Publishing; 2022.
4. Jena MK, Sharma NR, Petitt M, Maulik D, Nayak NR. Pathogenesis of preeclampsia and therapeutic approaches targeting the placenta. Biomolecules. 2020;10(6):953.
5. Chau K, Hennessy A, Makris A. Placental growth factor and pre-eclampsia. J Hum Hypertens. 2017;31(12):782–6.

6. Diguisto C, Piver E, Gouge AL, Eboue F, Vaillant CL, Maréchaud M, Goua V, Giraudeau B, Perrotin F. First trimester uterine artery Doppler, sFlt-1 and PIGF to predict preeclampsia in a high-risk population. J Maternal-Fetal Neonatal Med. 2017;30(13):1514–9.
7. Fialova L, Malbohan IM. Pregnancy-associated plasma protein A (PAPP-A): theoretical and clinical aspects. Bratisl Lek Listy. 2002;103:194–205.
8. Wells G, Bleicher K, Han X, McShane M, Chan YF, Bartlett A, et al. Maternal diabetes, large-for-gestational-age births, and first trimester pregnancy-associated plasma protein-A. J Clin Endocrinol Metab. 2015;100:2372–9.
9. Shah KH, Anjum A, Nair P, Bhat P, Bhat RG, Bhat S. Pregnancy associated plasma protein A: an indicator of adverse obstetric outcomes in a South India population. Turkish J Obstet Gynecol. 2020;17(1):40.
10. Hahn S, Rusterholz C, Hösli I, Lapaire O. Cell-free nucleic acids as potential markers for preeclampsia. Placenta. 2011;32:S17–20.
11. Lo YD, Corbetta N, Chamberlain PF, Rai V, Sargent IL, Redman CW, Wainscoat JS. Presence of fetal DNA in maternal plasma and serum. Lancet. 1997;350(9076):485–7.
12. Ashoor G, Syngelaki A, Poon LC, Rezende JC, Nicolaides KH. Fetal fraction in maternal plasma cell-free DNA at 11–13 weeks' gestation: relation to maternal and fetal characteristics. Ultrasound Obstet Gynecol. 2013;41(1):26–32.
13. American College of Obstetricians and Gynecologists. Committee opinion no. 640: cell-free DNA screening for fetal aneuploidy. Obstet Gynecol. 2015;126(3):e31–7.
14. Sifakis S, Koukou Z, Spandidos DA. Cell-free fetal DNA and pregnancy-related complications. Mol Med Rep. 2015;11(4):2367–72.
15. Sekizawa A, Jimbo M, Saito H, Iwasaki M, Sugito Y, Yukimoto Y, Otsuka J, Okai T. Increased cell-free fetal DNA in plasma of two women with invasive placenta. Clin Chem. 2002;48(2):353–4.
16. Lo YD, Leung TN, Tein MS, Sargent IL, Zhang J, Lau TK, Haines CJ, Redman CW. Quantitative abnormalities of fetal DNA in maternal serum in preeclampsia. Clin Chem. 1999;45(2):184–8.
17. Zhong XY, Holzgreve W, Hahn S. The levels of circulatory fetal DNA in maternal plasma are elevated prior to the onset of preeclampsia. Hypertens Pregnancy. 2002;21(1):77–83.
18. Levine RJ, Qian C, LeShane ES, Kai FY, England LJ, Schisterman EF, Wataganara T, Romero R, Bianchi DW. Two stage elevation of cell-free fetal DNA in maternal sera before onset of preeclampsia. Am J Obstet Gynecol. 2004;190(3):707–13.
19. Sifakis S, Zaravinos A, Maiz N, Spandidos DA, Nicolaides KH. First-trimester maternal plasma cell-free fetal DNA and preeclampsia. Am J Obstet Gynecol. 2009;201(5):472-e1.
20. Catalanotto C, Cogoni C, Zardo G. MicroRNA in control of gene expression: an overview of nuclear functions. Int J Mol Sci. 2016;17(10):1712.
21. Chan KA, Zhang J, Hui AB, Wong N, Lau TK, Leung TN, Lo KW, Huang DW, Lo YD. Size distributions of maternal and fetal DNA in maternal plasma. Clin Chem. 2004;50(1):88–92.
22. Kumar N, Singh AK. Cell-free fetal DNA: a novel biomarker for early prediction of pre-eclampsia and other obstetric complications. Curr Hypertens Rev. 2019;15(1):57–63.
23. Meddeb R, Pisareva E, Thierry AR. Guidelines for the preanalytical conditions for analyzing circulating cell-free DNA. Clin Chem. 2019;65(5):623–33.
24. Nakayama Y, Yamaguchi H, Einaga N, Esumi M. Pitfalls of DNA quantification using DNA-binding fluorescent dyes and suggested solutions. PLoS One. 2016;11(3):e0150528.

12 Noninvasive Prenatal Testing for Copy Number Variation and Sub-Chromosomal Variations

Martina Rincic

12.1 Introduction

Prenatal testing refers to identifying and diagnosing any medical conditions or diseases that an embryo or fetus may have before birth. Since their inception in the 1970s, prenatal diagnosis programs have predominantly concentrated on detecting Down syndrome and congenital anomalies, notably the most prevalent neural tube defects. However, as genomic technologies have progressed, we can gain insight into fetal genetics early in pregnancy without relying on invasive procedures like amniocentesis or chorionic villus sampling. Moreover, advancements in technology now enable the detection of even subtle genetic alterations that can profoundly impact development, expanding beyond the common trisomies typically screened for in traditional methods.

Copy number variations (CNVs) are genetic variations characterized by alterations in the number of copies of a specific DNA segment within an individual's genome. Individual variants of CNVs can range from short segments to thousands of bases, collectively detected across approximately 12–16% of the human genome [1, 2]. They can be defined as a gain (duplication) or loss (deletions) of chromosomal material and may or may not contain a gene(s). CNVs can arise through various mechanisms due to the intricate nature and inherent instability of genomic architecture [3]. On average, individuals possess 3–7 rare variant CNVs [3]. Approximately 5–10% of people possess CNVs larger than 500 kb, while 1–2% of the population carry CNVs exceeding 1 Mb in size. There is a strong negative correlation between the frequency of a CNV and its size and gene density [4]. Clinical significances or consequences of harboring CNVs in humans are variable. For

M. Rincic (✉)
Croatian Institute for Brain Research, School of Medicine University of Zagreb, Zagreb, Croatia
e-mail: martina.rincic@hiim.hr

R. A. Rather, S. C. Saha (eds.), *Non-invasive Prenatal Screening (NIPS) in Clinical Practice*, https://doi.org/10.1007/978-981-97-6402-0_12

example, there are pathogenic CNVs causing disease [5]. Conversely, CNVs without clinical outcomes significantly contribute to individuals' genome variability. This has been demonstrated by projects such as the 1000 Genomes Project, Exome Aggregation Consortium, and other large-scale efforts [1, 6–9]. Furthermore, they are now regarded as beneficiaries of human evolution and adaptation [10]. In the last several decades, a particular class, sub-chromosomal microdeletions, and microduplications, which alter more than 50 bp but <5 Mb, have been extensively associated mainly with developmental delay, intellectual disability, and other syndromes or diseases [3]. If recurrent microdeletions and microduplications are associated with the specific clinical presentation, they are called microdeletion/microduplication syndromes (MMS). The MMSs have been mostly associated with syndromic forms of developmental delay and intellectual disability since the 1980s. Over 200 recurrent microdeletion syndromes and over 80 recurrent microduplication syndromes have been identified [11–14]. Charcot–Marie–Tooth syndrome type 1A (CMT1A) and its reciprocal microdeletion syndrome, hereditary neuropathy with liability to pressure palsies (HNPP), serve as model systems for studying MMSs. These syndromes are caused by a duplication of the *PMP22* gene in the case of CMT1A and a microdeletion of the same gene in the case of HNPP [15]. An essential characteristic of sub-chromosomal CNVs is that their detection is limited to DNA-based copy number methods, such as PCR-based methods and fluorescent in situ hybridization for targeted regions or chromosomal microarray and sequencing-based CNV detection for genome-wide analysis. With advancements in genome technologies, our ability to detect smaller and smaller CNVs has significantly improved. At present, whole genome screening technology facilitates the identification of recurrent microdeletion and microduplication syndromes, enabling us to detect syndromic CNVs. Similar to syndromic CNVs, novel etiologies can be found using high-throughput genomic technologies called non-syndromic CNVs. Non-syndromic CNVs are copy number variations that do not show any particular syndromic features or signs but often can be associated with common diseases. Dosage-sensitive genes (haploinsufficient genes) are often implicated in non-syndromic CNVs related to human diseases, thus classifying such CNVs as pathogenic variants.

Around 3% of births in developed nations and 6% globally result in newborns with congenital disabilities, classifying them as individuals with congenital disorders, disease complexes, or syndromes [16]. EUROCAT, the European Surveillance of Congenital Anomalies, reports a total prevalence of significant congenital anomalies of 26.01 per 1000 births for 2010–2020 [17]. Congenital anomalies can be categorized into three primary groups based on their causes: The first group consists of genetic factors. The second group includes external teratogenic influences such as infections, nutrient deficiencies, substance abuse, toxic exposures, diabetes, and maternal age. Finally, the third group includes cases where the cause is unknown. Potentially genetically caused congenital anomalies can be found in ~40% of all significant congenital anomalies diagnosed before the age of 1 year [18]. Furthermore, chromosomal abnormalities are associated with 25% of perinatal deaths [19]. Most genetic abnormalities in congenital cases are multi-genetic diseases, followed by monogenic syndromes and chromosomal disorders. Additionally, pathogenic CNVs found in unselected, presumably healthy adult populations may be linked to unrecognized clinical consequences [20]. Pathogenic CNVs in the general population can reach 1.7%, much higher than the Down syndrome incidence (0.13–0.17%) [21]. In clinical cohorts, the incidence of pathogenic CNVs is even higher. Therefore, genomic testing for neuropsychiatric disorders, including autism spectrum disorder, schizophrenia, bipolar disorder, and other brain conditions like epilepsy, is increasingly utilized in pediatric settings. This approach has diagnostic yields for sequence variants and CNVs nearing 40% [22, 23]. Considering CNVs are recognized as one of the most common reasons for a wide range of human diseases, early diagnosis, and accurate interpretation is essential for timely intervention and targeted clinical management.

Comprehensive postnatal genetic testing for CNVs in the form of chromosomal microarray, multigene panel testing, exome sequencing, and genome sequencing is widely available and becoming more and more affordable, with the price for whole genome sequencing (WGS) dropping to the cost of chromosomal microarray analysis (CMA) [24]. Additionally, in 2021, the American College of Medical Genetics and Genomics (ACMG) recommended that ES/GS should be used as a first- or second-line test for pediatric patients presenting with one or more congenital anomalies with onset before 1 year of age or developmental delay or intellectual disability manifesting before 18 years of age [25]. Unsurprisingly, genetic testing for CNVs in postnatal clinical settings is broadly used.

Prenatally, CNV detection is more complex as different approaches for implementing prenatal tests have been used. Owing to the compassionate questions regarding pregnancy decisions and outcomes, prenatal CNV testing should be evaluated with extreme attention. In prenatal diagnostic testing, it is essential to distinguish between the general population and high-risk group pregnancies. Furthermore, it is necessary to consider whether prenatal detection was made by invasive or noninvasive means and which diagnostic test was applied. Above all, it is important to emphasize which sample is used for prenatal testing: cell-free vs. cell-based testing. Nowadays, we are witnessing increasingly pronounced pressure that noninvasive prenatal testing (NIPT) should be offered broadly to all pregnant women regardless of whether there is a predetermined risk. Along the same line, NIPT is offered at practically every gynecology clinic. Unfortunately, genetic counseling before and after NIPT is often not provided according to high standards based on a critical review of recent scientific literature and professional recommendations. In worst-case scenarios, genetic counseling is not provided at all. This chapter will cover the basic principles for noninvasive CNV detection to expand the knowledge. Besides technical aspects, this chapter will cover clinical application and ethical and legal considerations of NIPT for CNVs, all based on the most recent scientific literature.

12.2 Principles of NIPT for CNV and Sub-Chromosomal Variations Detection

The basis for all types of NIPT of any genetic aberrations, including CNVs and sub-chromosomal variations, lies in the availability of fetal cells and cell-free placental DNA (cfpDNA) circulating in maternal peripheral blood, which is intentionally and deceptively labeled as "cell-free fetal DNA" (cffDNA) in literature associated with NIPT, right from the start [26, 27]. Various types of fetal cells, including trophoblastic cells, lymphocytes, hematopoietic stem cells, and fetal nucleated erythrocytes, can be detected in maternal circulation [28]. First discovered circulating fetal cells in maternal blood were fetal lymphocytes with a male karyotype in 1969 [29]. In a normal pregnancy, the total number of fetal cells in maternal blood ranges from 2 to 6 cells per milliliter during the second trimester [30]. Increased numbers of various fetal cell types in maternal circulation can occur in cases of fetal and placental abnormalities, including fetal aneuploidy, preeclampsia, maternal-fetal isoimmunization, and other maternal obstetric complications such as gestational diabetes and bleeding [31–34]. Although fetal cells in maternal circulation accurately represent the genetic state of the fetus, their low numbers, as well as the difficulties in enriching and culturing them ex vivo, present significant challenges for their routine use in NIPT in clinical settings [35]. Another critical step forward in routine NIPT was done in 1997 by discovering circulating cell-free "fetal" cells in blood plasma and serum samples from pregnant women.

Since then, there has been significant interest in utilizing these fetal cells as a diagnostic tool to gain insights into fetal genetics [36]. Over the years, several technological methods have been advancing to offer comprehensive genetic data about the developing fetus, with next-generation sequencing (NGS) technology taking the lead over other techniques.

NGS technology was first introduced in the mid-2000s enabling high-throughput sequencing of DNA at a fraction of the time and cost of previous methods. They also enabled the sequencing of entire genomes and the sequencing of multiple samples simultaneously, making large-scale genomic studies possible. It is important to emphasize that NGS was a sequencing technology primarily used for DNA and RNA sequencing and was not intended for detecting structural variants, including CNVs. The basic workflow for NGS includes: (i) *Library preparation*: The DNA or RNA sample is typically fragmented to 300–600 bp, and adapters are added to the ends of the fragments to enable them to be sequenced on a specific platform. (ii) *Sequencing*: The library is loaded onto a sequencing instrument, and the DNA/RNA fragments are sequenced using one of several NGS technologies. (iii) *Data analysis*: The raw sequencing data is processed and analyzed using specialized software tools to align the reads to a reference genome or de novo assembly and identify variants. During NGS, millions of DNA fragments are sequenced simultaneously, generating massive amounts of data with a typical read length of 2 × 150 bp in case paired-end sequencing is performed. Paired-end sequencing is a high-throughput DNA sequencing method that generates sequence data from both ends of a DNA fragment. In paired-end sequencing, two reads are generated from each fragment, one from each end. The two reads are typically of equal length and are separated by a fixed distance, which is determined by the size of the DNA fragment and the sequencing technology used. The two reads are then used to assemble the original sequence of the DNA fragment. Paired-end sequencing has several advantages over single-end sequencing. First, it increases the accuracy of sequence alignment and assembly, especially in regions with complex or repetitive sequences. Second, it enables the detection of structural variations, such as insertions, deletions, and inversions, by comparing the distances between the paired reads. Sequencing-based computational approaches to detect CNVs using shotgun sequencing data (CNV-seq) were introduced in 2009 [37]. There are four strategies, all aimed at mapping sequence reads to the reference genome and subsequently identifying discordant signatures or patterns that indicate various types of structural variants.

Not long after CNV-seq was introduced, a new era in NIPT started with applying NGS to detect fetal aneuploidy by analyzing cfpDNA from a mother's blood [38]. This paved the way for modern aneuploidy screening targeting the most prevalent trisomies: trisomy 21 (Down syndrome), trisomy 18 (Edward syndrome), and trisomy 13 (Patau syndrome). In 2011, a breakthrough demonstration showed that noninvasive analysis of fetal DNA in maternal plasma could identify chromosomal microdeletions. The proof-of-concept involved a couple seeking prenatal genetic counseling due to a previous child with growth issues, developmental delays, and distinctive physical features attributed to a paternally inherited 4.2-Mb deletion on chromosome 12 (12p11.22–12p12.1). The father also had Asperger's syndrome, facial anomalies, brachydactyly, and short stature. At 21 weeks of pregnancy, amniocentesis confirmed the same heterozygous deletion in the male fetus using chromosomal microarray analysis. At 35 weeks, a maternal blood sample was taken, and plasma DNA, without additional enrichment, revealed a placental DNA prevalence of 5.7%. Using NGS, a 4-Mb depletion in DNA copy number on chromosome 12p (12p11.22–12p12.1) was detected [39]. Since this initial case report, the clinical application of NIPT for detecting copy number variations from sequence data has been rapidly advancing.

NGS technology is presently the most frequently utilized technique for NIPT for CNVs. The fundamental concept behind detecting CNVs noninvasively is to compare the amount and ratio of cfpDNA from mothers' blood serum to maternal DNA. However, well before the widespread adoption of NGS, many groups were already working on developing a cfDNA test for NIPT, such as real-time or digital PCR, microarray, and single nucleotide polymorphism (SNP)-based methods [40].

12.3 Technical Approaches for NIPD of CNV and Sub-Chromosomal Variations

Detecting CNVs from DNA samples can be done in several ways, including microarray analysis and CNV-seq and in principle, all these technologies could be used to detect CNVs from cfpDNA extracted from the plasma of a pregnant woman. Currently, due to technical issues, NIPT tests for CNV and sub-chromosomal variations can be divided according to the technology used on NGS and microarray-based technology. Additionally, based on the scope, there are two main approaches: whole genome analysis and targeted analysis.

12.3.1 Next-Generation Technology for CNV Detection—CNV-Seq

The fundamental approach for identifying CNVs from NGS reads involves creating consensus sequences through mapping to a reference genome or de novo assembly. This process enables the detection of anomalies caused by structural variations in patients' genomes. Once the reads are mapped, characteristics such as insertion size, orientation, and alignment length are analyzed to identify potential structural variation candidates. The capability to detect CNVs from sequencing data, whether prenatal or postnatal, relies on four main analytical approaches: read-pair analysis, read-depth analysis, split-read analysis, and assembly algorithms.

1. Read-pair methods analyze the span and orientation of paired-end reads, identifying "discordant" pairs where the mapping range and/or direction do not align with the reference genome. This approach is versatile and capable of detecting various types of genetic variations: widely spaced read pairs indicate deletions, closely positioned pairs suggest insertions, and discrepancies in orientation can reveal inversions and specific forms of tandem duplications. Read pairs in which only one end cluster and the others do not map to the reference have been used to flag variant sequences not included in the reference genome (novel insertions).
2. Read-depth approaches are based on the assumption of a random distribution in mapping depth, analyzing deviations from this pattern to identify duplications and deletions in the sequenced sample. The fundamental concept is that duplicated regions will exhibit significantly higher read depth, while deletions will show reduced read depth compared to diploid regions.
3. Split-read methods detect unmatched read pairs. The goal is to identify the breakpoint of a structural variant using a "split" sequence-read signature. The core principle is that the alignment to the genome is disrupted. A continuous stretch of gaps in the read suggests either a deletion in the patient's genome or an insertion in the reference genome.
4. Sequence assembly from NGS data theoretically enables accurate typing of all forms of structural variation, including copy number, content, and structure, provided the sequence reads are sufficiently long and accurate for de novo assembly. Despite the availability of various algorithms, this method is often too labor-intensive and expensive to be widely implemented.

Although conceptually simple, CNV detection from NGS data is computationally far from trivial, and every analytical approach for calling CNVs from NGS data has a specific computational algorithm. Over the past decade, over 100 short read-based mappers have been introduced, and numerous bioinformatics tools have been developed to detect CNVs in NGS data. First, development tools relied exclusively on limited number of information like the distance and orientation of paired-end reads but had a limited resolution leading to missed variations. Additional information can be added to increase the resolution, such as split reads and coverage. However, most CNV-seq tools are designed to detect large CNVs, typically in the size of megabases, making them suitable primarily for whole genome or whole exome sequencing data. Another challenge is handling sparser data from NGS gene panels frequently used in routine genetic testing. For instance, in diagnostic laboratories, targeted gene panels are commonly used, with the primary objective being to detect small (intragenic) disease-associated CNVs within partial, single, or a few small exons [41]. However, detecting smaller CNVs, particularly those involving partial or single exons or mosaic CNVs, remains challenging with CNV-seq. Achieving high sensitivity and specificity consistent with diagnostic standards is particularly difficult. This challenge arises from the inherent limitations of technology, such as short read lengths and bias toward GC content. In addition, CNV-seq methods are often blind to specific regions, including low complexity, highly repetitive, highly mutated regions of the human genome. It is important to emphasize that NGS is highly effective in identifying single-nucleotide variations, small insertions, and deletions (up to 50 bp) in genetic testing for Mendelian disorders. However, detecting structural rearrangements such as CNV and sub-chromosomal variations from NGS data still needs to be done by considering the limitations of the technique and CNV calling tools. Recently, a study assessed the performance of 50 CNV calling tools on whole WES and whole genome sequencing (WGS) data from various samples. It was found that many of these tools were either outdated, with their last updates over 5 years ago, or not widely utilized. The study concluded that when only WGS data is available, using multiple CNV calling tools is necessary for higher precision in CNV detection. It suggested combining tools with the best recall using consensus callers and prioritizing the CNV calls made by these tools [42]. Several factors contribute to the accuracy, sensitivity, and specificity of CNV-seq detection methods.

1. Read depth: The depth of sequencing coverage, which refers to the average number of reads that align to a specific genomic position, directly impacts CNV detection. Higher read depth increases the likelihood of capturing variations accurately and reduces false positives. Adequate read depth is crucial for detecting CNVs, especially those with low copy number changes or occurring in regions with low mappability [43].
2. Data quality: The quality of NGS data affects CNV detection. High-quality data with minimal sequencing errors, low background noise, and accurate base calling increase the reliability of CNV detection methods. Factors like sequencing platforms, library preparation protocols, and sequencing chemistry can influence data quality.
3. Library preparation biases: Library preparation methods can introduce biases, such as uneven coverage, GC bias, and amplification biases, which affect the accurate representation of the genome. These biases can lead to false positive or

false negative CNV calls. Careful consideration of library preparation protocols and the use of appropriate controls or normalization methods help mitigate these biases.

4. Read length: The length of sequencing reads impacts CNV detection. Longer reads provide better coverage across repetitive regions and enable more accurate alignment, reducing the ambiguity of mapping reads to repetitive elements. Longer reads also facilitate the detection of breakpoints and complex structural variations.
5. Algorithm and analysis method: The choice of CNV detection algorithm or analysis method plays a significant role. Different algorithms employ distinct statistical models, segmentation methods, or hidden Markov models to identify CNVs. Each algorithm has its sensitivity, specificity, and limitations, and the selection of an appropriate algorithm depends on the characteristics of the data and the specific research goals.
6. Reference genome quality: The quality and completeness of the reference genome used for read mapping and CNV calling are crucial. Inaccuracies, missing or misassembled regions, or genomic structural variations in the reference can affect the mapping and subsequent CNV detection.
7. Validation techniques: Independent validation of CNV calls using orthogonal methods, such as quantitative PCR (qPCR), microarrays, or other sequencing platforms, is essential to confirm the accuracy of CNV detection. Validation helps to reduce false positives and false negatives and increases confidence in the detected CNVs.

Regarding the scope of testing, CNV-seq can generally be applied in two ways: whole genome analysis and targeted analysis. WGS enables identifying CNVs across the entire genome. Conversely, targeted analyses include panel sequencing, which focuses on specific genes of interest; clinical exome sequencing, which examines genes linked to known clinical conditions (Mendeliom); and whole exome sequencing, which analyzes the protein-coding regions (exons) of all known genes, covering approximately 1–2% of the genome.

12.3.2 CNV-Seq for Prenatal cfpDNA Testing

Using NGS to detect genetic abnormalities in fetuses was more than logical. The basic principle involves counting the number of sequences reads mapped to each chromosome. The method detects over- or underrepresentation of any chromosome in maternal plasma DNA contributed by an aneuploid fetus. This approach does not necessitate distinguishing between placental and maternal DNA, and with sufficiently large read counts, it can be applied to even very small fractions of fetal DNA.

Soon after NGS was used to screen for common aneuploidies, CNV-seq was introduced in NIPT. The principle of using NGS data to detect CNVs and subchromosomal variations in prenatal settings follows the same protocol as CNV-seq but with special consideration regarding starting material. Following the initial demonstration of CNV-seq's ability to identify microdeletion syndrome in unborn babies [39], multiple research teams have successfully validated CNV-seq for NIPD of fetal chromosomal abnormalities. However, their focus has mainly been on a limited number of common syndromes, including 22q11.2 deletion, Prader-Willi syndrome, Angelman syndrome, 1p36 deletion, and Cri-du-chat syndrome [44, 45]. It is important to note that while these studies achieved an approximate detection rate of 90%, it should be emphasized that they utilized deep sequencing with high coverage making this an expensive approach for wide application in routine testing. To compensate for high cost and regional specificity, low-pass whole-genome sequencing was introduced for NIPD for structural variations in parallel.

Special consideration should be focused on technical factors contributing to the accuracy, sensitivity, and specificity of NIPD for structural variations. In addition to general factors, CNV-seq in context of NIPT for CNVs and sub-chromosomal variations has specific aspects that can impact the analysis and interpretation of NIPT results [46]. First is the proportion of cfpDNA within the maternal plasma, known as the fetal fraction [47]. Higher fetal DNA fraction is crucial for CNV detection in NIPT, increasing the sensitivity and accuracy of CNV detection. In addition, adequate fetal DNA fraction is required to distinguish fetal-specific CNVs from maternal background noise. Next in line is the size of structural variations. As discussed, CNV-seq has the best performance capturing longer structural variations. For example, the sensitivity for NIPT CNVs > 10 Mb was higher (91.67%) than for CNVs < 5 Mb (68.42%), with sensitivity for detecting CNVs < 2 Mb is only 21.43% [48]. Furthermore, sequencing depth affects CNV detection in NIPT. Sufficient sequencing depth is necessary to detect CNVs accurately, especially those with low-level mosaicism or minor copy number changes [43]. In addition, technical and biological noise could impact CNV detection in NIPT. Various technical and biological noise sources, such as sequencing artifacts, mapping biases, and biological factors affecting cfDNA release and clearance, must be considered in NIPT for structural variations [49]. Here are listed some of the most crucial biological factors that could directly affect the amount, integrity, and composition of cfpDNA:

1. Gestational age: The stage of pregnancy can influence the release and clearance of cfpDNA. During early pregnancy, the fetal fraction of cfDNA is generally lower compared to later stages. As gestational age progresses, the amount of fetal cfDNA increases, making it easier to detect fetal genetic information. The optimal timing for NIPT may vary depending on the specific analysis method and the desired accuracy of the test.
2. Placental health and function: The health and function of the placenta play a crucial role in the release of cfpDNA into the maternal bloodstream. Therefore, factors such as placental pathology, placental insufficiency, preeclampsia, or other placental abnormalities can impact the amount and quality of placental cfDNA. In addition, placenta abnormalities can affect fetal DNA representation in the maternal circulation, potentially leading to false-negative or false-positive NIPT results.

3. Placental turnover and apoptosis: The turnover and apoptosis of placental cells contribute to the release of cfpDNA into the maternal bloodstream. As placental cells undergo natural cell death and turnover, fragments of DNA are released and circulate in the maternal plasma. Therefore, factors that affect placental turnover and apoptosis, such as inflammation or oxidative stress, may influence the amount and composition of cfpDNA.
4. Fetal-maternal circulation and exchange: The exchange of cfDNA between the fetal and maternal circulations can impact the levels of fetal cfDNA in maternal blood. The efficiency of this exchange process can vary among pregnancies and individuals. In addition, factors that affect the placental barrier, such as placental abnormalities or vascular conditions, can influence the transfer of cfDNA and alter the representation of fetal DNA in the maternal bloodstream.
5. Maternal factors: Maternal factors, such as maternal body mass index (BMI), ethnicity, and medical conditions, can also influence the levels and characteristics of cfDNA. Maternal cfDNA, derived from the mother's cells, contributes to the total cfDNA pool and can affect the analysis of cfpDNA. Higher maternal BMI and certain medical conditions, like cancer or organ transplantation, may increase maternal cfDNA levels, potentially affecting the accuracy of NIPT results.

12.3.3 SNP-Sequencing for Prenatal Cell-Free DNA Testing

Compared to standard CNV-seq, in SNV-seq, in addition to quantitative (read count) information, qualitative (genotypic) information is added. The technique relies on the fact that SNPs are inherited from both parents. Multiplex PCR amplification of plasma DNA, which includes a mixture of maternal and placental DNA, along with the buffy coat (containing only maternal DNA), targets SNP sequences followed by direct sequencing to identify present amplified products [50]. Thousands of SNPs are analyzed for each chromosome of interest. Maternal SNPs are compared with SNP sequences derived from the aggregated maternal/placental cfDNA mixture in plasma. SNP differences from maternal SNPs indicate informative SNPs for the fetus. In cases where the mother's genotype is homozygous (A/A) at a specific SNP locus and the paternal genotype is unknown, the fetus could be either homozygous (A/A) or heterozygous (A/B), where "A" and "B" denote possible alleles (wild or mutant). If the mother is heterozygous (A/B), the fetus could have the genotypes A/A, B/B, or A/B. Differences between fetal and maternal genotypes result in a higher abundance of the paternal allele in plasma DNA (containing fetal DNA) compared to maternal DNA (lacking fetal DNA). This allele imbalance is detected by amplicon sequencing, which compares allele counts. Conversely, no allele imbalance occurs when fetal and maternal genotypes are identical. It is crucial to note that SNP testing cannot be performed in the presence of circulating cfDNA from a third source contributing additional unique SNPs, such as from organ transplants or egg donors.

12.3.4 Microarray Technology for Detecting CNV and Sub-Chromosomal Variations

CMA is a genomic testing method to detect CNVs throughout the genome. It involves using a microarray, a glass slide, or a chip containing thousands to millions of DNA probes that target specific genomic regions. The DNA on each spot is typically 25–80 bp long on most clinically based prenatal arrays. These probes are designed to hybridize with DNA fragments from the patient or control sample, allowing for the identification of genomic gains or losses. The diagnostic yield of CMA for detecting germline CNVs in patients with developmental delay, intellectual disability, autism, and/or multiple congenital anomalies has been extensively documented. CMA is endorsed by the ACMG as the primary test for specified indications [51]. Two techniques are commonly used in CMA to identify chromosomal imbalances: array comparative genomic hybridization (aCGH) and SNP array. The term CMA covers both types of testing methodologies.

aCGH determines the genomic DNA quantity in a patient's sample by comparing it to DNA from a standard control sample. This assay involves enzymatically or heat fragmenting both the patient and control DNA samples, which are then labeled with distinct fluorescent colors, typically green and red. The labeled DNAs, mixed equally, are applied to an array (a glass slide) that contains multiple probes representing various sequences throughout the human genome. The patient and control DNA molecules selectively bind (hybridize) to complementary sequences within the probe DNA on the array. The fluorescence signal intensity of each probe is measured using specialized software and normalized for direct comparison between the patient and control samples. The normalized intensity data is commonly represented as a Log2 ratio, where a ratio of 0 indicates a normal DNA copy number of 2. An increased Log2 ratio indicates a gain in DNA copy number in the patient, while a decreased Log2 ratio indicates a loss of DNA copy number. Clinical CGH arrays typically include several hundred thousand probes, whereas research CGH arrays may contain millions of probes.

SNP arrays utilize DNA probes from genomic regions containing SNPs, which are sites differing by a single base pair. Unlike aCGH, SNP arrays measure absolute fluorescence probe intensities of the patient sample compared to normalized intensities of multiple control samples combined to create a reference set (in silico comparison). Copy number data is typically displayed as a Log2 ratio plot. SNP arrays also provide additional clinically relevant information through genotype plots generated by SNPs. This includes details about mosaicism, twin–twin or maternal cell contamination, zygosity, consanguinity, limited outbreeding, triploidy, uniparental disomy, partial and complete hydatidiform moles, and parent of origin (requiring parental samples). It is important to note that triploidy cannot be solely detected from the Log2 ratio plots generated by SNP or aCGH arrays. The identification of triploidy relies on assessing the SNP allele patterns (genotypes) on an SNP array. Additionally, SNP arrays reveal long contiguous stretches of homozygosity (LCSH), which can guide focused investigations such as genotyping or sequencing. LCSH regions are particularly useful when investigating recessive gene disorders that align with the patient's phenotype or family history.

Several factors contribute to the accuracy, sensitivity, and specificity of CNV detection by CMA, including:

1. Probe Density. Higher probe density increases the resolution and sensitivity of CNV detection. Arrays with a larger number of probes covering the genome more comprehensively can better identify CNVs [52].
2. Probe Design. The design of probes plays a crucial role. Probes should be specific to the regions of interest and provide adequate coverage across the genome. In addition, they should have minimal cross-hybridization with similar genomic sequences to ensure accurate and reliable results.
3. Noise Reduction. Various noise reduction techniques are employed to minimize false-positive results. These include background subtraction, normalization, and statistical algorithms to distinguish true CNVs from noise or artifacts.
4. Quality Control. Rigorous quality control measures are essential to ensure the accuracy and reliability of the microarray results. This involves assessing starting sample quality, probe performance, signal-to-noise ratios, and overall data quality [53].
5. Data Analysis Algorithms. The use of advanced data analysis algorithms helps in accurately identifying and characterizing CNVs. These algorithms consider factors such as probe intensity, neighboring probes, and statistical methods to determine the presence and extent of CNVs.

As with any technology, CMA has several advantages and limitations. As per the latest ACMG technical standards from 2021 [51], the advantages of using chromosomal microarray analysis (CMA) include the capability to:

- Analyze any sample that provides DNA of sufficient quality and quantity.
- Detect abnormalities that are cytogenetically cryptic and not visible by standard G-banded chromosome analysis.
- Better define and characterize abnormalities identified by standard G-banded chromosome analysis.
- Customize the CMA platform to focus probes on specific areas of interest.
- Interpret objective data rather than relying on subjective visual assessments of band intensities.
- Detect regions of homozygosity (ROH), also known as copy-neutral loss of heterozygosity (CN-LOH), using CMA platforms incorporating SNP probes.
- Interface the data with genome browsers and databases for comprehensive analysis and interpretation.

According to the 2021 revision of ACMG technical standards [51], limitations of chromosomal microarray

- Inability to detect molecularly balanced chromosomal rearrangements that do not alter relative copy numbers, although CMA may reveal changes near breakpoint sites.
- Challenges in detecting low-level mosaicism for unbalanced rearrangements, constitutional aneuploidy, or tumor-specific changes in a small cell percentage, limiting its use in post-therapy follow-up or minimal residual disease detection in neoplastic settings.
- Requirement for conventional cytogenetic or FISH studies to characterize certain genetic imbalances like tandem duplications, unbalanced insertions, or marker chromosomes.
- Limitations in detecting tetraploidy or other ploidy levels, despite enhanced capabilities with CMA platforms integrating SNP probes.
- Inability to assess copy-number changes in genomic regions not covered by the CMA platform.
- Challenges in detecting duplications, deletions, single-nucleotide variants (SNVs), insertions/deletions (indels), gene expression changes, or epigenetic modifications not within the platform's coverage.
- Non-exclusivity of disease association with detected CNVs; absence of a CNV at a locus does not rule out a disease diagnosis.
- Partial characterization of clonal and subclonal cell populations, without capturing all significant clonal diversity.

12.3.5 CMA for NIPT of CNVs and Sub-Chromosomal Variations

Given that CMA is a well-established technology for CNV detection in a postnatal setting, it is common for CMA to be used to analyze preimplantation embryos and amniotic samples as it gives an increased detection rate of chromosomal imbalances, compared with conventional karyotyping [54]. The challenge lies in the fact that while CMA routinely examines pure samples of fetal DNA obtained from amniotic fluid and chorionic villus sampling, cfDNA typically comprises only 11% fetal DNA. In addition, the quality, size, and amount of cfDNA are usually inadequate to produce reliable CMA results.

For now, only several studies employed CMA for NIPT. One approach showed that a custom microarray is utilized instead of NGS as a counting method. Custom microarray was designed to target specifically chromosomes 13, 18, and 21 as a proof of principle. The findings of this study indicate that microarrays outperform NGS in addressing two key sources of data variability: (1) variability in nonpolymorphic assays across samples (assay variability), and (2) variability in measured fetal fraction using polymorphic assays (fetal fraction variability). In addition, this study showed improved turnaround time with microarrays compared to NGS, 7.5 h for microarray versus 56 h for NGS approaches [55]. When evaluating the clinical effectiveness of targeted cell-free DNA (cfDNA) analysis using microarray-based quantitation across a diverse pregnancy cohort encompassing over 23,000 pregnancies, including singleton and twin gestations as well as pregnancies achieved via in-vitro fertilization with donor oocyte, it was determined that the sensitivity exceeded 99% for trisomy 21, 97% for trisomy 18, and 94% for trisomy 13. Specificity for each trisomy was >99.9% [56]. Recently published meta-analysis found a direct equivalent for test accuracy comparing microarray- and sequencing-based cfpDNA for fetal trisomy 21, 18, and 13 [57]. This shows an excellent potential for microarrays to be used in NIPT for CNVs and sub-chromosomal variations by expanding the number of probes included. This possibility was employed for a targeted microarray-based cfpDNA test to detect 22q11.2 microdeletion. Array-based NIPT for common aneuploidies was enhanced with additional probes targeting the 3 Mb region of interest in the 22q11.2 region. Digital Analysis of Selected Regions (DANSR) assays were employed in a study involving 1953 plasma samples. The analytical sensitivity was 75.4% (95% CI: 67.1–82.2%) based on 122 samples with deletions ranging from 1.96 to 3.25 Mb. Specificity was determined to be at least 99.5% (95% CI: 99.0–99.7%) in 1614 presumed unaffected samples. Within the clinical cohort, 5 out of 7 samples from pregnancies affected with 22q11.2 deletion showed a high probability of deletion, with no false positive results observed in the 210 unaffected samples [58]. Despite its demonstrated effectiveness, microarrays are not widely employed in cfpDNA testing.

12.4 Scope of NIPD for CNVs and Sub-Chromosomal Detection

Based on the scope, the NIPT for CNVs can be divided into the whole genome and target analysis. In genome-wide NIPT for CNVs, all chromosomes are randomly analyzed using massively parallel sequencing, theoretically enabling the detection of CNVs across the entire genome. Therefore, whole genome NIPT for CNVs can be achieved exclusively by WGS [59]. On the other hand, targeted analysis can be accomplished by NGS and microarray technology. In targeted analysis, NGS can be used for chromosome-selective sequencing, targeted capture enrichment assay [60, 61], panel sequencing, whole exome sequencing, or clinical exome sequencing [62]. Targeted analysis of NIPT for CNVs with microarray technology includes a custom-targeted set of probes [63].

12.5 Placental Versus Maternal Cell-Free DNA in Plasma for NIPT of Structural Variation

In addition to all specifics of CNV-seq for NIPT for structural variations, it is essential to emphasize that during "classical" cell-free DNA sequencing, it is impossible to distinguish placental versus maternal DNA precisely. Instead, it is done based on fragment size analysis. Placental cfDNA fragments are generally shorter than maternal cfDNA fragments due to differences in nucleosome positioning. Therefore, analyzing the size distribution of cfDNA fragments can provide insights into their placental or maternal origin. However, due to the presence of fetal and maternal cfDNA fragments in maternal plasma, accurately determining the fetal genotype poses a significant challenge, primarily due to the predominant presence of maternal-derived fragments. This challenge becomes particularly pronounced when both parents carry the same disease-causing mutation. This is first recognized in monogenic disorders but extends to CNV detection. Therefore, the research is focused on finding the best cost-effective methods for differentiating placental versus maternal cell-free DNA fragments. In general, several approaches are used to accurately determine the placental and maternal origin of cfDNA, but they all demand that the maternal DNA be tested in parallel. Here are a few employed methods:

1. SNP Analysis. The SNP analysis compares the genetic variations between placental and maternal DNA at specific single nucleotide positions. By analyzing a panel of known SNPs, it is possible to identify SNPs particular to the placental or maternal DNA. For example, the presence of placental-specific SNPs in the cfDNA indicates a placental origin.
2. Methylation-based Analysis. DNA methylation patterns can differ between placental and maternal DNA. Methylation-based approaches involve analyzing the methylation status of specific genomic regions. Differences in methylation patterns can be used to distinguish between placental and maternal DNA fragments within the cfDNA population [64].
3. Haplotype Analysis. Haplotypes are combinations of genetic variants inherited on a chromosome. By analyzing haplotypes, it is possible to determine the parental origin of specific alleles in the cfDNA. Comparing haplotypes between placental and maternal DNA can help differentiate their origins.
4. Y Chromosome Analysis. If the fetus is male, Y chromosome-specific sequences in the cfDNA indicate a placental origin. Therefore, detecting Y chromosome sequences makes it possible to determine whether the cfDNA is derived from the fetus.

12.6 Cell-Based NIPT for CNVs

For the last two decades, there has been ongoing research in capturing and examining intact fetal cells without invasive methods. However, this progress has faced significant technical difficulties due to various factors, such as the limited availability of these cells, the need for more reliable and scalable methods for capturing them, and the variations observed in individual samples [65]. Hence, there is significant interest in noninvasively capturing intact fetal cells and interrogating them, as this approach could provide precise and comprehensive profiling of the fetal genome for microdeletions and microduplications. Recent years have seen substantial advancements in developing more efficient workflows for isolating circulating fetal trophoblasts and performing single-cell genomic profiling to detect submicroscopic variants. The development of whole-genome amplification and single-cell analysis technologies potentiated this. A typical workflow for cell-based noninvasive prenatal testing for CNVs involves several key steps: (1) collecting the maternal sample,

(2) initial processing of the sample and enriching fetal cells using markers or size, optionally depleting maternal cells, (3) identifying fetal cells, (4) isolating the fetal cells, and (5) performing downstream analysis including low-coverage NGS. Besides, whole genome amplification is essential in single circulating fetal cell analysis to ensure sufficient DNA input for the downstream analysis.

Literature regarding clinical validation of cell-based NIPT for CNVs is still scarce, including a smaller number of samples and employing various extraction methodologies. As a result, several papers use cell-based NIPT for CNV. In a validation study involving 95 cases, after isolating circulating trophoblasts, 55% were found to be of high quality and suitable for both aneuploidy and CNV testing [66]. This cell-based NIPT method has demonstrated reliability in diagnosing fetal chromosomal abnormalities down to 1 Mb in size [67]. In another study, DNA from individual trophoblasts underwent low-coverage NGS for genome-wide CNV analysis, successfully identifying fetal aneuploidy in 11 cases and known CNVs in four cases, ranging from 1.1 Mb to 19 Mb in size [68].

In 2022, the automated and high-throughput procedure was introduced as a multi-step workflow for the semi-automatic isolation and analysis of circulating extravillous trophoblasts (cEVTs) as a proof of concept. In addition, the fetal cell origin's genetic conformation was performed using a proprietary integrated algorithm based on SNP analysis [69]. In this study involving a cohort of 61 cases undergoing fetal CMA analysis, two cases were identified with pathogenic CNVs, while the remaining 59 cases showed normal CMA results. Single-cell analysis of cEVTs demonstrated complete concordance with clinical fetal CMA results. Among the group with normal CMA results, six (likely) benign CNVs were identified, all involving copy number gains ranging from 900 Kb to 3 Mb in size, as detected by single-cell analysis in cEVTs. Furthermore, this study demonstrated that the limit for single-cell cEVTs CNV detection using low-coverage NGS is 800 Kb [69].

One challenge in CNV detection in single-cell analysis is the potential recovery of cells in low-quality states during apoptosis or in the mitotic S-phase, which can result in DNA that is not suitable for analysis. Nevertheless, the S-phase cells were successfully used to identify aneuploidy. These observations highlight the significance of recovering multiple cells from each sample to reduce inconclusive results or limit the assessment solely to aneuploidy. To address these situations, a potential strategy could involve increasing the volume of blood collected or processed initially.

12.7 Clinical Applications of NIPT for CNV and Sub-Chromosomal Variations Detection

Generally, today cell-free NIPT is propagated to be offered to all pregnant women regardless of the predisposing risk. Although this can justify common aneuploidy testing, offering cell-free NIPT for CNVs and sub-chromosomal variation should be carefully considered. Therefore, several questions regarding NIPT for CNV and sub-chromosomal variations in clinical settings must be contemplated.

12.7.1 To Whom NIPT for CNV and Sub-Chromosomal Variations Should Be Offered?

When considering NIPT for CNVs, it should be noted that there were no large-scale randomized control studies conducted to assess clinical validity or clinical utility for cell-free NIPT for CNVs and cub-chromosomal variations. On the other hand, multiple studies have been undertaken to determine the sensitivity and specificity of NIPT for common aneuploidy, but with an emphasis on high-risk patient populations with singleton pregnancies. Therefore, there is an ongoing effort to make a statement offering NIPT in general, but NIPT for CNVs especially, should be considered regarding the population of general risk individuals. This implies that obstetrical practice encounters a population spanning a spectrum of risks, encompassing low-risk, intermediate-risk, and high-risk patients. In this way, before offering the NIPT for CNVs, clinicians should consider the range of risks in every case.

Traditionally, maternal age and increased nuchal translucency (above 3.5 mm) are the most common referral reason for prenatal testing [70]. In contrast to the common trisomy, the prevalence of CNVs is independent of maternal age (68). This can work in favor of NIPT for CNVs, but on the other hand prevalence of CNVs in low-risk pregnancies is extremely low. It is difficult to estimate the prevalence of clinically significant CNVs in general-risk pregnancies as most students are biased toward including intermediate-risk and high-risk patients. Based on a systematic review and meta-analysis involving ten publications with cohorts exceeding 300 cases each, pooled estimates reveal that 0.84% of fetuses undergoing invasive testing due to advanced maternal age or parental anxiety harbored pathogenic clinically significant submicroscopic aberrations [71]. Conversely, among tested samples from fetuses with suspected growth or structural anomalies, 2.8% showed clinically relevant CNVs detected by microarray that were not identified by karyotyping [21].

Another factor that may contribute to performing prenatal tests is to assess recurrence risk (RR). Research indicates that women who experienced a fetal structural anomaly in their initial pregnancy face a 2% absolute recurrence risk of a similar anomaly in subsequent pregnancies, contributing to an overall 4.08% risk of congenital anomalies when factoring in an additional 2% risk for dissimilar abnormalities [70]. The recurrence risk for a similar anomaly is notably elevated (RR = 24) [70]. This risk escalates significantly when molecular analysis confirms an autosomal recessive or X-linked disease.

According to European guidelines for constitutional cytogenomic analysis, prenatal diagnosis is recommended in cases of abnormal fetal ultrasound findings, high-risk maternal serum screening, or NIPT results indicating increased risk of chromosomal abnormalities, parental chromosome rearrangements, mosaicism, or previous aneuploidy. It is also indicated after previous live births or stillbirths with chromosome abnormalities, detection of possible fetal mosaicism in prior prenatal studies, and in families with monogenic disorders such as cystic fibrosis or Noonan syndrome. These guidelines help identify genetic abnormalities early in pregnancy, enabling informed decision-making and appropriate management strategies [72].

12.7.2 How to Address the Clinical Significance of Test Results? What About Variants of Uncertain Significance?

Determining whether CNVs are pathogenic or not is of utmost importance in clinical settings, particularly in prenatal diagnosis, as it can have significant implications. Identifying whether a CNV is pathogenic or benign is crucial in guiding decisions on further diagnostic testing, pregnancy management, and counseling prospective parents. Ideally, CNV discovery, genotyping, and clinical significance assessment require accurately detecting three main features: copy, content, and structure.

Currently, available technical standards for the interpretation and reporting of constitutional copy number variants in the prenatal and postnatal setting as a joint consensus of the ACMG and the Clinical Genome Resource (ClinGen) recommend using a 5-tiered system. This recommendation includes five categories of clinical significance ranging from pathogenic (P), likely pathogenic (LP), variants of uncertain significance (VUS), likely benign (LB), to benign (B) [73]. CNVs can impact phenotypes through several molecular mechanisms, including the dosage sensitivity of genes within the CNV, disruption or fusion of genes at breakpoint junctions, deletion of regulatory elements, and the unmasking of recessive alleles or functional polymorphisms. Additionally, CNVs can influence noncoding regulatory elements such as promoters or enhancers, further complicating their functional implications in genomic disorders [74]. In recent years, awareness of the limitations of NGS and the pitfalls of genome-wide investigations has been raised. For now, this has been an issue in postnatal genetic testing, but these findings are also significant in the context of prenatal testing. Determining the clinical significance of genetic findings and predicting their impact on health outcomes, including potential medical or cognitive issues, is crucial for guiding decisions during pregnancy. An important factor to consider in interpreting genetic variants is the use of alternative transcripts. Multiple transcript databases exist for annotating variants, leading to significant variability in transcript selection among sequencing laboratories [75]. This variability was underscored by a comparison study involving 80 million genomic variants using RefSeq and Ensembl transcript sets, which showed agreement in annotation for only 44% of putative loss of function (LoF) variants [76]. In 2022, a unified transcript set by NCBI and EMBL-EBI for clinical genomics and research, known as MANE, was introduced. There is a recommendation for international adoption of MANE in best practice guidelines to ensure consistency in reporting the primary consequence of specific genetic changes across clinical laboratories and online databases. This standardization aims to provide clear, stable, and accurate reports on actionable genetic variants for both specialist and nonspecialist practitioners. However, integrating tissue-specific transcripts remains an ongoing challenge.

WGS was in the past years favored and advertised as "*the holy grail,*" which provides complete insight into our genome. However, from today's perspective, based on scientific findings and critical analysis, we know this is far from the truth. Regrettably, there is a scarcity of studies that have comprehensively compared the diagnostic efficacy of GS versus ES within identical patient cohorts. Nevertheless, existing data suggest that GS generally yields approximately 9–16% higher diagnostic rates compared to ES in postnatal clinical scenarios [77, 78]. However, this must be taken into consideration that this is biased data, including the diseased population.

The issue of misclassifying the pathogenicity of genetic variants is widely recognized. For example, one study compared the ClinGen Dosage Map, a resource documenting known haploinsufficiency (HI) and triplosensitivity (TS) genes/regions, with germline clinical CNV interpretations in the ClinVar database and identified 251 CNVs requiring re-evaluation [79]. Of the 246 re-evaluated CNVs, 157 (63.8%) needed an updated clinical classification, 79 (32.1%) retained their original classification, and 10 (4.1%) led to other updates in ClinVar records [79]. Additionally, a recent study found that 40% of variants in the ClinVar database labeled "Pathogenic" or "Likely Pathogenic," without conflicting interpretations and with an allele frequency above 0.5% in healthy individuals of at least one ancestry, were reclassified as "Benign," "Likely Benign," or "Variant of Uncertain Significance" when manually reviewed using ACMG criteria [80]. Furthermore, investigations into variants listed as disease-causing in the Human Gene Mutation Database revealed that about 3.5% were misclassified. Various factors contribute to labeling a variant not associated with disease as disease-causing in a fetus without a compatible phenotype.

Generally, in prenatal whole genome/exome sequencing, it is possible to detect variants in genes with a relatively limited amount of evidence supporting their association with diseases. These genes are frequently termed candidate genes, genes of uncertain significance, or genes with limited evidence. Their potential disease association may be supported by factors such as internal laboratory data, animal studies, or a few published case reports. Therefore, regardless of the criteria for CNV interpretation, any variant identified in a candidate gene should be classified as no higher than a VUS.

Finally, addressing the clinical significance of detected CNVs should be considered regarding confined placental mosaicism (CPM). Notwithstanding that during NIPT, detected CNVs can sometimes be attributed to CPM. Patients with CPM of rare structural variations are linked to pregnancy complications and adverse outcomes, such as fetal growth restriction, preeclampsia, and preterm birth [81–83]. Moreover, since cell-free DNA tested in NIPT predominantly originates from placental trophoblasts, placental mosaicism can lead to false-positive and false-negative results [84].

12.7.3 Which Variant to Report?

Generally, pathogenic and likely pathogenic variants should be considered for reporting in prenatal testing. On the other hand, reporting CNVs associated with variable phenotypes and low penetrance, like 15q11.2 breakpoints 1–2 deletions and duplications, is still a point of contention [85]. Most recently, reporting secondary findings (SF) and incidental findings (IF) in prenatal sequencing came into focus. Unfortunately, professional bodies issue few guidelines or position statements that address the return of secondary findings, specifically in the context of prenatal testing. On the contrary, whether to return IF and/or actively search for SF has been discussed extensively for postnatal testing, and several well-established guidelines for postnatal testing are updated regularly [86, 87]. The highly complex situation regarding IF and SF in prenatal diagnosis includes practical consideration regarding test design, which genes to report, counseling, consent, and laboratory workload. In the context of the test, employing a trio design that includes the fetus and both parents necessitates obtaining additional consent from both parents. This is due to the potential detection of incidental or secondary findings in the genotypes of both the parents and the fetus, as well as the possible identification of consanguinity and nonpaternity [88]. The question of which genes should be reported as IF and SF in prenatal testing can be addressed by an alternative approach. In contrast to SF lists in postnatal testing, specialized gene lists developed for newborn genomic screening could be a good starting point [89–91]. This customized gene list excludes disorders that manifest in adulthood but includes information about carrier status. It also encompasses childhood-onset conditions for which no treatment currently exists. The rationale behind this inclusion is that early detection might obviate the need for extensive clinical tests, shorten the diagnostic process, and enable the implementation of nonspecific early interventions that could potentially enhance outcomes. However, a crucial distinction between newborn screening and prenatal testing is that in the latter case, a test result may lead to a decision to terminate the pregnancy. This aspect contradicts the initial notion of "actionability," which primarily aimed to benefit the health of the tested individual. When considering the clinical significance of detected CNVs, one has to remember that changes in the interpretation of clinical genomic test results are expected and that reevaluation and reanalysis of genomic test results over time are already well established for postnatal genomic testing [92].

12.7.4 Which Type of NIPT for CNVs and Sub-Chromosomal Variations Should Be Offered?

As discussed, NIPT for CNVs and sub-chromosomal variations can be done by microarray or NGS technology. NGS offers whole genome CNV detection but with lower resolution than microarray. On the other hand, microarray can detect significantly smaller CNVs but has limited scope, offering targeted analysis. Therefore, depending on the referral indication, the practitioner should consider which test is more suitable to meet the testing needs.

1. Does NIPT for CNVs offer advantages compared to traditional prenatal diagnosis methods for CNVs?

   Most commercially available prenatal tests include testing for CNVs >7 Mb. Notably, the approximate resolution of classical karyotyping by G-banding at 500–850 bands is 3–5 Mb. However, it is essential to acknowledge that sequencing does not provide insight into the structural characteristics of the genome, unlike classical cytogenetic analysis, which can present the genome structure.

Generally, comparing "*traditional*" prenatal diagnosis methods for CNV detection with NIPT is a challenging task. "*Traditional*" prenatal diagnosis methods for detecting CNVs include chromosomal microarray analysis and sequencing, similar to those used in NIPT. However, prenatal samples are sourced from chorionic villi, amniotic fluid, and cord blood. It is well established that even low-pass genome sequencing in invasive prenatal testing provides additional clinically significant information with higher resolution and increased sensitivity in detecting mosaicism compared to chromosomal microarrays [93, 94]. In evaluating the advantages and limitations of NIPT for CNVs compared to traditional methods, it is crucial to note that the sensitivity of NIPT, as reported in the literature, ranges from 20% to 100%. Specificity ranges from 81.62% to 100%, and the positive predictive value ranges from 3% to 100% for cases with diagnostic or clinical follow-up information [95].

2. Does NIPT for CNVs offer comparable performance in singleton and twin pregnancies?

   Regarding twin pregnancies and NIPT for CNVs, it is essential to consider zygosity and chorionicity. Concerning chorionicity, all dizygotic pregnancies and approximately one-third of monozygotic pregnancies are dichorionic [96]. Only a limited number of studies have investigated the performance of genome-wide NIPT in large cohorts of twin samples. For example, a report from a commercial laboratory. Over half of the enrolled individuals selected the "*Enhanced Sequencing*" option to test for additional aneuploidies (except T21, T18, and T13) and microdeletion syndromes. Regarding the analysis of NIPT results for microdeletions in twin pregnancies, six samples had a positive outcome. Out of these, four samples underwent invasive diagnostic testing: three were confirmed true positives, and one was a false positive. The remaining two cases, although not confirmed diagnostically, were reported to be clinically consistent with the suspected microdeletion syndrome. Notably, all samples with positive microdeletion results were from higher-risk pregnancies, such as those with ultrasound findings or other high-risk factors [97]. In a more recent study involving 1244 twin pregnancies that underwent NIPT, four dichorionic samples were identified as high risk for CNVs without presenting ultrasound anomalies. Following amniocentesis, two of these samples were determined to be false positives, and the remaining case was lost to follow-up [98].

3. What about the mother's genetic variations in cell-free testing?

   The existence of maternal CNVs presents a challenge when screening for fetal CNVs. In theory, identifying maternal CNVs is relatively straightforward because approximately 90% of the cfDNA mixture originates from the mother. Surprisingly, statistically speaking, maternal CNVs are more common than anticipated. This was demonstrated in a study using parental data from 653 affected probands. Among the 66 CNVs, 18 were found to be exclusive de novo, while the majority (48/66) were inherited [99]. Another instance illustrating this phenomenon is documented in a study where 55 microdeletions were identified among a randomly sampled population of 175,393 pregnancies, with 25 cases attributed to maternal origins [100]. Although CNV-seq NIPT can technically detect maternal CNVs, it cannot distinguish whether a CNV is solely maternal or if it affects both the mother and the fetus. Consequently, when a maternal CNV is suspected, prenatal diagnosis becomes necessary to determine the status of the fetus, leading to more invasive testing. One way to overcome invasive testing to determine CNV origin is parent-offspring trio analysis, or at least a maternal sample should be included [101]. When CNVs are identified as maternal, NIPT for CNVs demonstrates high sensitivity, specificity, and positive predictive value across both single and multiple pregnancies, regardless of risk status, with excellent performance observed genome-wide [102].

Regarding reporting maternal outcomes, there are only several studies available. Commonly reported outcomes included maternal neoplasms and abnormalities involving the maternal X-chromosome. Additionally, findings encompassed actionable maternal CNVs, identification of CNVs in the Duchenne muscular dystrophy gene, and various structural chromosomal abnormalities like mosaicism involving interstitial deletions and unbalanced translocations [103].

Although maternal malignancy is relatively uncommon, with an incidence ranging from 1 in 1000 to 1 in 5000, most reported maternal outcomes are incidental detection of maternal cancer during NIPT [104, 105]. Recent retrospective studies have compiled data on these incidental findings, where maternal malignancies were discovered following a "false positive" NIPT. These studies have reported positive predictive values for cancer detection exceeding 50% [106, 107]. Considering the substantial positive predictive value observed in detecting maternal malignancies during routine NIPT, the advantages of reporting these findings significantly outweigh the associated risks.

12.8 Ethical and Legal Considerations of NIPT for CNVs and Sub-Chromosomal Variations Detection

During the last 5 years, there has been a substantial discussion regarding legal and ethical questions considering NIPT in general. Recently, several countries have introduced publicly funded NIPT as a second-tier test offered between the 11th and 14th week of pregnancy for common chromosomal aneuploidies such as trisomy 21, 18, and/or 13 [108, 109]. Regulations in some countries stipulate specific criteria for accessing free NIPT for these conditions, while others lack legal frameworks governing its use. Legal regulations regarding decisions on pregnancy and unborn children often intersect with reproductive rights and access to termination of pregnancy (TOP). Laws regarding reproductive rights and TOP vary widely across countries, and no universal regulation or direction can be applied universally. To date, according to our knowledge, there is no legal regulation regarding NIPT for CNVs and sub-chromosomal variations. Commercial companies mainly offer NIPT for CNVs, and in most countries, no legal regulation states that NIPT may or may not be shown, which means that it is not defined whether it is legal or illegal to offer NIPT for CNVs and sub-chromosomal variations.

Profound ethical issues arise when considering the permissibility of screening for specific traits, the provision and accessibility of such information, the allocation of public funds for particular purposes, and the need for regulations on the commercial direct-to-consumer sector. The discourse surrounding the inclusion of specific elements in prenatal screening should not be confined solely to accepting or rejecting what is currently feasible but rather should delve into the broader considerations at hand. Beauchamp and Childress outline four ethical principles that guide medical practice: beneficence, nonmaleficence, autonomy, and justice [110]. In contrast, the Wilson and Jungner principles, established nearly 50 years ago, continue to influence decision-making for screening [111]. Originally formulated as guidelines for population screening, these principles emphasize the importance of understanding the health issue's significance, disease progression, test characteristics, treatment options, and cost-effectiveness. However, prenatal screening programs differ in their objectives compared to other population screening initiatives. The thesis prompted extensive NIPT debates, particularly on detecting CNVs and sub-chromosomal variations. A recent review of the literature on NIPT's ethical considerations in England, France, and Germany has identified common ethical concerns among these countries [108]. Besides, this is not a particularity of these three countries, and the same ethical issues are discussed mainly in international literature. Several ethical aspects of NIPT, in general, could extend to CNVs. Firstly, concerns arise regarding the potential for increased stigmatization and discrimination of individuals with autosomal aneuploidies if NIPT becomes a routine part of antenatal care. These concerns could negatively impact the support available to women who choose to raise a child with a trisomy condition. This could be extrapolated to women who decide to raise a child with known pathogenic/likely pathogenic CNVs, which cause microdeletion/microduplication syndromes. Another significant issue raised in the literature is the possibility of NIPT becoming a standardized procedure. This development has the potential to exacerbate the existing complexities surrounding prenatal testing. There are concerns that routine implementation of NIPT may pressure women to undergo testing, compromising their ability to make fully informed decisions and diminishing their reproductive autonomy [112]. Additional arguments revolve around the potential dangers of prenatal sex selection [113] and the risks associated with screening for conditions considered "less severe" and those related to adult-onset conditions and carrier status. Ethical concerns also arise regarding the disclosure of results, the management of information in situations involving misattributed paternity, and the identification of secondary or incidental findings. These findings may have implications for the fetus's health and the well-being of the mother and other family members.

Interestingly and intriguing, the same literature overview offers insight into charges of eugenics to denounce NIPT [108]. This perspective should be carefully considered in the context of NIPD for CNVs and sub-chromosomal variations. The risk of a drift toward 'eugenics' appears in different forms, and in Germany and France, it is already used as an argument against the implementation of NIPT [114, 115]. But not to be forgotten that these countries have endorsed NIPT for common chromosomal aneuploidies (trisomy 21, 18, and/or 13). This fact emphasizes how complicated are the ethical issues surrounding NIPT.

Another critical ethical concern regarding NIPT involves examining whether there has been any shift in birth rates since its introduction into general practice and whether it has contributed to a reduction in the barrier to TOP. Although this is a fundamental question, there is little evidence for or against this thesis in the literature. Studies reporting pregnancy outcomes following NIPT were mainly based on aneuploidy screening in the past. The rates of selective TOP may not differ much from other prenatal screening methods [116]. Only a single study reported a decreased rate of trisomy 21 from 2014 to 2018, from 0.06% of live births to 0.04%, respectively [117]. Data on NIPT for CNVs and still limited. A recent study indicated that pregnancies with inherited variants of uncertain significance were typically continued based on primary ultrasound findings combined with noninvasive prenatal screening results, whereas pregnancies involving de novo variants of uncertain significance were often terminated [118]. Examining "non-medical" traits that can potentially impact cognitive ability or conditions deemed less severe is another ethical consideration of NIPT for CNVs and sub-chromosomal variations. For instance, the 15q11.2 microdeletion involving NIPA1 has been shown to reduce IQ by an average of 4.3 points, suggesting a sub-clinical effect on cognitive ability while also slightly increasing the risk of intellectual disability [85, 119]. Whole genome analysis provides extensive predictive information, including the use of polygenic scores. A polygenic risk score is derived from the cumulative impact of multiple genetic variations. It holds the potential for predicting various non-medical attributes such as cognitive aptitude, artistic ingenuity, physical stature, and outward appearance. In the clinical domain, polygenic scores for diabetes and heart disease are often deliberated upon as they contribute to advancing "personalized medicine" approaches. Ethical issues raised here are should we offer such screening just because it is possible?

12.8.1 Informed Consent, Counseling, and Reporting

Appropriate pre- and post-testing counseling should be integral to informed decision-making in NIPT, as with all prenatal screening and diagnostic testing. It is crucial to recognize that even when utilizing comprehensive techniques like karyotyping, chromosomal microarray, and NGS, there is no guarantee of identifying the specific genetic variant in all cases. As a result, it is essential to provide counseling that covers various aspects of the procedure and obtain informed consent, considering clinical suspicion. Accurate and precise reports should be issued to referral practitioners to facilitate post-test counseling and informed decision-making. In the context of general antenatal care, referral to genetic counseling is indicated for several reasons. These include a family history that includes a known or suspected genetic disorder, being a carrier of a genetic condition, or consanguinity within the family (such as between parents, first cousins, or closer relatives). Detection of fetal structural anomalies during prenatal ultrasound, abnormal results from traditional serum screening or noninvasive prenatal testing (NIPT), and known exposure to teratogens during pregnancy also warrant referral. Additionally, recurrent pregnancy loss (defined as three or more first-trimester losses, stillbirths, or neonatal/fetal losses associated with structural anomalies) prompts referral, as does patient-initiated interest in obtaining more information about genetic testing options. These indications highlight the importance of genetic counseling in assessing genetic risks and informing decision-making during pregnancy.

Genetic counseling considerations for NIPT for CNVs include a broad range of potential indications for testing. Advances in prenatal sequencing have boosted the effectiveness of genetic diagnostics, yet they have also heightened the likelihood of encountering ambiguous results. Several aspects of uncertainties associated with prenatal whole genome analysis are discussed regarding pre- and post-testing counseling. Still, most overlap, thus emphasizing the complexity of counseling in NIPT for CNVs. It is essential to emphasize that some of the guidelines for prenatal counseling are based on older literature where chromosomal microarrays are used for CNV detection. However, most of them can be applied in NIPT for CNVs and sub-chromosomal variations, bearing in mind all technical aspects of technology used in NIPT. The latest literature review summarizes the most common uncertainties practitioners may encounter during routine clinical prenatal diagnostics [120].

12.8.2 Uncertainties Related to the Technology

If a finding is based on insufficient read depth or potentially unreliable CNVs, validating the results using a different technique is recommended, and it is essential to discuss this during pre-test counseling. This validation step helps minimize the impact of false positives. It is crucial to thoroughly evaluate and re-evaluate candidate variants by referring to relevant databases. In the laboratory report provided to the clinician, an estimated diagnostic specificity can be included to indicate the risk of a false positive if feasible. The limitations of enrichment methods and sequencing platforms should guide the decision on whether additional testing is necessary and what type of testing should be pursued. Communicating disclaimers and limitations regarding coverage in the laboratory report is essential. Finally, disclosing testing characteristics is crucial to provide an unbiased perspective for patients about NIPT for CNVs. This includes, when possible, reporting and explaining false positives and false negative predictive values, test sensitivity and specificity, and positive predictive values for specific screening tests.

12.8.3 Uncertainty Related to Clinical Effectiveness—Diagnostic Yield

Uncertainty related to clinical effectiveness is associated with the question what is the likelihood of providing a diagnosis. The pre-test genetic counseling and informed consent procedures should cover a thorough discussion about realistic anticipations regarding the probability of a diagnosis, the chance of not receiving a result before birth, and the extent and accuracy of the test. Furthermore, test failure (or "*no-call*" results) should be explained and discussed in pre-test genetic counseling. During pre- and post-test counseling, it is crucial to inform parents that the absence of a causative variant for the primary indication is a potential outcome, and it should not be interpreted as conclusive evidence ruling out a genetic cause. Post-test counseling should explain the remaining risk if a genetic reason for the prenatal abnormalities cannot be identified.

Addressing and clarifying the analysis resolution is crucial in managing expectations regarding diagnostic outcomes. Laboratory reports and informed consent documents should explicitly outline the analysis resolution used during testing. Recommendations for resolution typically range from 200 kb to 400 kb, aiming to reduce the likelihood of identifying variants of uncertain significance while optimizing diagnostic yield. Additionally, discussing trio analysis during pre-test counseling is beneficial as it can enhance the accuracy of diagnoses by comparing fetal and parental genomes simultaneously, thereby expediting the testing process and improving diagnostic success rates.

12.8.4 Uncertainties Related to Incomplete Knowledge—Prenatal Phenotype

The correlation between prenatal genotype and phenotype is often regarded as uncertain due to limited knowledge about the prenatal manifestations of specific genetic variations. Many microdeletion/microduplication syndromes have been identified postnatally, with inadequate information about their prenatal presentation or a biased focus on more severely affected fetuses. To address this challenge effectively, it is essential to ensure standardized clinical information is provided. This includes comprehensive details such as thorough fetal imaging reports, results from previous prenatal tests, parental medical histories, ethnicity, reproductive backgrounds, family histories (with attention to consanguinity), and any other pertinent data [121]. Collecting phenotypic data enables better connections between genetic information and potential diseases, even though it can be challenging when comprehensive or specific clinical characterization of the fetus is impossible. Laboratories should establish systems for the submission of this clinical information. In cases where genetic data are the only available information, post-test counseling is particularly important. Analyzing and interpreting variants remains challenging, especially in prenatal settings where the interpretation often relies on limited or nonspecific fetal phenotypic information. Compared to the postnatal setting, the prenatal variant analysis may be more challenging due to including an often incomplete or ambiguous fetal phenotype. The phenotype may still be incomplete or only partially detectable depending on the gestational age, as anomalies can manifest sequentially or later in pregnancy.

In a recent paper, it was proposed to address disorders-related and variant-related questions in variant interpretation. In the first section, disorders-related questions emphasize the well-established gene-disease correlations. In prenatal settings, exome sequencing is recommended due to its focus on coding sequences of known disease-causing genes. This approach is particularly useful when ultrasound anomalies are present and standard microarray testing has not provided a definitive diagnosis [121]. Accurate classification of copy number variants (CNVs) is crucial for proper reporting and should adhere to established standards, typically utilizing a 5-tiered system supported by clinical information and empirical data [73]. Laboratory reports should include several key elements: (1) determination of *cis* or trans configurations for potentially multiple pathogenic variants, (2) supporting evidence from literature, prediction programs, or databases to justify pathogenicity assignments, (3) integration of results from multiple analyses on a single sample, (4) appropriate references used for result interpretation, especially for rare or unclassified variants, (5) consideration of additional information that clinicians may provide to aid interpretation accuracy, and (6) clear delineation of the significance of the findings. These practices ensure thorough and informative reporting to guide clinical decision-making in prenatal care [122]. Re-evaluation of clinical significance for detected CNVs should not be routinely performed in prenatal settings [123, 124]. However, it is advised to provide a revised report to the clinician when there is a significant change in the categorization of a variant, such as a shift from being considered (likely) pathogenic to (likely) benign. During post-test counseling, parents should be informed about this possibility. Patients can also request further re-analyses through their clinician, particularly in the case of future pregnancies or after a substantial amount of time has passed (e.g., more than 12 months) if the initial exome sequencing report did not include an extended phenotype or if the phenotype has evolved postnatally. Moreover, new associations between genes and diseases may have been established, and the fetal phenotype could now be linked to the genotype [121].

Clinical implications of variants of uncertain significance should be discussed during pre-test counseling. VUS constitutes a distinct category of chromosomal CNVs where reliable population data is generally lacking, making it challenging to determine their pathogenicity definitively. Nonetheless, several studies have shown that parent-of-origin investigations can offer valuable insights in counseling such cases, aiding in the clinical interpretation of fetal CNVs and potential pregnancy outcomes. These studies have indicated that most VUS are inherited and do not exhibit apparent indications of abnormal birth outcomes. However, declining parent-of-origin testing complicates genetic and general prenatal counseling, particularly in decisions regarding continuation or termination of the pregnancy. Guidelines occasionally provide clarity on managing variants of VUS but frequently rely on local practices for resolution. Some guidelines discourage the reporting of variants of uncertain significance (VUS) [125, 126], while others recommend including VUS in the report if they might become clinically significant in the future [122, 127]. There is ongoing debate regarding whether to report VUS that could potentially contribute to abnormal fetal phenotypes [121]. Despite these inconsistencies, genetic healthcare professionals should gather clinical information and consent during pre-test counseling, where the types of variants to be reported are discussed. Parents should have the option to decline receiving variants in non-disease genes. It is crucial to explain to parents that different laboratories may have varying policies on reporting VUS during prenatal testing. Furthermore, it should be emphasized that VUS may not be reported in the future, particularly after targeted testing has been performed.

What an aid of genomic sequencing; findings unrelated to the primary clinical question can be incidentally discovered or actively searched for. SF and IF, counseling can be offered as an opt-in or opt-out option. Still, it has to be discussed during pre-test counseling, clearly stating that it refers to both mother and child, and father in the case of trio analysis. This includes reporting for a mother's conditions like malignancy and late-onset disorders. In recent recommendations concerning the fetus, it has been advised to establish clear policies regarding (1) whether to report only SFs in the fetus or also SFs detected in the parents, and (2) whether the analysis, filtering, and variant calling of SFs in parents should be restricted to those identified in the fetus [121]. Some guidelines do not advocate actively searching for SFs but instead recommend targeted analysis of genes relevant to the primary indication [123, 124]. Currently, there are no established criteria for SF, but every laboratory should determine if and which SF is to be reported. This should be determined by their ability to provide accurate information, and clinicians should discuss which SFs are routinely analyzed and reported during the informed consent process. Clinical data is crucial in supporting variant interpretation, as an SF may not always be unexpected. There is a growing interest in SF, and one way to overcome this is to perform a search for SF separately and with separate informed consent from the patient [123]. In general, patients should be provided with information regarding the following aspects:

1. Whether the analysis includes genes unrelated to the phenotype, such as those listed in the ACMG secondary findings gene list [86].
2. Patients can opt-in or opt-out of receiving secondary findings.
3. How the reporting of secondary findings differs from reporting variants associated with the primary indication.
4. In the case of trio analysis, separate consent should be obtained from each individual (except the proband) who undergoes sequencing.

At the time of pre-test counseling and the informed consent process, discussions should cover the likelihood of identifying secondary findings and whether they are included or excluded from the analysis. It is recommended to offer an opt-out option for receiving secondary findings from the ACMG list. The multidisciplinary team should decide to opt-in for receiving (likely) pathogenic secondary findings and communicate to the laboratory. If treatable pathogenic secondary findings are reported, it should be explicitly explained that they are unrelated to the primary indication.

It is advisable to minimize the likelihood of IFs by focusing on the gene panel under investigation or by targeted testing aligned with the referral reason and informed consent. When conducting trio analysis, only variants identified in the fetus should be analyzed in parental samples. A clear protocol or policy should dictate whether IFs are reported and whether an opt-in or opt-out option is provided. Generally, IFs lacking health implications or of uncertain significance should not be reported. Guidelines for prenatal exome sequencing recommend: (1) refraining from reporting variants lacking a known fetal or childhood phenotype, (2) withholding reports of heterozygous variants for autosomal recessive or X-linked disorders, and (3) reporting highly penetrant pathogenic IFs potentially causing moderate to severe early onset disorders [121, 125, 128]. While some suggest reporting all relevant results if requested by parents, this practice raises ethical concerns. It is considered more ethical to report actionable known and likely pathogenic IFs only. Early onset disorders are typically reported, whereas late-onset disorders may be considered for an opt-in if treatable but not for non-treatable conditions [129]. Deletions in genes linked to recessive disorders not aligning with the fetal phenotype should only be reported if the carrier frequency exceeds 1 in 50 [126]. When reporting treatable pathogenic IFs, it should be explicitly stated that these findings are unrelated to the initial test indication.

12.9 Conclusion

The American College of Obstetrics and Gynecology advises against routine cell-free DNA screening for microdeletion syndromes [130]. Similarly, the American Society of Human Genetics and the European Society of Human Genetics concur that such screening should not be routinely performed at present due to concerns over reduced diagnostic specificity, particularly with the inclusion of sub-chromosomal aneuploidies in NIPT [131]. However, the ACMG provides conditional recommendations specifically for NIPT targeting 22q11.2 deletion syndrome, acknowledging moderate certainty of evidence supporting its use. This suggests that NIPT for 22q11.2 deletion syndrome may be offered under conditions where patients are well-informed about the benefits and limitations, and decision-making is shared [132]. Conversely, for CNVs other than 22q11.2 deletion, there is currently no recommendation due to insufficient clinically relevant evidence and validity, aligning with previous statements from 2016 regarding genome-wide CNV screening in NIPT [132, 133].

Notwithstanding, NIPT for CNVs is not recommended, at its current state, testing for the more common microdeletion syndromes like the DiGeorge, Cri-du-chat, Prader–Willi/Angelman, 1p36 deletion, and Wolf–Hirschhorn syndromes will most likely be more and more advertised and propagated to pregnant women around the world regardless of the predisposing risk. There is some solid background to discuss NIPT for CNVs if there is an indication for one of the stated microdeletion syndromes (i.e., ultrasound finding or other high-risk factors). Offering prenatal genome-wide CNV testing should be approached cautiously, particularly for fetuses with multiple anomalies or those selected following genetic counseling, as these cases may yield higher diagnostic rates. Particular emphasis must be placed on pretest counseling explaining two main points. First, any positive results from the screening should be verified through invasive diagnostic procedures. Second, it is essential to understand that the absence of a detected CNV does not guarantee its absence altogether. It is crucial to recognize that even when utilizing comprehensive techniques like karyotyping, chromosomal microarray, and NGS, there is no guarantee of identifying the specific genetic variant in all cases.

Moreover, it is widely accepted that employing a trio test design is optimal in prenatal settings. This approach facilitates the rapid identification of de novo variants, segregation analysis of recessive variants, and minimizes the reporting of uncertain variants, which can have significant psychosocial implications.

Additional research is necessary to address various aspects of implementing NIPT for CNVs and sub-chromosomal variations to ensure its accuracy and reliability. Standardized and regularly updated education and training programs for healthcare providers offering NIPT should be established to inform them about new field developments. It is essential to develop and validate unbiased educational materials, free from marketing interests, and provide accurate information about the benefits and limitations of NIPT in general, particularly for CNVs.

The potential harms associated with NIPT require further investigation, particularly regarding the specific conditions included in different NIPT assays, like maternal conditions, variants of uncertain clinical significance, and the need for an invasive confirmation test. Understanding and assessing these potential risks are crucial for NIPT's responsible and informed use. NIPT for CNVs and sub-chromosomal variations are currently a subject of intense deliberation in both scientific and public discourse, lacking definitive resolutions.

NIPT has undoubtedly brought significant advancements to prenatal care, but expanding NIPT for CNVs into general-risk populations should be approached carefully. It is essential to ensure that NIPT for CNVs is implemented responsibly and with thorough consideration of its implications and limitations. NIPT is poised to become a permanent fixture in the prenatal field. Yet, healthcare providers must uphold the utmost standards and quality by employing rigorous, impartial evaluation of scientific literature. It is up to them to provide unbiased information to the patients to facilitate informed decision-making. This approach will enable them to deliver the best possible prenatal care to expectant mothers. Unfortunately, despite the appealing concept of NIPT for CNVs and sub-chromosomal variations, it gives rise to more unanswered questions than we currently have solutions for, and future work is needed to form reliable guidelines.

References

1. Redon R, Ishikawa S, Fitch KR, Feuk L, Perry GH, Andrews TD, et al. Global variation in copy number in the human genome. Nature. 2006;444:444–54.
2. Itsara A, Cooper GM, Baker C, Girirajan S, Li J, Absher D, et al. Population analysis of large copy number variants and hotspots of human genetic disease. Am J Hum Genet. 2009;84:148–61.
3. Harel T, Lupski JR. Genomic disorders 20 years on—mechanisms for clinical manifestations. Clin Genet. 2018;93:439–49.

4. Rice AM, McLysaght A. Dosage sensitivity is a major determinant of human copy number variant pathogenicity. Nat Commun. 2017;8:14366.
5. Zhang F, Gu W, Hurles ME, Lupski JR. Copy number variation in human health, disease, and evolution. Annu Rev Genomics Hum Genet. 2009;10:451–81.
6. Sebat J, Lakshmi B, Troge J, Alexander J, Young J, Lundin P, et al. Large-scale copy number polymorphism in the human genome. Science. 2004;305:525–8.
7. Mills RE, Walter K, Stewart C, Handsaker RE, Chen K, Alkan C, et al. Mapping copy number variation by population-scale genome sequencing. Nature. 2011;470:59–65.
8. Sudmant PH, Rausch T, Gardner EJ, Handsaker RE, Abyzov A, Huddleston J, et al. An integrated map of structural variation in 2504 human genomes. Nature. 2015;526:75–81.
9. Lek M, Karczewski KJ, Minikel EV, Samocha KE, Banks E, Fennell T, et al. Analysis of protein-coding genetic variation in 60,706 humans. Nature. 2016;536:285–91.
10. Soto DC, Uribe-Salazar JM, Shew CJ, Sekar A, McGinty SP, Dennis MY. Genomic structural variation: a complex but important driver of human evolution. Am J Biol Anthropol. 2023;181:118.
11. Weise A, Mrasek K, Klein E, Mulatinho M, Llerena JC, Hardekopf D, et al. Microdeletion and microduplication syndromes. J Histochem Cytochem Off J Histochem Soc. 2012;60:346–58.
12. Goldenberg P. An update on common chromosome microdeletion and microduplication syndromes. Pediatr Ann. 2018;47:e198–203.
13. Nevado J, Mergener R, Palomares-Bralo M, Souza KR, Vallespín E, Mena R, et al. New microdeletion and microduplication syndromes: a comprehensive review. Genet Mol Biol. 2014;37:210–9.
14. Wetzel AS, Darbro BW. A comprehensive list of human microdeletion and microduplication syndromes. BMC Genomic Data. 2022;23:82.
15. Lupski JR. Charcot-Marie-tooth polyneuropathy: duplication, gene dosage, and genetic heterogeneity. Pediatr Res. 1999;45:159–65.
16. Scharf A. First trimester screening with biochemical markers and ultrasound in relation to non-invasive prenatal testing (NIPT). J Perinat Med. 2021;49:990–7.
17. European Platform on Rare Disease Registration. Available from: https://eu-rd-platform.jrc.ec.europa.eu. [cited 2023 May 11].
18. Liehr T. False-positives and false-negatives in non-invasive prenatal testing (NIPT): what can we learn from a meta-analyses on > 750,000 tests? Mol Cytogenet. 2022;15:36.
19. Zeitlin J, Mohangoo A, Cuttini M, EUROPERISTAT Report Writing Committee, Alexander S, Barros H, et al. The European perinatal health Report: comparing the health and care of pregnant women and newborn babies in Europe. J Epidemiol Community Health. 2009;63:681–2.
20. Mefford HC. Copy number variants in the population: unselected does not mean unaffected. Epilepsy Curr. 2016;16:91–3.
21. Wapner RJ, Martin CL, Levy B, Ballif BC, Eng CM, Zachary JM, et al. Chromosomal microarray versus karyotyping for prenatal diagnosis. N Engl J Med. 2012;367:2175–84.
22. Martin CL, Wain KE, Oetjens MT, Tolwinski K, Palen E, Hare-Harris A, et al. Identification of neuropsychiatric copy number variants in a health care system population. JAMA Psychiatry. 2020;77:1276–85.
23. Stefanski A, Calle-López Y, Leu C, Pérez-Palma E, Pestana-Knight E, Lal D. Clinical sequencing yield in epilepsy, autism spectrum disorder, and intellectual disability: a systematic review and meta-analysis. Epilepsia. 2021;62:143–51.
24. Runheim H, Pettersson M, Hammarsjö A, Nordgren A, Henriksson M, Lindstrand A, et al. The cost-effectiveness of WGS in neurodevelopmental disorders. Sci Rep. 2023;13:6904.
25. Manickam K, McClain MR, Demmer LA, Biswas S, Kearney HM, Malinowski J, et al. Exome and genome sequencing for pediatric patients with congenital anomalies or intellectual disability: an evidence-based clinical guideline of the American College of Medical Genetics and Genomics (ACMG). Genet Med. 2021;23:2029–37.
26. Levine RJ, Qian C, Leshane ES, Yu KF, England LJ, Schisterman EF, et al. Two-stage elevation of cell-free fetal DNA in maternal sera before onset of preeclampsia. Am J Obstet Gynecol. 2004;190:707–13.

27. Liehr T. Non-invasive prenatal testing, what patients do not learn, may be due to lack of specialist genetic training by gynecologists and obstetricians? Front Genet. 2021;12:682980.
28. Sabbatinelli G, Fantasia D, Palka C, Morizio E, Alfonsi M, Calabrese G. Isolation and enrichment of circulating fetal cells for NIPD: an overview. Diagnostics. 2021;11:2239.
29. Walknowska J, Conte Felix A, Grumbach MM. Practical and theoretical implications of fetal/maternal lymphocyte transfer. Lancet. 1969;293:1119–22.
30. Krabchi K, Gros-Louis F, Yan J, Bronsard M, Massé J, Forest J-C, et al. Quantification of all fetal nucleated cells in maternal blood between the 18th and 22nd weeks of pregnancy using molecular cytogenetic techniques. Clin Genet. 2001;60:145–50.
31. Nelson JL, Furst DE, Maloney S, Gooley T, Evans PC, Smith A, et al. Microchimerism and HLA-compatible relationships of pregnancy in scleroderma. Lancet. 1998;351:559–62.
32. Bianchi DW, Simpson JL, Jackson LG, Elias S, Holzgreve W, Evans MI, et al. Fetal gender and aneuploidy detection using fetal cells in maternal blood: analysis of NIFTY I data. National Institute of child health and development fetal cell isolation study. Prenat Diagn. 2002;22:609–15.
33. Krabchi K, Gadji M, Forest J-C, Drouin R. Quantification of all fetal nucleated cells in maternal blood in different cases of aneuploidies. Clin Genet. 2006;69:145–54.
34. Krabchi K, Gadji M, Samassekou O, Grégoire M-C, Forest J-C, Drouin R. Quantification of fetal nucleated cells in maternal blood of pregnant women with a male trisomy 21 fetus using molecular cytogenetic techniques. Prenat Diagn. 2006;26:28–34.
35. Hatt L, Brinch M, Singh R, Møller K, Lauridsen RH, Uldbjerg N, et al. Characterization of fetal cells from the maternal circulation by microarray gene expression analysis—could the extravillous trophoblasts be a target for future cell-based non-invasive prenatal diagnosis? Fetal Diagn Ther. 2014;35:218–27.
36. Lo YM, Corbetta N, Chamberlain PF, Rai V, Sargent IL, Redman CW, et al. Presence of fetal DNA in maternal plasma and serum. Lancet Lond Engl. 1997;350:485–7.
37. Xie C, Tammi MT. CNV-seq, a new method to detect copy number variation using high-throughput sequencing. BMC Bioinform. 2009;10:80.
38. Fan HC, Blumenfeld YJ, Chitkara U, Hudgins L, Quake SR. Noninvasive diagnosis of fetal aneuploidy by shotgun sequencing DNA from maternal blood. Proc Natl Acad Sci U S A. 2008;105:16266–71.
39. Peters D, Chu T, Yatsenko SA, Hendrix N, Hogge WA, Surti U, et al. Noninvasive prenatal diagnosis of a fetal microdeletion syndrome. N Engl J Med. 2011;365:1847–8.
40. Gil MM, Nicolaides KH. Implementation of maternal blood cell-free DNA testing in early screening for aneuploidies. Med Genet. 2019;31:283–8.
41. Truty R, Paul J, Kennemer M, Lincoln SE, Olivares E, Nussbaum RL, et al. Prevalence and properties of intragenic copy-number variation in Mendelian disease genes. Genet Med. 2019;21:114–23.
42. Moreno-Cabrera JM, del Valle J, Castellanos E, Feliubadaló L, Pineda M, Brunet J, et al. Evaluation of CNV detection tools for NGS panel data in genetic diagnostics. Eur J Hum Genet. 2020;28:1645–55.
43. Chen S, Zhang L, Gao J, Li S, Chang C, Chen Y, et al. Expanding the scope of non-invasive prenatal testing to detect fetal chromosomal copy number variations. Front Mol Biosci. 2021;8:8. https://doi.org/10.3389/fmolb.2021.649169. [cited 2023 May 23].
44. Jensen TJ, Dzakula Z, Deciu C, van den Boom D, Ehrich M. Detection of microdeletion 22q11.2 in a fetus by next-generation sequencing of maternal plasma. Clin Chem. 2012;58:1148–51.
45. Wapner RJ, Babiarz JE, Levy B, Stosic M, Zimmermann B, Sigurjonsson S, et al. Expanding the scope of noninvasive prenatal testing: detection of fetal microdeletion syndromes. Am J Obstet Gynecol. 2015;212(332):e1–9.
46. Zhao C, Tynan J, Ehrich M, Hannum G, McCullough R, Saldivar J-S, et al. Detection of fetal subchromosomal abnormalities by sequencing circulating cell-free DNA from maternal plasma. Clin Chem. 2015;61:608–16.

47. Canick JA, Palomaki GE, Kloza EM, Lambert-Messerlian GM, Haddow JE. The impact of maternal plasma DNA fetal fraction on next generation sequencing tests for common fetal aneuploidies. Prenat Diagn. 2013;33:667–74.
48. Chen Y, Yu Q, Mao X, Lei W, He M, Lu W. Noninvasive prenatal testing for chromosome aneuploidies and subchromosomal microdeletions/microduplications in a cohort of 42,910 single pregnancies with different clinical features. Hum Genomics. 2019;13:60.
49. Xiang J, Peng Z. Applications of noninvasive prenatal testing for subchromosomal copy number variations using cell-free DNA. Clin Lab Med. 2022;42:613–25.
50. Kim K. Advantages of the single nucleotide polymorphism-based noninvasive prenatal test. J Genet Med. 2015;12:66–71.
51. Shao L, Akkari Y, Cooley LD, Miller DT, Seifert BA, Wolff DJ, et al. Chromosomal microarray analysis, including constitutional and neoplastic disease applications, 2021 revision: a technical standard of the American College of Medical Genetics and Genomics (ACMG). Genet Med. 2021;23:1818–29.
52. Coe BP, Ylstra B, Carvalho B, Meijer GA, MacAulay C, Lam WL. Resolving the resolution of array CGH. Genomics. 2007;89:647–53.
53. Craig JM, Vena N, Ramkissoon S, Idbaih A, Fouse SD, Ozek M, et al. DNA fragmentation simulation method (FSM) and fragment size matching improve aCGH performance of FFPE tissues. PLoS One. 2012;7:e38881.
54. Hillman SC, Pretlove S, Coomarasamy A, McMullan DJ, Davison EV, Maher ER, et al. Additional information from array comparative genomic hybridization technology over conventional karyotyping in prenatal diagnosis: a systematic review and meta-analysis. Ultrasound Obstet Gynecol. 2011;37:6–14.
55. Juneau K, Bogard PE, Huang S, Mohseni M, Wang ET, Ryvkin P, et al. Microarray-based cell-free DNA analysis improves noninvasive prenatal testing. Fetal Diagn Ther. 2014;36:282–6.
56. Stokowski R, Wang E, White K, Batey A, Jacobsson B, Brar H, et al. Clinical performance of non-invasive prenatal testing (NIPT) using targeted cell-free DNA analysis in maternal plasma with microarrays or next generation sequencing (NGS) is consistent across multiple controlled clinical studies. Prenat Diagn. 2015;35(12):1243–6.
57. Geppert J, Stinton C, Johnson S, Clarke A, Grammatopoulos D, Taylor-Phillips S. Antenatal screening for fetal trisomies using microarray-based cell-free DNA testing: a systematic review and meta-analysis. Prenat Diagn. 2020;40:454–62.
58. Schmid M, Wang E, Bogard PE, Bevilacqua E, Hacker C, Wang S, et al. Prenatal screening for 22q11.2 deletion using a targeted microarray-based cell-free DNA test. Fetal Diagn Ther. 2018;44:299–304.
59. Kypri E, Ioannides M, Achilleos A, Koumbaris G, Patsalis P, Stumm M. Non-invasive prenatal screening tests—update 2022. J Lab Med. 2022;46:311–20.
60. Koumbaris G, Achilleos A, Nicolaou M, Loizides C, Tsangaras K, Kypri E, et al. Targeted capture enrichment followed by NGS: development and validation of a single comprehensive NIPT for chromosomal aneuploidies, microdeletion syndromes and monogenic diseases. Mol Cytogenet. 2019;12:48.
61. Neofytou MC, Tsangaras K, Kypri E, Loizides C, Ioannides M, Achilleos A, et al. Targeted capture enrichment assay for non-invasive prenatal testing of large and small size sub-chromosomal deletions and duplications. PLoS One. 2017;12:e0171319.
62. Liehr T, Harutyunyan T, Williams H, Weise A. Non-invasive prenatal testing in Germany. Diagn Basel Switz. 2022;12:2816.
63. Bevilacqua E, Jani JC, Chaoui R, Suk E-KA, Palma-Dias R, Ko T-M, et al. Performance of a targeted cell-free DNA prenatal test for 22q11.2 deletion in a large clinical cohort. Ultrasound Obstet Gynecol. 2021;58:597–602.
64. Hatt L, Aagaard MM, Graakjaer J, Bach C, Sommer S, Agerholm IE, et al. Microarray-based analysis of methylation status of CpGs in placental DNA and maternal blood DNA—potential new epigenetic biomarkers for cell free fetal DNA-based diagnosis. PLoS One. 2015;10:e0128918.

65. Vossaert L, Chakchouk I, Zemet R, Van den Veyver IB. Overview and recent developments in cell-based noninvasive prenatal testing. Prenat Diagn. 2021;41:1202–14.
66. Vossaert L, Wang Q, Salman R, McCombs AK, Patel V, Qu C, et al. Validation studies for single circulating trophoblast genetic testing as a form of noninvasive prenatal diagnosis. Am J Hum Genet. 2019;105:1262–73.
67. Vossaert L, Wang Q, Salman R, Zhuo X, Qu C, Henke D, et al. Reliable detection of subchromosomal deletions and duplications using cell-based noninvasive prenatal testing. Prenat Diagn. 2018;38:1069–78.
68. Zhuo X, Wang Q, Vossaert L, Salman R, Kim A, den Veyver IV, et al. Use of amplicon-based sequencing for testing fetal identity and monogenic traits with single circulating trophoblast (SCT) as one form of cell-based NIPT. PLoS One. 2021;16:e0249695.
69. Doffini A, Forcato C, Mangano C, Lattuada D, Aversa R, Maranta C, et al. Isolation of single circulating trophoblasts from maternal circulation for noninvasive fetal copy number variant profiling. Prenat Diagn. 2023;43:14–27.
70. Bunnell ME, Adams S, Pelletier A, Hoffman SY. Increased use of diagnostic testing after increased nuchal translucency: the influence of non-invasive prenatal testing and chromosomal microarray. Prenat Diagn. 2022;42:1606–11.
71. Srebniak MI, Joosten M, Knapen MFCM, Arends LR, Polak M, van Veen S, et al. Frequency of submicroscopic chromosomal aberrations in pregnancies without increased risk for structural chromosomal aberrations: systematic review and meta-analysis. Ultrasound Obstet Gynecol. 2018;51:445–52.
72. Silva M, de Leeuw N, Mann K, Schuring-Blom H, Morgan S, Giardino D, et al. European guidelines for constitutional cytogenomic analysis. Eur J Hum Genet. 2019;27:1–16.
73. Riggs ER, Andersen EF, Cherry AM, Kantarci S, Kearney H, Patel A, et al. Technical standards for the interpretation and reporting of constitutional copy-number variants: a joint consensus recommendation of the American College of Medical Genetics and Genomics (ACMG) and the clinical genome resource (ClinGen). Genet Med. 2020;22:245–57.
74. Hehir-Kwa JY, Pfundt R, Veltman JA, de Leeuw N. Pathogenic or not? Assessing the clinical relevance of copy number variants. Clin Genet. 2013;84:415–21.
75. Bodian DL, Kothiyal P, Hauser NS. Pitfalls of clinical exome and gene panel testing: alternative transcripts. Genet Med Off J Am Coll Med Genet. 2019;21:1240–5.
76. McCarthy DJ, Humburg P, Kanapin A, Rivas MA, Gaulton K, Cazier J-B, et al. Choice of transcripts and software has a large effect on variant annotation. Genome Med. 2014;6:26.
77. Alfares A, Aloraini T, Subaie LA, Alissa A, Qudsi AA, Alahmad A, et al. Whole-genome sequencing offers additional but limited clinical utility compared with reanalysis of whole-exome sequencing. Genet Med Off J Am Coll Med Genet. 2018;20:1328–33.
78. Shashi V, Schoch K, Spillmann R, Cope H, Tan QK-G, Walley N, et al. A comprehensive iterative approach is highly effective in diagnosing individuals who are exome negative. Genet Med Off J Am Coll Med Genet. 2019;21:161–72.
79. Riggs ER, Nelson T, Merz A, Ackley T, Bunke B, Collins CD, et al. Copy number variant discrepancy resolution using the ClinGen dosage sensitivity map results in updated clinical interpretations in ClinVar. Hum Mutat. 2018;39:1650–9.
80. Xiang J, Yang J, Chen L, Chen Q, Yang H, Sun C, et al. Reinterpretation of common pathogenic variants in ClinVar revealed a high proportion of downgrades. Sci Rep. 2020;10(1):331.
81. van Prooyen SL, Sistermans EA, Van Opstal D, Henneman L, Bekker MN, Bax CJ, et al. Clinical impact of additional findings detected by genome-wide non-invasive prenatal testing: follow-up results of the TRIDENT-2 study. Am J Hum Genet. 2022;109:1140–52.
82. Eggenhuizen GM, Go A, Koster MPH, Baart EB, Galjaard RJ. Confined placental mosaicism and the association with pregnancy outcome and fetal growth: a review of the literature. Hum Reprod Update. 2021;27:885–903.
83. Mossfield T, Soster E, Menezes M, Agenbag G, Dubois M-L, Gekas J, et al. Multisite assessment of the impact of cell-free DNA-based screening for rare autosomal aneuploidies on pregnancy management and outcomes. Front Genet. 2022;13:975987. https://doi.org/10.3389/fgene.2022.975987. [cited 2023 May 26]

84. Grati FR, Malvestiti F, Ferreira JCPB, Bajaj K, Gaetani E, Agrati C, et al. Fetoplacental mosaicism: potential implications for false-positive and false-negative noninvasive prenatal screening results. Genet Med Off J Am Coll Med Genet. 2014;16:620–4.
85. Maya I, Perlman S, Shohat M, Kahana S, Yacobson S, Tenne T, et al. Should we report 15q11.2 BP1-BP2 deletions and duplications in the prenatal setting? J Clin Med. 2020;9(8):2602.
86. Miller DT, Lee K, Chung WK, Gordon AS, Herman GE, Klein TE, et al. ACMG SF v3.0 list for reporting of secondary findings in clinical exome and genome sequencing: a policy statement of the American College of Medical Genetics and Genomics (ACMG). Genet Med Off J Am Coll Med Genet. 2021;23:1381–90.
87. Matsui K, Yamamoto K, Tashiro S, Ibuki T. A systematic approach to the disclosure of genomic findings in clinical practice and research: a proposed framework with colored matrix and decision-making pathways. BMC Med Ethics. 2021;22:168.
88. Best S, Wou K, Vora N, Van der Veyver IB, Wapner R, Chitty LS. Promises, pitfalls and practicalities of prenatal whole exome sequencing. Prenat Diagn. 2018;38:10–9.
89. Downie L, Halliday JL, Burt RA, Lunke S, Lynch E, Martyn M, et al. A protocol for whole-exome sequencing in newborns with congenital deafness: a prospective population-based cohort. BMJ Paediatr Open. 2017;1:e000119.
90. DeCristo DM, Milko LV, O'Daniel JM, Foreman AKM, Mollison LF, Powell BC, et al. Actionability of commercial laboratory sequencing panels for newborn screening and the importance of transparency for parental decision-making. Genome Med. 2021;13:50.
91. Ceyhan-Birsoy O, Machini K, Lebo MS, Yu TW, Agrawal PB, Parad RB, et al. A curated gene list for reporting results of newborn genomic sequencing. Genet Med Off J Am Coll Med Genet. 2017;19:809–18.
92. Deignan JL, Chung WK, Kearney HM, Monaghan KG, Rehder CW, Chao EC. Points to consider in the reevaluation and reanalysis of genomic test results: a statement of the American College of Medical Genetics and Genomics (ACMG). Genet Med Off J Am Coll Med Genet. 2019;21:1267–70.
93. Wang H, Dong Z, Zhang R, Chau MHK, Yang Z, Tsang KYC, et al. Low-pass genome sequencing versus chromosomal microarray analysis: implementation in prenatal diagnosis. Genet Med Off J Am Coll Med Genet. 2020;22:500–10.
94. Pauta M, Martinez-Portilla RJ, Borrell A. Prenatal exome sequencing in recurrent fetal structural anomalies: systematic review and meta-analysis. J Clin Med. 2021;10(20):4739.
95. Zaninović L, Bašković M, Ježek D, Katušić BA. Validity and utility of non-invasive prenatal testing for copy number variations and microdeletions: a systematic review. J Clin Med. 2022;11:3350.
96. Audibert F, Gagnon A. No. 262-prenatal screening for and diagnosis of aneuploidy in twin pregnancies. J Obstet Gynaecol Can. 2017;39:e347–61.
97. Dyr B, Boomer T, Almasri EA, Wardrop JL, Rafalko J, Chibuk J, et al. A new era in aneuploidy screening: cfDNA testing in >30,000 multifetal gestations: experience at one clinical laboratory. PLoS One. 2019;14:e0220979.
98. De Falco L, Savarese G, Savarese P, Petrillo N, Ianniello M, Ruggiero R, et al. Clinical experience with genome-wide noninvasive prenatal screening in a large cohort of twin pregnancies. Genes. 2023;14:982.
99. Girirajan S, Rosenfeld JA, Coe BP, Parikh S, Friedman N, Goldstein A, et al. Phenotypic heterogeneity of genomic disorders and rare copy-number variants. N Engl J Med. 2012;367:1321–31.
100. Helgeson J, Wardrop J, Boomer T, Almasri E, Paxton WB, Saldivar JS, et al. Clinical outcome of subchromosomal events detected by whole-genome noninvasive prenatal testing. Prenat Diagn. 2015;35:999–1004.
101. Gabriel H, Korinth D, Ritthaler M, Schulte B, Battke F, von Kaisenberg C, et al. Trio exome sequencing is highly relevant in prenatal diagnostics. Prenat Diagn. 2022;42:845–51.
102. Basel-Salmon L, Sukenik-Halevy R. Challenges in variant interpretation in prenatal exome sequencing. Eur J Med Genet. 2022;65:104410.

103. Rose NC, Barrie ES, Malinowski J, Jenkins GP, McClain MR, LaGrave D, et al. Systematic evidence-based review: the application of noninvasive prenatal screening using cell-free DNA in general-risk pregnancies. Genet Med. 2022;24:1379–91.
104. Ji X, Chen F, Zhou Y, Li J, Yuan Y, Mo Y, et al. Copy number variation profile in noninvasive prenatal testing (NIPT) can identify co-existing maternal malignancies: case reports and a literature review. Taiwan J Obstet Gynecol. 2018;57:871–7.
105. Paolucci SA, Ma KK, Paulson V, Gadi VK, Lockwood CM. Multiple copy number variants detected by noninvasive prenatal testing. Clin Chem. 2022;68:627–32.
106. Lenaerts L, Jatsenko T, Amant F, Robert VJ. Noninvasive prenatal testing and detection of occult maternal malignancies. Clin Chem. 2019;65:1484–6.
107. Lenaerts L, Brison N, Maggen C, Vancoillie L, Che H, Vandenberghe P, et al. Comprehensive genome-wide analysis of routine non-invasive test data allows cancer prediction: a single-center retrospective analysis of over 85,000 pregnancies. EClinicalMedicine. 2021;35:100856.
108. Perrot A, Horn R. The ethical landscape(s) of non-invasive prenatal testing in England, France and Germany: findings from a comparative literature review. Eur J Hum Genet. 2022;30:676–81.
109. Ellison J, Wang C, Yarrington C, Connors P, Hanchate A. Insurance and geographic variations in non-invasive prenatal testing. Prenat Diagn. 2022;42:1004–7.
110. Beauchamp TL, Childress JF. Principles of biomedical ethics. 7th ed. New York: Oxford University Press; 2013.
111. Wilson JMG, Jungner G, Organization WH. Principles and practice of screening for disease. 1968;
112. van Schendel RV, Kleinveld JH, Dondorp WJ, Pajkrt E, Timmermans DRM, Holtkamp KCA, et al. Attitudes of pregnant women and male partners towards non-invasive prenatal testing and widening the scope of prenatal screening. Eur J Hum Genet. 2014;22:1345–50.
113. Chapman AR, Benn PA. Noninvasive prenatal testing for early sex identification: a few benefits and many concerns. Perspect Biol Med. 2013;56:530–47.
114. Gaille M, Viot G. Prenatal diagnosis as a tool and support for eugenics: myth or reality in contemporary French society? Med Health Care Philos. 2013;16:83–91.
115. Bowman-Smart H, Wiesemann C, Horn R. Non-invasive prenatal testing in Germany: a unique ethical and policy landscape. Eur J Hum Genet. 2023;31:562–7.
116. Hill M, Barrett A, Choolani M, Lewis C, Fisher J, Chitty LS. Has noninvasive prenatal testing impacted termination of pregnancy and live birth rates of infants with down syndrome? Prenat Diagn. 2017;37:1281–90.
117. Van Den Bogaert K, Lannoo L, Brison N, Gatinois V, Baetens M, Blaumeiser B, et al. Outcome of publicly funded nationwide first-tier noninvasive prenatal screening. Genet Med. 2021;23:1137–42.
118. Shi P, Liang H, Hou Y, Chen D, Ren H, Wang C, et al. The uncertainty of copy number variants: pregnancy decisions and clinical follow up. Am J Obstet Gynecol. 2023;S0002-9378(23):00062–5.
119. Jønch AE, Douard E, Moreau C, Van Dijck A, Passeggeri M, Kooy F, et al. Estimating the effect size of the 15Q11.2 BP1-BP2 deletion and its contribution to neurodevelopmental symptoms: recommendations for practice. J Med Genet. 2019;56:701–10.
120. Klapwijk JE, Srebniak MI, Go ATJI, Govaerts LCP, Lewis C, Hammond J, et al. How to deal with uncertainty in prenatal genomics: a systematic review of guidelines and policies. Clin Genet. 2021;100:647–58.
121. Monaghan KG, Leach NT, Pekarek D, Prasad P, Rose NC, ACMG Professional Practice and Guidelines Committee. The use of fetal exome sequencing in prenatal diagnosis: a points to consider document of the American College of Medical Genetics and Genomics (ACMG). Genet Med. 2020;22:675–80.
122. Claustres M, Kožich V, Dequeker E, Fowler B, Hehir-Kwa JY, Miller K, et al. Recommendations for reporting results of diagnostic genetic testing (biochemical, cytogenetic and molecular genetic). Eur J Hum Genet. 2014;22:160–70.

123. Vears DF, Sénécal K, Clarke AJ, Jackson L, Laberge AM, Lovrecic L, et al. Points to consider for laboratories reporting results from diagnostic genomic sequencing. Eur J Hum Genet. 2018;26:36–43.
124. Matthijs G, Souche E, Alders M, Corveleyn A, Eck S, Feenstra I, et al. Guidelines for diagnostic next-generation sequencing. Eur J Hum Genet. 2016;24:2–5.
125. Armour CM, Dougan SD, Brock J-A, Chari R, Chodirker BN, DeBie I, et al. Practice guideline: joint CCMG-SOGC recommendations for the use of chromosomal microarray analysis for prenatal diagnosis and assessment of fetal loss in Canada. J Med Genet. 2018;55:215–21.
126. Vanakker O, Vilain C, Janssens K, Van der Aa N, Smits G, Bandelier C, et al. Implementation of genomic arrays in prenatal diagnosis: the Belgian approach to meet the challenges. Eur J Med Genet. 2014;57:151–6.
127. Deans ZC, Ahn JW, Carreira IM, Dequeker E, Henderson M, Lovrecic L, et al. Recommendations for reporting results of diagnostic genomic testing. Eur J Hum Genet. 2022;30:1011–6.
128. Vears D, Amor DJ. A framework for reporting secondary and incidental findings in prenatal sequencing: when and for whom? Prenat Diagn. 2022;42:697–704.
129. Vetro A, Bouman K, Hastings R, McMullan DJ, Vermeesch JR, Miller K, et al. The introduction of arrays in prenatal diagnosis: a special challenge. Hum Mutat. 2012;33:923–9.
130. Rose NC, Kaimal AJ, Dugoff L, Norton ME, Medicin AC. Of O and GC on PB-O on G for M-F. Screening for fetal chromosomal abnormalities: ACOG Practice Bulletin, number 226. Obstet Gynecol. 2020;136:e48.
131. Dondorp W, de Wert G, Bombard Y, Bianchi DW, Bergmann C, Borry P, et al. Non-invasive prenatal testing for aneuploidy and beyond: challenges of responsible innovation in prenatal screening. Eur J Hum Genet. 2015;23:1438–50.
132. Dungan JS, Klugman S, Darilek S, Malinowski J, Akkari YMN, Monaghan KG, Erwin A, Best RG, ACMG Board of Directors. Noninvasive prenatal screening (NIPS) for fetal chromosome abnormalities in a general-risk population: An evidence-based clinical guideline of the American College of Medical Genetics and Genomics (ACMG). Genet Med. 2023;25(2):100336.
133. Gregg AR, Skotko BG, Benkendorf JL, Monaghan KG, Bajaj K, Best RG, et al. Noninvasive prenatal screening for fetal aneuploidy, 2016 update: a position statement of the American College of Medical Genetics and Genomics. Genet Med. 2016;18:1056–65.

Circulation Transcriptome of Maternal Plasma and Its Use in Noninvasive Prenatal Screening

13

Om Kumari and Latika Chawla

13.1 Introduction

Noninvasive prenatal screening (NIPS) involves analyzing cell-free fetal nucleic acids in maternal plasma or serum to detect genetic conditions in the fetus. Traditionally, diagnosing fetal chromosomal or genetic abnormalities necessitated obtaining fetal or placental tissues through procedures like amniocentesis or chorionic villus sampling. These invasive methods, however, come with a minor but significant risk of fetal loss and maternal complications. Currently, NIPS is mainly utilized for screening high-risk pregnancies for conditions like trisomy 21, trisomy 18, and trisomy 13. However, it is not a diagnostic test and cannot replace invasive diagnostic procedures such as chorionic villus sampling and amniocentesis. Prenatal screening for fetal chromosomal anomalies has undergone a paradigm shift as a result of cell-free fetal DNA (cf-fDNA) detection in maternal plasma. These nucleic acids are different from the intact genomic DNA in their size and characteristics. Cell-free fetal nucleic acids exist in form of small fragments (50–150 bp) in maternal plasma. These cell-free DNAs or cell-free RNAs (cfRNA) of fetal origin exist in minority in maternal plasma admixed with majority of cell-free DNA of maternal origin and are generally smaller in size as compared to maternal cell-free DNA and contribute approximately 10–20% of whole cell-free DNA of maternal plasma. The development of massively parallel sequencing (MPS) has significantly enhanced the robustness of NIPS for detecting chromosomal aneuploidies, making these tests widely accessible for clinical use. Through MPS of cell-free fetal nucleic acids (DNA and RNA) in maternal plasma, it is possible to decode the fetal genome at chromosomal, subchromosomal, and single-gene levels, as well as the methylome and transcriptome. Analyzing the fetal transcriptome and methylome can provide

O. Kumari (✉) · L. Chawla
Department of Obstetrics and Gynaecology, All India Institute of Medical Sciences, Rishikesh, Uttarakhand, India

R. A. Rather, S. C. Saha (eds.), *Non-invasive Prenatal Screening (NIPS) in Clinical Practice*, https://doi.org/10.1007/978-981-97-6402-0_13

valuable insights into both fetal and maternal health. There are now more clinical uses for cell-free fetal nucleic acid analysis, including the assessment of preeclampsia, the investigation of early or recurrent pregnancy losses, and the prenatal diagnosis of single gene abnormalities. As sequencing technology continues to advance, the analysis of cell-free nucleic acids will become increasingly important in prenatal screening, diagnosis, monitoring, and risk stratification for both fetal and maternal conditions.

13.1.1 What Is Transcriptome?

The transcriptome encompasses the entire collection of ribonucleic acid (RNA) molecules produced in an organism and expressed in a specific entity, such as a cell, tissue, or the organism itself [1]. RNAs are important macromolecules that are produced by the cellular process of transcription based on the genomic template. The human genome includes templates for around 21,000 protein-coding genes as well as numerous noncoding RNA genes [2].

Transcriptomics technologies are the methods used to analyze an individual's transcriptome, which includes all RNA transcripts within the organism. The organism's genetic information is stored in its DNA and expressed via transcription, where messenger RNAs (mRNAs) act as temporary intermediaries, while noncoding RNAs have various other functions. A transcriptome provides a snapshot of all transcripts present in a cell at a specific moment in time. Transcriptomics involves all types of RNA transcripts, including mRNAs, microRNAs (miRNAs), and various long noncoding RNAs (lncRNAs). It examines their transcription and expression levels, functions, locations, trafficking, and degradation. These RNAs are involved in a wide array of cellular activities. Mature mRNA transcripts contain the coding regions that direct protein translation, while tRNA plays a role in this translation process. Long and short noncoding RNA genes can regulate other RNAs, and miRNAs specifically regulate many protein-coding genes. However, the detailed functions of miRNAs within a living cell remain largely unknown.

The advent of high-throughput sequencing technologies, including RNA sequencing (RNA-seq) and microarray technology, offers valuable insights into differential gene and transcript expression patterns in specific cell types. Microarray quantifies a set of predetermined sequences while RNA sequencing is high thorough output sequencing to capture all sequences. Unlike the relatively stable DNA molecule, whose primary function is to transmit genetic information across cell lineages, RNAs serve dual roles as carriers of genetic information and catalysts within cells. RNA molecules are dynamic entities, undergoing synthesis and degradation regulated by various factors, thereby contributing significantly to the cellular dynamics. RNAs have distinct, time-dependent, and tissue-specific patterns of expression. RNA and protein abundance do not always exhibit tight correlation due to the presence of multiple layers of post-transcriptional regulation and varying rates of degradation. Consequently, establishing a straightforward correlation between mRNA and protein expressions can be challenging.

13.2 Overview of Maternal Plasma Transcriptome

The presence of fetal RNA in maternal plasma was first verified by Poon et al. 2000 [3]. The specific circulating cfRNA transcriptome during pregnancy primarily originates from the placenta. This is attributed to placental cell turnover and active exosome secretion from the placenta. Whole blood transcriptome profiling of maternal plasma involves the sequencing of cfRNA which may allow minimally invasive assessment of fetal tissues. Both the fetus and the mother contribute to the cfRNA found in maternal circulation, and their respective contributions can be differentiated using exonic single nucleotide variants. CfRNA profiles reflect diverse gene expression patterns across various tissues and cell types. The detection of transcripts specific to fetal tissues circulating in maternal blood can vary throughout gestation. In the early first trimester, fetal-specific contributions are typically <1%, increasing to approximately 15% by the second and third trimesters [4]. CfRNA primarily comprises ribosomal RNA (rRNA), which accounts for over 95% of transcripts, along with mitochondrial RNA (mtRNA), while mRNA constitutes about 2% of the composition, mirroring the transcript profile found in most cell types [5].

The utilization of cfRNA for detection of fetal genetic disorders has limitations as compared to cf-fDNA due to its highly unstable and fragmented nature along with variable fetal tissue-specific transcripts circulating in maternal blood. However, with advancing technologies, collection and extraction can be optimised preventing degradation of cfRNA. It can offer a snapshot of gene expression, as placental transcripts clear within 2 hours of delivery, akin to cfDNA.

13.3 Evolution of NIPT

The primary goal of prenatal screening and diagnosis is to decrease the occurrence and prevalence of genetic conditions, which significantly affect the psychosocial and economic aspects of people's lives. Over the past decade, prenatal screening using sequencing of circulating cell-free DNA has revolutionized obstetric care, markedly decreasing the need for invasive diagnostic procedures such as amniocentesis for genetic disorders. The effectiveness of noninvasive screening for common fetal aneuploidies has prompted the expansion of fetal genetic screening to include rare autosomal trisomies and microdeletions. cf-fDNA can be analyzed through a blood sample from the pregnant woman. In 1997, Lo et al. pioneered the detection of cf-fDNA in maternal plasma using real-time PCR, enabling the determination of fetal sex and RhD blood group. They also found that the concentration of cf-fDNA in maternal blood increases with gestational age, making it suitable for pregnancy tests due to its rapid clearance after the pregnancy ends [6]. In 2008, Fan et al. [7] and Chiu et al. [8] demonstrated the feasibility of screening for Trisomy 21 (Down syndrome) by sequencing cf-fDNA in maternal plasma with a very low false positive rate (FPR). A significant advancement occurred in 2011 when the first report was published on the detection of trisomy 21 using MPS [9]. Subsequently, newer genetic diseases have been identified prenatally. However, major obstetrical syndromes such as preterm birth and preeclampsia, leading causes of fetal and maternal complications and death, respectively, remain under-researched in terms of proactive clinical care and precise diagnosis before the onset of clinical symptoms. For these complications, emergency care often remains the obstetrician's primary option. Recently, the concept of liquid biopsy tests represents a step toward addressing this gap by identifying at-risk pregnancies early and fulfilling the potential of precision medicine. Liquid biopsies, which analyze macromolecules or whole cells present in blood—such as DNA, RNA, white blood cells, proteins, lipids, and metabolites—offer a promising solution to predict risk and monitor progression. While advances in NIPT have predominantly focused on cell-free nucleic acids, with more than half of prenatal genetic tests involving cf-fDNA, clinical research is now increasingly exploring the role of cf-fRNAs. These molecules hold potential for detecting pregnancy-associated diseases like preeclampsia, fetal growth restriction, congenital heart diseases, gestational diabetes, and preterm births. Determining the biological function and therapeutic utility of cell-free nucleic acids is a major undertaking. These molecules may consist of nucleic acids such as DNA, mtDNA, mRNA, miRNA, lncRNA, and circRNA. They can be used to diagnose a variety of disorders because they are present in several bodily fluids (Table 13.1).

Cell-free nucleic acids play crucial roles in normal physiological processes and disease development. With the advent of next-generation sequencing, cf-fDNA is already widely employed clinically for prenatal detection of genetic diseases. However, the clinical applications of various cfRNAs as diagnostic or prognostic markers are still in the experimental phase. These RNAs are expressed in all cell types, and changes in their expression patterns could serve as potential biomarkers for detecting pathological conditions. Recently, cfRNAs have become a focus of clinical research aimed at detecting common pregnancy-associated diseases such as preeclampsia, fetal growth restriction, congenital heart diseases, and gestational diabetes (Table 13.2).

Table 13.1 Cell-free nucleic acids having possible prenatal screening potential

Name of transcriptome	Fetus/pregnant associated	Prenatal/clinical application
Cf-fDNA	Fetus	Trisomy/copy number variants, mutations, microdeletions, or duplications
MicroRNA	Antenatal woman/fetus	Preeclampsia, gestational diabetes Congenital heart disease (fetus)
Circular RNA	Antenatal woman/fetus	Gestational diabetes, congenital heart disease
Long noncoding RNA	Fetus	Congenital heart disease
Messenger RNA	Unknown	Not used
Mitochondrial DNA	Antenatal woman	Preterm prelabour rupture of membranes

Table 13.2 Comparision of cf-fDNA versus cf-fRNA

Cf-fDNA	Cf-fRNA
Derived from syncytiotrophoblast (apoptosis of placenta)	Diverse sources (fetus, placenta, various maternal tissues)
Fetal fraction(FF) approximately 10–20%	Fetal fraction variable throughout gestation
FF decreases with increase in BMI	FF not affected by BMI
Short half-life	Short half-life
Disappears from maternal circulation shortly after delivery (2 h)	Disappears from maternal plasma shortly after delivery (2 h)

13.4 Cell-Free RNA (cfRNA)

The human genome contains approximately 23,000 genes encoding various types of RNAs, with mRNA serving as a type of coding RNA. These molecules are detectable in serum and other biological fluids and exhibit tissue-specific profiles. During pregnancy, the placenta expresses specific mRNAs that can be identified in maternal circulation. Go et al. demonstrated the use of single nucleotide polymorphisms (SNPs) to detect trisomies by analyzing the allelic ratio of specific heterozygous SNPs on cfRNA derived from maternal blood [10]. Currently, mRNAs are not utilized in prenatal screening for genetic diseases. However, due to their diverse origins during pregnancy, cfRNA can offer insights into the interactions between the pregnant individual, fetus, and placenta. The levels of gene expression measured using cfRNA change predictably during gestation, closely reflecting gene expression patterns in the placenta and fetus, thus providing a snapshot of physiological changes over time.

13.5 MicroRNA (miRNA)

MicroRNAs (miRNAs) are a subset of noncoding RNA molecules derived from longer precursors, typically ranging in size from 18 to 25 base pairs. They are found in diverse biological fluids such as plasma, amniotic fluid, and saliva, where they exert significant influence over gene regulation. During placental apoptosis and necrosis, miRNAs are packaged within encapsulated extracellular vesicles in maternal plasma, which confers stability and enables their dissemination throughout the body. These miRNAs demonstrate robust stability even after undergoing freezing and thawing cycles, maintaining their quality and concentration. Genomic research has identified an extensive array of miRNAs in the placenta, categorized into placenta-specific, placenta-associated, and placenta-derived miRNAs, each exhibiting distinct expression profiles across different placental regions [11, 12]. Their expression dynamics during pregnancy are intricately regulated by signal transduction pathways and environmental factors such as hypoxia and oxidative stress [13]. Cell-free miRNAs hold promising potential for diagnosing, prognosticating, and therapeutically addressing pregnancy-related conditions including preeclampsia, congenital heart diseases, gestational diabetes mellitus, and fetal growth restriction.

13.6 Role of miRNA in Preeclampsia

Preeclampsia represents a significant global health challenge, affecting approximately 3–5% of pregnancies worldwide. It contributes significantly to maternal and fetal morbidity and mortality, accounting for 10–15% of maternal deaths and 15–20% of preterm births. This multiorgan syndrome is characterized by abnormal placentation and an inflammatory response during early pregnancy, followed by endothelial dysfunction and subsequent end-organ damage. Severe cases can lead to complications such as stroke and even death. Although the precise biological mechanisms underlying the development of preeclampsia are not fully elucidated, early prediction of this condition is crucial to mitigate the need for emergency medical interventions. Numerous research teams have documented aberrant miRNA expression in the disease's pathophysiological process. These have an impact on immunological response, cell adhesion, metabolic modifications, cardiovascular development, etc. There is upregulation of miRNA-210 in pregnancies with preeclampsia [14, 15]. A review by Skallis et al. on miRNAs in preeclampsia categorized their effects into distinct groups: miRNAs contributing to impaired angiogenesis (including miR-210, miR-21, and miR-22), those affecting defective trophoblast migration and invasion (such as miR-195, miR-276C, miR-278a-5p, and miR-210), and miRNAs involved in the dysregulation of the maternal immune system (including miR-223, miR-148a, and miR-152) [16].

13.7 Role of miRNA in Congenital Heart Diseases (CHD)

CHD represents the most prevalent congenital anomalies, affecting approximately 1% of live births. These defects contribute significantly to health issues, causing 30–50% of mortality among newborns and infants. Prenatal detection rates vary widely, particularly since most cases occur in pregnancies considered low risk. Early prenatal diagnosis facilitates delivery planning through interdisciplinary collaboration, thereby reducing postnatal morbidity and mortality. Zhu et al. utilized SOLiD sequencing to compare miRNA profiles in pregnant women carrying fetuses with CHD, including Tetralogy of Fallot, atrial septal defect, and ventricular septal defect, against those with normal fetuses. Their findings revealed significantly elevated levels of miR-19b, miR-22, miR-29c, and miR-375 in the patient group [17]. Kehler et al. reported elevated level of miR-99a in maternal plasma as a possible biomarker for detection of CHD [18]. Although CHD is the most prevalent cause of birth abnormalities, high-risk cases cannot be reliably identified by current prenatal screening techniques. miRNAs in the mother's circulation may enhance diagnostic accuracy and open up new avenues for the study and detection of congenital heart disease.

13.8 Infections During Pregnancy

Pregnant women are vulnerable to various infections that can traverse the placenta, directly affecting the fetus, or indirectly impairing placental function through inflammation and sepsis. Infections acquired in utero significantly contribute to fetal and neonatal mortality and are major drivers of childhood morbidity. Infections during pregnancy can lead to complications such as preterm rupture of membranes, preterm birth, preeclampsia, and fetal abnormalities such as growth restriction, developmental delay, anomalies, or fetal demise.

Recent developments in cfRNA sequencing offer the capability to detect and identify nucleic acids associated with pathogens through the analysis of reads that do not match the human genome. This may have limitations of providing false positive reports in case sample is contaminated during sampling process. Both RNA- and DNA-based methods can be used for prenatal detection of infections using standard NIPT cfDNA and cfRNA sequencing. Furthermore, detecting viral nucleic acid sequences through cfDNA-based NIPT does not compromise the accuracy of aneuploidy screening [19, 20]. For cfRNA-based NIPT, identifying pathogenic RNA can be combined with measurements of human immune-specific RNA to reflect the maternal immune response [21].

The most frequent infections during pregnancy are well recognized TORCH acronym. It includes; toxoplasmosis, rubella, cytomegalovirus, herpes simplex virus, and others (syphilis, varicella-zoster, zika virus). Currently, clinician use maternal serum antibodies to identify or detect acute and past infections in antenatal women while fetal affection can be confirmed by invasive tests (amniocentesis/cordocentesis) followed by PCR technique. All organisms in the TORCH group can be identified using NIPT cfDNA sequencing, while rubella can be detected using a parallel cfRNA sequencing approach. However, the clinical utility of NIPT-based nucleic acid sequencing needs to be validated by further research.

13.9 Cancer Detection During Pregnancy

Recent advancements in NIPT methodology have enhanced its capability to analyze chromosomal abnormalities, copy number variants, and microdeletions. Techniques such as MPS, chromosome-selective sequencing, and SNP analysis via multiplex PCR demonstrate the feasibility of detecting micro-deletions/duplications with a resolution comparable to that of microarray analysis. In pregnant women with undiagnosed cancer, tumor-derived cfDNA contributes to the total cfDNA sequenced from maternal plasma. Sequencing all cfDNA, which includes tumor-derived, placental, and other background DNA, can create the impression of multiple aneuploidies when mapped to the human genome. Several studies have reported a correlation between multiple aneuploidies detected by cfDNA screening and cancer in pregnant women [22–25]. The presence of tumor-derived cfDNA can distort NIPT results and confound the fetal risk assessment for aneuploidy. Multiple aneuploidies detected in NIPT must be followed by diagnostic invasive tests, chorionic villous sampling or amniocentesis for fetal diagnosis. Currently, there are no evidence-based guidelines for reporting and management of incidental detection of parental malignancy by cfDNA-based NIPT. Hence, clinical validation studies are required to improve outcomes by early detection of silent malignancy in pregnant women.

13.10 Gestational Diabetes Mellitus (GDM)

GDM has been linked to circulating miRNAs that express themselves irregularly. They might act as prospective early indicators. The first identified miRNA to exhibit altered expression in GDM is miR-518d, which is a member of the C19MC cluster [26]. Wander et al. recently reported that GDM was associated with increased plasma levels of 10 microRNAs, including miR-155-5p and miR21-3p, whose expression was changed [27]. MicroRNAs require more in-depth study before they can be used as biomarkers in the early detection of GDM.

13.11 Circular RNAs (circRNAS)

These are unique noncoding RNAs with tissue specificity, structural stability, and evolutionary conservation. They control the expression of several genes and function similarly to miRNA sponges [28]. CircRNAs are found in the placenta and may play a role in pathological events associated with pregnancy [29]. Wang et al. assessed three lncRNA levels in GDM and paired with 99 miRNAs' expression; however, the levels of circ_5824, circ_3636, and circ_0395 were significantly lower in GDM [30]. CircRNA can serve as additional intriguing molecules as biomarker for GDM.

13.12 Long Noncoding RNA (LncRNA)

Long noncoding RNAs are non-translating RNA molecules longer than 200 nucleotides. They are stable in plasma and other biological fluids and exhibit disease and tissue specificity. Over 1000 lncRNAs participate in various biological processes. LncRNAs play a role in heart development, and CHD-related lncRNAs can be detected in placental tissues and maternal circulation [31].

cfRNAs are expressed in all cell types, and variations in how they are expressed may indicate the presence of pathogenic diseases. Less is known about circRNAs and lncRNAs, although they may be useful in medicine in the coming years. The research focuses on a number of conditions related to pregnancy, including congenital cardiac disorders, gestational diabetes, and preeclampsia.

13.13 Cell-Free Fetal DNA

Most contemporary clinical NIPT screens for aneuploidy by quantifying the fetal contribution of cf-fDNA. This cf-fDNA is primarily derived from the turnover of placental trophoblasts and can be detected in maternal plasma as early as 5–7 weeks of gestation. Its concentration increases as the pregnancy progresses. To ensure an adequate fetal fraction (FF), NIPT should not be performed before 10 weeks. The FF is the proportion of cf-fDNA relative to the total cfDNA. The higher the FF in maternal blood, there is less false negative and more reliable result. FF fluctuates based on environmental and maternal factors. The FF of cf-fDNA is influenced by various factors. It is typically lower in twin pregnancies, Afro-Caribbean and South Asian women, and in pregnancies conceived via in vitro fertilization (IVF). Additionally, a high pulsatility index of uterine arteries at the first-trimester scan is associated with a lower FF. Conversely, FF tends to be higher in East Asian women and smokers, primarily due to a decrease in maternal cfDNA. On the other hand, increased maternal age and body mass index (BMI) are associated with a lower FF, mainly because of a relative increase in maternal cfDNA rather than a decrease in cffDNA [32, 33].

13.14 Reliability of Cf-fDNA Test

As cf-fDNA comes from placental cytotrophoblast, low FF could result in test failure or false negative tests. Important contributors to test failure are IVF conception, high BMI, increasing age, parity, and some specific ethnic groups. Studies have shown lower FF in cases of aneuploidies and triploidies [34, 35]. When NIPT results are not reported or are uninterpretable due to low FF, the possibility of fetal aneuploidy may be higher. Patients should be informed of this concern, and it is recommended to follow up with a thorough ultrasound, genetic counseling, and diagnostic testing. Potential reasons for FPR include placental mosaicism, maternal chromosomal abnormalities, fetal mosaicism, pathogenic copy number variants, or the presence of a vanishing twin. Unknown maternal cancer can potentially cause a false positive. NIPT has high detection rate (DR) for common chromosome abnormalities, but the positive predictive value is not always as high which also depends on the prevalence in the population. Cf-fDNA is also utilized for detecting sex chromosome aneuploidies, such as monosomy X (Turner syndrome), Klinefelter syndrome (47,XXY or 48,XXYY), triple X syndrome (47,XXX), and 47,XYY. According to a recent meta-analysis, DR and FPR in singletons are 99.7% and 0.04% for trisomy 21 (T21), 97.9% and 0.04% for trisomy 18 (T18), and 99.0% and 0.04% for trisomy 13 (T13), respectively. For monosomy X (Turner syndrome), the DR and FPR in singletons have been reported as 95.8% and 0.14%, respectively [36]. The rise of prenatal screening test DR has reduced the need for invasive tests but also increased the revelation of sex chromosome aneuploidies. The high incidence of mosaicisms for sex chromosomal aneuploidies (both maternal and/or fetal) and low prevalence in population pose a limitation for the performance of NIPT. Hence, the results should be confirmed by amniocentesis. It is recommended that patients with a high nuchal translucency in the first trimester, or those presenting with cystic hygroma or hydrops in the second trimester, should undergo an invasive diagnostic test rather than relying on screening by cf-fDNA analysis.

13.15 Clinical Implications of NIPT

13.15.1 Chromosomal Aneuploidy Detection

T21 is the most prevalent chromosomal aneuploidy and a significant contributor to intellectual disability. Other notable aneuploidy syndromes include trisomy 18, trisomy 13, and monosomy X. NIPT through cell-free DNA analysis is tailored to identify abnormal chromosome counts, leveraging the principle that in T21 pregnancies, there is a disproportionate amount of cf-fDNA molecules from chromosome 21 compared to euploid pregnancies.

13.15.2 Subchromosomal Aberration

Subchromosomal aberration syndromes are frequent causes of congenital physical and intellectual disabilities. The approach used to detect chromosomal aneuploidies can similarly be applied to identify subchromosomal deletions or duplications down to a resolution of 1 Mb. Unlike MPS-NIPT for chromosomal aneuploidies, detecting subchromosomal deletions or duplications typically necessitates random MPS with increased sequencing depth or deep-targeted sequencing. Although several commercial laboratories already offer MPS-NIPT for specific subchromosomal deletion syndromes (such as 22q11.2 deletion, 5p deletion, 15q deletion, and 4p deletion), further clinical validation studies are essential before these methods can be widely adopted for prenatal testing.

13.15.3 Diagnosis of Monogenic Disorders

NIPT has the capability to determine fetal sex and RhD genotyping, as well as to detect paternally inherited or de novo mutations for autosomal dominant diseases in maternal plasma, indicating whether the fetus is affected. In cases of autosomal recessive diseases where both parents carry different mutations, the absence of the paternal mutation in cell-free DNA testing effectively rules out an affected fetus. Lun et al. (2008) pioneered a digital PCR-based method known as relative mutation dosage (RMD) analysis for NIPT of β-thalassemia [37]. Subsequent studies have demonstrated the feasibility of digital PCR-RMD analysis in NIPT for sickle cell anemia (autosomal recessive) and hemophilia (X-linked) [38, 39].

13.15.4 Noninvasive Whole/Targeted-Fetal-Genome Sequencing

Advancements in MPS now allow for comprehensive scanning of the entire fetal genome for genetic diseases. The approach typically begins with genome-wide genotyping of both parents using SNP microarrays or whole-genome sequencing. By detecting paternal-specific alleles in maternal plasma, it becomes possible to infer paternal inheritance patterns. Concurrently, determining the maternal haplotype enables the application of the relative haplotype dosage (RHDO) method to deduce fetal inheritance of maternal haplotypes. As sequencing and haplotyping technologies continue to improve, it is anticipated that more NIPTs will be developed for monogenic diseases, leveraging noninvasive targeted or whole-fetal-genome sequencing with haplotype-assisted analysis.

13.16 Prenatal Plasma Transcriptomic Analysis

Analyzing fetal RNA in maternal plasma is inherently more complex than analyzing cf-fDNA for several reasons. First, cell-free mRNA, which constitutes a major RNA species, reflects gene transcription that occurs intracellularly and is released extracellularly. Unlike the genome, which remains relatively stable, gene transcription is dynamic and subject to tissue-specific and temporal variations over time. In 2014, Tsui et al. used high-throughput RNA sequencing (RNA-seq) to profile the plasma transcriptome during pregnancy, identifying 131 transcripts associated with pregnancy [40]. RNA-seq, an adaptation of DNA-based MPS (DNA-seq), includes additional steps like rRNA depletion/mRNA selection and reverse transcription before sequencing. The FF of transcripts detected in maternal plasma was observed to rise from 3.70% in early pregnancy to 11.28% in late pregnancy [40]. This underscores the potential of RNA-seq as a noninvasive method for monitoring placental and fetal health in the future.

13.17 Summary

Analysis of nucleic acids in maternal plasma by MPS has changed noninvasive detection of genetic information related to the fetus, including chromosomal abnormalities and single-gene disorders. NIPT has the potential to extend its scope to include subchromosomal deletions and duplications; however, further clinical validation studies are needed for widespread adoption in obstetric practice. The advent of deep sequencing technologies enables comprehensive exploration of the whole fetal genome, transcriptome, and methylome from maternal plasma, offering novel avenues for research. Examining the circulating cfRNA transcriptome during pregnancy promises real-time insights into biological changes across gestation and early detection of obstetric disorders before clinical symptoms emerge. RNA sequencing technology facilitates quantifying fetal-derived transcripts and identifying pregnancy-associated genes in maternal plasma. Maternal plasma transcriptomic profiling holds promise for identifying biomarkers implicated in pregnancy-related diseases, offering potential avenues for early diagnosis and intervention.

References

1. Lowe R, Shirley N, Bleackley M, Dolan S, Shafee T. Transcriptomics technologies. PLoS Comput Biol. 2017;13:5.
2. Nagano T, Fraser P. No-nonsense functions for long noncoding RNAs. Cell. 2011;145(2):178–81.
3. Poon LL, Leung TN, Lau TK, Lo YM. Presence of fetal RNA in maternal plasma. Clin Chem. 2000;46:1832–4.
4. Moufarrej MN, Wong RJ, Shaw GM, Stevenson DK, Quake SR. Investigating pregnancy and its complications using circulating cell-free RNA in women's blood during gestation. Front Pediatr. 2020;8:605219.
5. Larson MH, Pan W, Kim HJ, Mauntz RE, Stuart SM, Pimentel M, et al. A comprehensive characterization of the cell-free transcriptome reveals tissue- and subtype-specific biomarkers for cancer detection. Nat Commun. 2021;12:2357.
6. Lo YM, Corbetta N, Chamberlain PF, Rai V, Sargent IL, Redman CW, et al. Presence of fetal DNA in maternal plasma and serum. Lancet. 1997;350:485–7.
7. Fan HC, Blumenfeld YJ, Chitkara U, Hudgins L, Quake SR. Noninvasive diagnosis of fetal aneuploidy by shotgun sequencing DNA from maternal blood. Proc Natl Acad Sci U S A. 2008;105:16266–71.
8. Chiu RW, Chan KA, Gao Y, Lau VY, Zheng W, Leung TY, Foo CH, Xie B, Tsui NB, Lun FM, Zee BC. Noninvasive prenatal diagnosis of fetal chromosomal aneuploidy by massively parallel genomic sequencing of DNA in maternal plasma. Proc Natl Acad Sci U S A. 2008;105:20458–63.
9. Palomaki GE, Kloza EM, Lambert-Messerlian GM. DNA sequencing of maternal plasma to detect down syndrome: an international clinical validation study. Genet Med. 2011;13(11):913–20.
10. Go AT, van Vugt JM, Oudejans CB. Non-invasive aneuploidy detection using free fetal DNA and RNA in maternal plasma: recent progress and future possibilities. Hum Reprod Update. 2011;17(3):372–82.
11. Mouillet J-F, Ouyang Y, Coyne CB, Sadovsky Y. MicroRNAs in placental health and disease. Am J Obstet Gynecol. 2015;213(4):S163–72.
12. Cai M, Kolluru GK, Ahmed A. Small molecule, big prospects: microRNA in pregnancy and its complications. J Pregn. 2017;2017:6972732.
13. Fu G, Brkić J, Hayder H, Peng C. MicroRNAs in human placental development and pregnancy complications. Int J Mol Sci. 2013;14(3):5519–44.
14. Malnou EC, Umlauf D, Mouysset M, Cavaillé J. Imprinted microRNA gene clusters in the evolution, development, and functions of mammalian placenta. Front Genet. 2019;9:706.
15. Lv Y, Lu C, Ji X, Miao Z, Long W, Ding H, Lv M. Roles of microRNAs in preeclampsia. J Cell Physiol. 2019;234(2):1052–61.
16. Skalis G, Katsi V, Miliou A, Georgiopoulos G, Papazachou O, Vamvakou G, et al. MicroRNAs in preeclampsia. MicroRNA. 2019;8(1):28–35.
17. Zhu S, Cao L, Zhu J, Li K, Jin J, Qian L, et al. Identification of maternal serum microRNAs as novel non-invasive biomarkers for prenatal detection of fetal congenital heart defects. Clin Chim Acta. 2013;424:66–72.
18. Kehler L, Biro O, Lazar L, Rigo J Jr, Nagy B. Elevated hsa-miR-99a levels in maternal plasma may indicate congenital heart defects. Biomed Rep. 2015;3(6):869–73.
19. Linthorst J, Baksi MMM, Welkers MRA, Sistermans EA. The cell-free DNA virome of 108,349 Dutch pregnant women. Prenat Diagn. 2023;43(4):448–56.
20. Chesnais V, Ott A, Chaplais E, Gabillard S, Pallares D, Vauloup-Fellous C, et al. Using massively parallel shotgun sequencing of maternal plasmatic cell-free DNA for cytomegalovirus DNA detection during pregnancy: a proof of concept study. Sci Rep. 2018;8(1):4321.
21. Pan W, Ngo TTM, Camunas-Soler J, Song CX, Kowarsky M, Blumenfeld YJ, et al. Simultaneously monitoring immune response and microbial infections during pregnancy through plasma cfRNA sequencing. Clin Chem. 2017;63(11):1695–704.

22. van der Meij KRM, Sistermans EA, Macville MVE, Stevens SJC, Bax CJ, et al. TRIDENT-2: national implementation of genome-wide non-invasive prenatal testing as a first-tier screening test in The Netherlands. Am J Hum Genet. 2019;105(6):1091–101.
23. Ji X, Li J, Huang Y, Sung P-L, Yuan Y, Liu Q, Chen Y, et al. Identifying occult maternal malignancies from 1.93 million pregnant women undergoing noninvasive prenatal screening tests. Genet Med. 2019;21(10):2293–302.
24. Lenaerts L, Brison N, Maggen C, Vancoillie L, Che H, Vandenberghe P, et al. Comprehensive genome-wide analysis of routine non-invasive test data allows cancer prediction: a single-center retrospective analysis of over 85,000 pregnancies. EClinicalMedicine. 2021;35:100856.
25. Heesterbeek CJ, Aukema SM, Galjaard R-JH, Boon EMJ, Srebniak MI, Bouman K, et al. Noninvasive prenatal test results indicative of maternal malignancies: a nationwide genetic and clinical follow-up study. J Clin Oncol. 2022;40(22):2426–35.
26. Barchitta M, Maugeri A, Quattrocchi A, Agrifoglio O, Agodi A. The role of miRNAs as biomarkers for pregnancy outcomes: comprehensive review. In J Genomics. 2017;11:8067972.
27. Wander PL, Boyko EJ, Hevner K, Parikh VJ, Tadesse MG, Sorensen TK, et al. Circulating early- and mid-pregnancy microRNAs and risk of gestational diabetes. Diabetes Res Clin Pract. 2017;132:1–9.
28. Geng HH, Li R, Su YM, Xiao J, Pan M, Cai XX, et al. The circular RNA Cdr1 as promotes myocardial infarction by mediating the regulation of miR-7a on its target genes expression. PLoS One. 2016;11(3):e0151753.
29. Maass PG, Glazar P, Memczak S, Dittmar G, Hollfinger I, Schryer L, et al. A map of human circular RNAs in clinically relevant tissues. J Mol Med (Berlin). 2017;95:1179–89.
30. Wang H, She G, Zhou W, Liu K, Miao J, Yu B. Expression profile of circular RNAs in placentas of women with gestational diabetes mellitus. Endocr J. 2019;66(5):431–41.
31. Wang Y, Li B, Zhao Y. Inflammation in preeclampsia: genetic biomarkers, mechanisms, and therapeutic strategies. Front Immunol. 2022;13:883404.
32. Poon LC, Musci T, Song K, Syngelaki A, Nicolaides KH. Maternal plasma cell-free fetal and maternal DNA at 11–13 weeks' gestation: relation to fetal and maternal characteristics and pregnancy outcomes. Fetal Diagn Ther. 2013;33(4):215–23.
33. Sarno L, Revello R, Hanson E, Akolekar R, Nicolaides KH. Prospective first-trimester screening for trisomies by cell-free DNA testing of maternal blood in twin pregnancy. Ultrasound Obstet Gynecol. 2016;47(6):705–11.
34. Rava RP, Srinivasan A, Sehnert AJ, Bianchi DW. Circulating fetal cell-free DNA fractions differ in autosomal aneuploidies and monosomy X. Clin Chem. 2014;60(1):243–50.
35. Palomaki GE, Kloza EM, Lambert-Messerlian GM, van den Boom D, Ehrich M, et al. Circulating cell free DNA testing: are some test failures informative? Prenat Diagn. 2015;35(3):289–93.
36. Gil MM, Accurti V, Santacruz B, Plana MN, Nicolaides KH. Analysis of cell-free DNA in maternal blood in screening for aneuploidies: updated meta-analysis. Ultrasound Obstet Gynecol. 2017;50(3):302–14.
37. Lun FM, Tsui NB, Chan KC, Leung TY, Lau TK, Charoenkwan P, et al. Noninvasive prenatal diagnosis of monogenic diseases by digital size selection and relative mutation dosage on DNA in maternal plasma. Proc Natl Acad Sci U S A. 2008;105(50):19920–5.
38. Barrett AN, McDonnell TC, Chan KC, Chitty LS. Digital PCR analysis of maternal plasma for noninvasive detection of sickle cell anemia. Clin Chem. 2012;58:1026–32.
39. Tsui NB, Kadir RA, Chan KC, Chi C, Mellars G, Tuddenham EG, et al. Noninvasive prenatal diagnosis of hemophilia by microfluidics digital PCR analysis of maternal plasma DNA. Blood. 2011;117(13):3684–91.
40. Tsui NB, Jiang P, Wong YF, Leung TY, Chan KC, Chiu RW, et al. Maternal plasma RNA sequencing for genome-wide transcriptomic profiling and identification of pregnancy-associated transcripts. Clin Chem. 2014;60(7):954–62.

Part III

Challenges and Future Prospect

Challenges and Ethical Issues Surrounding Noninvasive Prenatal Screening (NIPS)

14

Thomas Liehr

14.1 Introduction

Since the 1970s, what was previously only science fiction, something has been possible for the first time in the history of mankind: for pregnant women, it has now become a matter of course that they can receive images and statements about the health and genetic state of their unborn child [1, 2]. In parallel to the remarkable progress regarding noninvasive sonographic visualization [3], (cyto-)genetics provided the tools to obtain reliable information about the genetic status of a given fetus [1]. Invasive or noninvasive approaches can be used in prenatal diagnostics [4]. In invasive approaches, placenta or fetal tissues are analyzed; clear results are obtained as to whether the expected baby suffers from a specific genetic problem that it has been tested for. Still, it is well-known that abnormal results obtained from placenta by chorionic villi sampling (CVS) should be checked by amniocentesis [1–4]. Invasive tests also include a certain risk for the unborn: while e.g., the abortion risk connected with amniocentesis was 1–3% in the 1990s, currently it is as low as 0.1–0.3% [2]. Nonetheless, the abortion risk of invasive approaches stimulated the development of noninvasive tests, like (i) sonography, (ii) protein marker testing from maternal blood, (iii) first-trimester screening (FTS, i.e. combining I and ii), and (iv), as a most recent tool the noninvasive prenatal screening (NIPS) [2]. All these noninvasive approaches have in common that they (a) do not study fetal cells, (b) are pure risk estimation tests, (c) just give risk figures, and (d) cannot diagnose or rule out a specific genetic disorder with 100% security.

NIPS was developed between 1997 [5] and 2010 [6] and is alternatively called noninvasive prenatal testing (NIPT) or even noninvasive prenatal diagnostics

T. Liehr (✉)
Jena University Hospital, Friedrich Schiller University, Institute of Human Genetics, Jena, Germany
e-mail: Thomas.Liehr@med.uni-jena.de

R. A. Rather, S. C. Saha (eds.), *Non-invasive Prenatal Screening (NIPS) in Clinical Practice*, https://doi.org/10.1007/978-981-97-6402-0_14

(NIPD). The latter designation should not be used, as NIPS is not a diagnostic but a screening test [2]. The so-called "cell-free fetal DNA = cffDNA" from maternal serum is studied in NIPS approaches [5], being detectable reliably there from the tenth week of gestation onwards. cffDNA is particularly suitable to check the actual product of conception because it disappears from the maternal blood shortly after the birth of the actual child [2]. The unique advantages of NIPS, as we recently noted, include that it: "(I) can be utilized earlier in pregnancy compared to other tests; (II) has the capability to reach populations without access to centers offering invasive diagnostics; (III) can most accurately exclude and detect trisomy 21 among non-invasive methods—with a positive predictive value (PPV) greater than 99% for trisomy 21 cases indicated by NIPS; (IV) provides psychological relief to pregnant women with a normal NIPS result; and (V) facilitates early pregnancy termination following the early detection of adverse genetic conditions through NIPS" [2].

However, NIPS was introduced remarkably fast worldwide and is connected with an unusual high financial impact for providers of the test [7]. Thus, it seems to be necessary to discuss what are the challenges and ethical issues of NIPS for users, being gynecologists/obstetricians, and pregnant women.

14.2 Challenges

It is strongly recommended by NIPS providers and human genetic societies to provide NIPS only together with the offer of pretest- and posttest-genetic counseling [2]. During these consultations, the couple shall be informed on how the test works, its advantages, possible problems, and consequences of a positive and a negative result. This includes also informing the pregnant woman and her partner on the points mentioned below in this chapter. Talking to dozens of MDs in Germany working as gynecologists or obstetricians, the author of this chapter found that many of them were not aware of all peculiarities listed below. Thus, the major challenge of NIPS is to teach and inform specialists, as well as public, as best possible on the points treated here. Specific challenges are included and highlighted after each corresponding paragraph and examples are given, where problems arose in connection with the respective NIPS-related challenge.

14.2.1 Technical Challenges

NIPS-test is based on next-generation sequencing (NGS), a relatively new approach. Understanding NGS in detail may not be necessary to explain how NIPS works. Still, there are several points as listed below, which can lead to problems for MDs, consultants, and public, if not considered at least.

14.2.1.1 What Does It Mean That There Are Many NIPS Variants?

Technical, one needs to understand at first, that in NIPS literature dozens of variants of the test are reported, but not always easy to distinguish [2, 8, 9]:

(a) NIPS can be done based on whole genomic sequencing (WGS), which itself has at least three technical subtypes and/or can be combined with other approaches [2] Consequently, triploidy can only be characterized by one NIPS type, which means exclusively when a single nucleotide polymorphism (SNP) based WGS platform is applied [10].

Challenge 1: *MDs need to understand that there are different kinds of NIPS-test platforms. They need to know which test they offer, as they are obliged to tell the couple about the technical shortcuts of the approach used.*

(b) NIPS can be performed using commercial kits or in-house proceedings—in both cases, technical details to reproduce the test or how cut-off rates were determined are sometimes not available to public or are at least hard to access [2, 8, 10].

Challenge 2: *NIPS tests should be offered only by MDs if their providers give clear cut-off rates, and the data collection is comprehensible.*

(c) NIPS can include testing the whole genome for (partial) chromosomal aneuploidies, or only checking for most common trisomies of the second trimenon (i.e., #13, #18, #21), which may include or not include a test for gonosomes (most frequent or selected), microdeletion syndromes, or other genetic alterations [2].

Challenge 3: *The scope covered by the NIPS test must be clear to the MD and carefully discussed with the pregnant couple, as well as adapted to their individual needs.*

Example 1

At the Institute of Human Genetics Jena (Germany), a pregnant woman (19th week of gestation = w.o.g.) showed up in the counseling unit. Her fetus had severe sonographic signs, including omphalocele and *hygroma colli*; she had a NIPS test in 11th w.o.g., which excluded trisomy 13, 18, and 21. She wanted to discuss the result of invasive diagnostics (amniocentesis) done in 16th w.o.g., which was 69,XXX, i.e., a triploidy. She did not understand, why trisomies of three chromosomes were excluded and at the same time a triploidy (= trisomy of all chromosomes) was present. Obviously, the MD offering the test to this woman did not explain that it was no SNP-WGS-based test; thus, a triploidy could not be accessed for technical reasons.

14.2.1.2 What Is cffDNA?

As mentioned in Introduction part of this chapter, cffDNA stands for "cell free fetal DNA." This term is at least misleading, as it was clear from the moment of its first description, that this kind of cell-free DNA (cfDNA) is derived from trophoblast, being part of placenta [5]. Accordingly, it is clearly not derived from the fetus.

Placenta and embryo develop from the same fertilized egg cell, the zygote; however, the separation of future placenta and future embryo cells is extremely early during pregnancy—only a few days after fertilization. Thus, it is not that surprising that placenta and fetus do not have in all cases the identical genetic constitution. From invasive diagnostics using CVS, it can be suggested that in tenth w.o.g. 2–4% of pregnancies have a discordant karyotype in placenta and fetus [10].

Challenge 4: MDs must be able to explain the mislabeling of cell-free placental DNA as cell-free fetal DNA in maternal serum. Only if this is properly done, consultants are able to understand later the biological challenge treated below, called "confined placental mosaicism" (see Sect. 2.2.3).

Example 2
In a BLOG of Nuffield Council on Bioethics with over 400 posts of women with experiences and doubts about NIPS [11], there is not a single contribution that would make the reader think that any of the persons/pregnant women is or was aware of the fact that placental DNA is accessed in this test. They all state that "fetal DNA" was studied. This is distressing and shows that this point is not known to, or at least not communicated by the MDs offering the test.

14.2.1.3 What Does Fetal Fraction Mean?

cffDNA, derived from placenta, is only one of two parts of the cfDNA being accessible in the blood serum of a pregnant woman. Major part of the cfDNA is derived from apoptotic cells of the pregnant woman herself. Placenta-derived "cffDNA" represents between 3% of cfDNA in fifth w.o.g. to ~30% in 40th w.o.g. [12]. and is called "fetal fraction." To perform an informative NIPS test, the fetal fraction must be at least 4% [13]. Consequence of a low fetal fraction may be a "no-call result" in NIPS, which means that no informative result could be obtained. In this case, NIPS providers normally offer a free repetition of the test.

No-call results can be a hint on a small placenta of the fetus, preeclampsia, intrauterine growth retardation and even a to be expected early abortion [2]. In Western countries, low fetal fraction is also often due to obesity of the pregnant woman—here the rate of cffDNA to maternal cfDNA is shifted to the disadvantage of cffDNA—in this population of women with high BMI NIPS failure rate is up to 24.3% [14].

Challenge: The chance to obtain a no-call result must be discussed in pretest counseling to avoid later scaring of pregnant woman and her partner.

Example 3
In this post taken from [11] (02 March 2021 by S.R.L.), avoidable misunderstanding of a no-call result is reported as follows:

> I underwent NIPT around 10 weeks and received 'high risk' results for a trisomy, but my fetal fraction was only 2.7%. The 'high risk' designation was based solely on an equation involving my advanced maternal age and other factors, not on DNA analysis. Initially, I didn't understand this, and my doctor didn't explain it, leaving me devastated and fearing the worst. A second blood draw and test two weeks later yielded the same results. A maternal fetal medicine specialist later clarified that the results were misleading in my case. They explained that low fetal fraction is common in women with high body mass, which applies to me. An ultrasound at nearly 14 weeks showed normal neck measurements [11].

14.2.1.4 What Does Positive Predictive Value Mean?

The reliability of a test result can be expressed in different ways; for NIPS tests, it was introduced to apply therefore the positive predictive value (PPV). According to Ref. [15], a PPV "is the ratio of patients truly diagnosed as positive to all those who had positive test results (including healthy subjects who were incorrectly diagnosed as positive). This characteristic predicts how likely it is for someone to truly be a patient if they have a positive test result." NIPS tests for trisomy 21 in general are given to have a PPV of 0.99. However, in a review from 2022, the PPV for trisomy 21 was determined to be only 0.95 [8], which means that 9.8% of pregnancies tested to have a trisomy 21 indeed are false positive (Table 14.1). PPVs for trisomy 13 and 18 are 0.69 and 0.84, respectively; for other autosomal trisomies PPV is 0.54 and sex chromosome aberrations 0.64 (Table 14.1). The PPV is even worse for microdeletion syndromes [16]. In Table 14.1, it is detailed that for example a PPV of 0.54 for rare autosomal trisomies means that 84.5% of women receiving a positive NIPS test outcome indeed get a false positive result. The meaning of, and how to calculate PPVs, is even unclear to specialists (see [17]).

Challenge: When offering a NIPS, the MD must be able to explain what is the PPV, what it means, and in how many cases the test may lead to false positive results.

Example 4

The following post taken from [11] (15 February 2021 by D.T.) highlights in sentence one, that PPVs are falsely explained even in printed patient information, and in sentence 2, that PPVs lead to confusion and wrong conclusions: "We received a leaflet from our doctor about NIPT, which indicated a PPV of 50% for Edwards' syndrome if the mother is under 35. This means that 1 in 2 positive screen results would be a false positive. In hindsight, we should have realized that a PPV of 50% is quite low, and we would likely need an amniocentesis to confirm the result."

Table 14.1 PPVs and underlying data taken from [13]

	NIPS-positive cases	NIPS-false-positive cases	PPV calculated acc. to [12]	Women getting a false positive result (%)
Trisomy 13	687	304	0.69	44.3
Trisomy 18	1447	268	0.84	18.5
Trisomy 21	5283	517	0.95	9.8
Rare autosomal trisomies	342	289	0.54	84.5
Sex-chromosome aberrations	1192	661	0.64	55.5

14.2.2 Biological Challenges

NIPS-test is based on cell-free placental DNA (mislabeled as "cffDNA") floating in the blood serum of a pregnant woman together with her own cfDNA. However, there can be more than just these two populations of cfDNA detected in the blood serum of a pregnant woman. There can be (A) a twin pregnancy (see Sect. 2.3.3), (B) a remainder singleton pregnancy, but an already deceased twin (plus placenta) can be in apoptosis, and flood the blood serum with its cfDNA (see Sect. 2.2.1—vanishing twin), (C) there can be extra maternal cfDNA with trisomies and/or monosomies derived from a yet undetected malignancy (see Sect. 2.2.2), and (D) the mother can have a cryptic mosaic in her blood cells (see Sect. 2.2.3).

Besides, one placenta can have two or more different cell populations with or without chromosomal imbalances, which must not represent the karyotype of the fetus (see Sect. 2.2.4). All these biological conditions must be known to the consumers of NIPS to be able to understand the meaning of a normal and an abnormal test result properly, as outlined below.

14.2.2.1 What Are the Consequences of a Vanishing Twin for NIPS Test?

Twins occur naturally in about one out of 250 births [18]. However, in about one-third of early twin pregnancies, one of the two embryos experiences intrauterine fetal death [19]. This dead embryo/fetus (and its placenta) may be absorbed by the maternal body, without the pregnant woman noticing. Still, this apoptotic process of the vanishing twin produces high rates of "cffDNA," and this embryo/fetus may have not survived, due to a chromosomal imbalance [20]. The latter can now be detected in NIPS test, and falsely attributed to the surviving singleton [10].

Challenge: Pretest counseling should include detailed explanations and why a false-positive NIPS result may happen—and that this may also be due to the presence of a vanishing twin. Furthermore, an abnormal NIPS test result must be an indication for a detailed sonography scan and the search for signs of a potential vanishing twin.

Example 5

Mrs. M.M. reports on [11] the following on 28 May 2020 after she just received a call from her MD that NIPS test resulted in a 50/50 chance of expecting a Turner syndrome baby: "I've done some research and I'm feeling a bit relieved. I had a vanishing twin that made it to about 4 weeks, 2 weeks behind the other one when it was found during my 6 weeks 6 days sonogram. It vanished and was dismissed. I've read that one reason for a false positive can be due to a vanishing twin, and its DNA can remain for up to 16 weeks."

This example shows that explanations on her test result had to be searched by the pregnant woman herself and was not provided by the NIPS-providing medical practice.

14.2.2.2 Why Can NIPS Detect a Maternal Malignancy?

Cancer is typically observed in elderly people. Still, also younger individuals can suffer from a malignancy [21]. As NIPS checks for chromosomal imbalances in cfDNA of pregnant women, there is also a (minor) chance to detect by this approach chromosomal imbalances derived from a yet not diagnosed maternal tumor. According to a recent study, the chance is at 0.02% [21].

Challenge: It is to be discussed if pretest counseling should include also this rare possibility of detecting a maternal malignancy. However, it is indisputable that each MD dealing with NIPS tests must know about this possibility, as it is just a variant of liquid tumor biopsy [22].

Example 6

Due to its rareness, here no patient report can be provided for a NIPS that detected a malignancy—however, the following piece translated from a Swiss Newspaper published in 2015 is still valid for detection of a cancer by NIPS: "How incidental findings are communicated has not been regulated until now. 'As a medical geneticist, I often receive calls and emails from gynecologists asking, 'What should I tell the patient?' " writes Bianchi in a commentary in Nature. Most clinicians are still struggling with the practical problems of introducing a new prenatal test into clinical practice, she says. Few are familiar with genome sequencing or trained to discuss the management of pregnancies complicated by the discovery of a maternal condition [23].

14.2.2.3 Why Are NIPS Detection Rates for Sex Chromosome Aberrations So Low?

In Table 14.1, it was highlighted that 55.5% of sex-chromosome aberrations picked up in NIPS are false positive. One reason therefore is that a subset of pregnant women are mosaic for their sex-chromosome constitution. According to [24], 35–40% of false-positive sex chromosome aberrations can be attributed to women who do not have a karyotype 46,XX in all their cells, but besides also such cells with only one (45, X) and/or three X-chromosomes (47,XXX). As NIPS compares the chromosome-specific amount of DNA in the mother with that of the placenta-derived DNA, such a condition is misinterpreted as a numerical sex-chromosome aberration of the developing child [24].

Challenge: Concerning testing of the gender in NIPS, at first country-specific regulations have to be respected. In many countries, sex is not allowed to be reported before a certain w.o.g. [2]. *Furthermore, the biologically caused lower reliability to detect sex chromosome aberrations than trisomies 13, 18, and 21 should be clarified in pretest counseling.*

Example 7

I.S. wrote on 16 February 2022 in [11]: "My NIPS test showed an increased risk for Triple X syndrome (47,XXX) at 12 weeks of gestation. I had never heard of this condition before, and my doctor told me these tests are 99% accurate. A few days later, I consulted with a genetic counselor who used a program to calculate that there is a 74% chance the NIPS test result for trisomy X is a false positive. My OBGYN suggested I shouldn't get my hopes up, but the genetic counselor's calculation seems to offer a different perspective. Has anyone else had experiences with Triple X syndrome detected through NIPS?"

This example shows that no proper pretest counseling was done, and that there is no clear idea about reliability of the NIPS test concerning the different tested conditions in MDs.

14.2.2.4 What Is Confined Placental Mosaicism?

From invasive prenatal diagnostics, we know that for cytogenetic results obtained in a CVS (placenta cells) are in 2% of the cases discrepant from those in amniocytes (fetal cells) [2, 4]. This observation has been denominated as confined placental mosaicism (CPM). CPM simply means that for example a trisomy 21 found in placenta can be restricted to placenta and not present in the fetus. All imaginable variations have been described between (a) no aberration in fetus and the aberration in all cells of the placenta or (b) vice versa. Thus, at least the same range of discrepancy is to be expected in NIPS and amniocentesis results. In addition, according to the tested condition (like trisomy 21 or trisomy 18), the false positive rates have to be different, as highlighted in Table 14.1.

It has been shown that the genome in early pregnancy is highly flexible; mosaicism of chromosomal aberration in these early embryonal cells, before they develop toward placenta and embryo, is extremely high [25]. Besides, there seem to be yet not completely understood mechanisms by which aberrant cells are developing predominantly toward placenta rather than fetus. Still, the latter can explain the relatively high false positive rate for rare autosomal trisomies in NIPS settings (Table 14.1).

Overall, the existence of CPM is responsible for major parts of false-positive and false-negative NIPS-test results [2]. This biological cause clearly explains why those are to be expected and must be observed in a certain subset of pregnant women.

Challenge: The possible discrepancy of NIPS results from placenta-derived cells and the real genetic status of the fetus (CPM) must be explained in pre-test counseling.

Example 8

In the laboratory of the author, several cases were studied like the following one: NIPS at 11th w.o.g. showed a trisomy 21 in cell-free placenta-derived DNA of a pregnant woman. Sonography was normal before and when amniocentesis was done in 18th w.o.g. The karyotype was determined as mos 47,XN,+mar/46,XN, and the small supernumerary marker chromosome (sSMC) was characterized as an inv dup(21)(q11.2).

Overall, this means most likely that there was a trisomy 21 in zygote; during first cell divisions, a trisomic rescue event occurred, which eliminated one of the three chromosomes 21 either completely or to an sSMC 21. Obviously only cells with normal karyotype or with the sSMC developed further to the fetus, while major parts of placenta still kept the karyotype 47,XN,+21. The latter was detected in NIPS. In several comparable cases, later on normal children were born, while the mentioned case was lost during follow-up. Similar cases are also reported in the literature [26].

14.2.3 Medical Challenges

Medical challenges are partially based on the before-treated technical and biological circumstances. Questions raised below are to be treated in pre-test counseling and after the test result came back.

14.2.3.1 Why a Normal NIPS Result Does Not Mean the Future Baby Will Be Healthy?

As highlighted previously in [2], women taking the NIPS test are often under the impression, that when they get back a negative test result the future baby is definitely healthy. This is the first thing medical specialists need to explain to the advised person(s): Even if a NIPS comes back with normal results this covers maximally 5–10% of possible events, which may lead to a baby with health problems and/ or inborn abnormalities [2]. Approximately 3–6% of newborns are born with significant physical abnormalities. Within this group, around 6% have a chromosomal disorder, 7% suffer from teratogenic damage, and 8% and 25% are affected by monogenetic and multigenetic diseases, respectively. For the remaining 54%, the condition is typically diagnosed as an "idiopathic disorder," meaning the cause remains unknown and leads to lifelong challenges [2].

Besides, there are other problems to consider, like above mentioned ones. An example is given from the literature below for CPM (see Sect. 2.3.1.1).

Challenge: It must be made clear to an advised couple before the test that a normal NIPS is not a stand-alone test. Further tests are advised according to the individual case and possible familial indications.

Example 9

Yang et al. [27] describe a case of a 32-year-old pregnant woman taking the NIPS test in 15th w.o.g. with normal result. Still there were abnormal biochemical tests and severely abnormal sonographic signs. The latter were confirmed in 24th and 30th w.o.g. Finally in 33rd w.o.g. chromosome analyses from umbilical cord blood revealed a karyotype 47,XN,+18[61]/46,XN[39], and pregnancy was terminated in 34th w.o.g. Placenta was studied afterward and showed mosaic rates of 33–72% of trisomy 18 in six different studied areas. Thus, here the NIPS was negative, due to a too-low percentage of cells with trisomy 18 in placenta to be detected. Overall, here the late termination could have been avoided easily by an earlier amniocentesis and earlier mistrusting in NIPS result.

14.2.3.2 Why an Abnormal NIPS Result Does Not Mean the Future Baby Will Be Affected?

As shown in Table 14.1, between 9.8% and 84.5% of aberrant NIPS test results are false positive—according to the tested condition these rates are different. This is the main reason why an abnormal NIPS result does not mean the future baby will be affected necessarily. Besides, recently we calculated the rate of real positive NIPS results and the rate of newborns to be expected with the corresponding conditions [8]. Surprisingly this calculation showed that 90% of those fetuses, which were terminated by *interruptio* would have been naturally aborted. They would have not survived until birth and induced abortion would not have been necessary.

Challenge: As shown by others and summarized by us [2, 8] *(see also* [11]*), a positive NIPS result means in the following for the MD a lot of necessary care for the pregnant woman. Apart from detailed pre- and post-test counseling for possible positive test results, further counseling dates would be advisable to answer in detail all questions and doubts of the tested woman.*

Example 10
On 13 December 2021, K.N. reports on [11] the following on a false positive result she got: "I was 38 and thus considered high risk. My NIPS test indicated a 15q deletion (Angelman/Prader Willi syndrome). I was not informed about the high rate of false positives. The doctor advised abortion as a viable option, amniocentesis, and joining support groups to 'understand what I am getting myself into.' After researching the miscarriage rates for amniocentesis (0.5%), I firmly told the doctor that I would take my chances, as a 1 in 200 risk was too significant for me. My daughter was born perfectly healthy. When the doctors came for her DNA evaluation at 2 months, they observed our bond and suggested the results would likely be negative. They were correct; she did not have the deletion. I cried during my pregnancy, constantly researching but never finding information about the high rate of false positives for deletions. I was furious that my doctor was unaware. How many women, seeing the 99% accuracy rate, might choose abortion for a perfectly healthy child? I decided that my daughter was perfect in whatever form. These companies can do more harm than good, especially if women are advised to consider abortion based on high rates of false positives. This should be made clear to all obstetricians and expectant mothers."

Example 11
On 21 June 2020, J.D. made a comment on [11] clearly highlighting that waiting for results is really distressing for women—here further genetic counseling appointments could help to reduce the stress.

She wrote: "I received an inconclusive NIPS test the first time. Now, at 19 weeks of gestation, my doctor discovered a choroid plexus cyst. She suggested retaking the NIPS since the initial test was inconclusive due to low fetal DNA. I'm extremely worried and have to wait at least a week to hear back. Has anyone else experienced this?"

14.2.3.3 Why Not to Test Twin Pregnancies?

As mentioned before naturally twins occur in 1/250 births. Due to increased application of artificial insemination in the USA, this rate increased to 1/30 births [28]. Thus, NIPS is applied in those cases more and more, too. Still, a recent study confirmed that PPVs, false positive rates, and inconclusive results are in the range from previous studies and thus, as stated in [16], NIPS should not be offered in twin pregnancies.

Challenge: MDs should have a clear idea about the low reliability of NIPS test results in twin pregnancies. Accordingly, they should do the counseling to couples.

Example 12

Mrs. S. wrote on 16 June 2020 about testing her twin pregnancy by NIPS [11]: "I'm currently 13 weeks pregnant with a natural twin pregnancy, and my NIPS test results came back inconclusive for all accounts. They suspect it's related to the 21st chromosome, and they don't believe redoing the test will change the outcome. My enhanced first trimester ultrasound and blood work were normal, which has left my doctors puzzled. It's unsettling to hear that such a situation is <1% likely to occur. I wasn't even considered high risk, and my doctor didn't initially recommend the NIPT test. I thought having the test done, given its purported 99% accuracy, would provide more conclusive results than other screenings. At this point, I've undergone additional genetic testing for myself, and I have an amniocentesis scheduled for early July. It's difficult to imagine how I'll manage until then without worrying myself sick."

On July 20, 2020, she added: "I'm really happy to say that after all the testing, our amniocentesis results came back completely normal for both babies. There's no trisomy 21 as the NIPS test initially suggested. What a rollercoaster ride it's been. I hope this brings peace of mind to all the moms who have received 'abnormal' NIPS test results (although I know that's unlikely—I've been there). This technology is still evolving, and there's still a lot of potential for errors. If anyone receives an 'abnormal' NIPS result, I strongly encourage invasive testing to confirm whether that's the case."

14.2.3.4 Proper Counseling Is Really Important for NIPS Testing

From the before listed points and examples the statement that well informed counselling really matters for NIPS-testing is consistent—see also [29]. For sure, majority of consumers of the test will have a negative result and feel relieved afterward. Still, a subset of them may have a rude awakening when nonetheless an impaired child is born. So, it must be clearly stated which conditions can and are finally tested in the individual NIPS test by the caring MD. Furthermore, 9.8–84.5% of the positive cases are false-positive ones, which also needs to be communicated. Besides, the latter cases need special attention from the attending physician.

Challenge: NIPS has the potential to bring relieve or excitement to pregnant women—both conditions have the potential to cause more work for the gynecologists and obstetricians. This is mainly because the NIPS test has not found its proper place in prenatal care yet.

Example 13

The following long post from L.M.P. from Australia (posted April 25, 2020) [11] highlights how eager tested women are for information, which normally should be available from the MD offering the test: "We opted to go ahead with the NIPS test, and a week later, the doctor called us urgently to share high-risk results for trisomy 13. We were informed that the test was highly accurate, and the hospital would reach out within 24 h for further testing and next steps. The news was devastating as we learned about the severe prognosis associated with trisomy 13. For 4 days, we anxiously called the hospital every day, seeking answers. Eventually, I connected with a sympathetic genetic counselor who scheduled an appointment for us 2 weeks later (since the referral hadn't been processed) and suggested having another ultrasound to provide more information and help pass the waiting period. During the ultrasound, everything appeared 'perfect' with no signs of trisomy 13 markers. The ultrasound technician warned that despite the reassuring scan, the NIPS test results were deemed more reliable. This gave us a small glimmer of hope. However, upon returning to the doctor, we were informed that the scan's accuracy might be questionable because I was only 11 weeks pregnant at the time, necessitating invasive diagnostic tests to prepare for all potential outcomes. At our initial hospital appointment, a professor, a doctor, and two midwives were present. I asked about the reliability of the NIPS test and the statistics supporting its advertised accuracy. They explained that the test is relatively new, and not enough is known about it. From their experience, in cases like mine (high-risk result but no markers on the scan), less than 50% of the babies actually have a genetic disorder.

At my second hospital appointment, I underwent another scan and amniocentesis. The team of doctors and nurses were optimistic. Two days later, they called to confirm that the baby does not have trisomy 13, Down syndrome, or Edwards syndrome. Saying we were relieved is an understatement. As my doctor learned more about the test, she was shocked by the information. I feel we were not adequately informed and should have been offered counseling before taking the Harmony test. It was only through the genetic counselor and the hospital that we learned the NIPS test is a 'screening test,' not a 'diagnosis.' My doctor never used such terms and assured us of the 99% accuracy. Before speaking with the genetic counselor, my husband and I were considering termination to avoid carrying a non-viable pregnancy. Now, at 17/18 weeks, I realize I spent half of my pregnancy extremely stressed, which significantly impacted my mental health, something I had never struggled with before. I know my doctor was doing her best with the information she had, but more training and information need to be made available for doctors offering this test."

14.3 Ethical Issues

NIPS and above mentioned level of uninformedness about possibilities, shortcuts, possible problems, technical, biological, and medical challenges provide already a number of ethical issues. Especially, it seems questionable that a test obviously not only really understood in public and by consumers (= pregnant women and their partners), but also by providers (here MDs in specific) is used in such a wide range worldwide. Misunderstandings about reliability (keyword: use of misleading PPV instead of giving percentage of false positives) is only one ethically questionable aspect here. Overall, there is a clear trend in NIPS-related publications to report as positive as possible on the test. Leaving out information, like the fact that a PPV of

0.99 applies only for testing of trisomy 21 unfortunately is not only present in advertising NIPS, but also in scientific papers. In 2019 D'Ambrosio et al. [30] wrote: "NIPT has changed the prenatal diagnosis landscape and remains the most superior screening method for detecting common autosomal aneuploidies"—if at all this is only valid for detection of trisomy 21. The same way Yang et al. stated in 2020 [31]: "Our findings indicate that NIPT is increasingly being utilized as a first-line screening tool in the general obstetric population. It is also gaining prevalence in screening for sex chromosome abnormalities and copy number variations." Is that really ethically justified when false positive rates are in a range of >50 and 95%? Besides these, only rarely considered ethical problems of applying a somehow immature, not really understood test to public, there are other ethical issues discussed in literature, as follows.

14.3.1 What About Reliability of Signed Informed Consent?

All the above-mentioned points on complexity of the NIPS test lead also to the question in as far a valid informed consent really can be achieved. This was best summarized by Thomas et al. [32] as follows:

"Respect for autonomy necessitates obtaining informed consent, which must be given voluntarily. Medico-legally, individuals must be sufficiently informed about a test or procedure, including its risks and benefits. The extent of information required is typically determined by what a reasonable person in the individual's circumstances would expect to receive. Ethically, it is crucial that the person comprehends the information, not just that they were provided with the legally required details. Given the complexity of extended panels and WGS NIPS, ensuring understanding necessitates a significant time investment. While it may be tempting to respect autonomy by avoiding paternalism and providing all information from prenatal testing to parents, this approach transfers the burden of uncertain results and the associated anxiety onto the parents. Research, both qualitative and quantitative, indicates higher levels of decisional regret among parents whose results revealed variations of uncertain significance. Many parents would not have consented to the test if they had known the potential implications. Additionally, the uncertainty faced by clinicians regarding the meaning of these results for the child's future exacerbates parental distress."

14.3.2 What About Rights of the Unborn and Selective Reproduction?

Overall, each prenatal test, which has been done since decades, has and had an ethical dilemma. The right of the unborn to be born in each biologically possible case is defined to be less important than the right of the mother to decide which baby (according to its health status) she wants to give live to and she feels comfortable to provide further care in later live [33–35]. Accordingly, the basic human right for life and integrity is trumped for an unborn here by other values.

14.3.3 What About Selective Reproduction?

NIPS enables an early access to the sex of the future child. As historical cultures of many countries still favor birth of a boy rather than a girl, most countries introduced legal boundaries to hamper sex selection [2]. Nonetheless, use of NIPS alerted disability rights organizations that this new test may increase the trends toward (i) selective reproduction, (ii) the devaluation of lives of persons with trisomy 21 or even (iii) eugenics [33]. Such apprehensions are further stired up by papers like that of Shang et al. [36], which declare that the use of NIPS is the most cost-efficient strategy for the society to avoid babies with Down syndrome (see as well Ref. [37]). Furthermore, Deans et al. [34] argued that NIPS may just be a first entrance door for later testing of unborn, newborn, and all adult for all possible conditions. As shown here and previously [2], this is not a realistic scenario, yet, considering all the problems, limitations, and shortcuts of the NIPS and its underlying techniques.

14.3.4 What About Social Pressure?

Already before NIPS was introduced, parents with a Down syndrome could suffer from attacks from their own family and/or society such as: Could this (person with trisomy 21) not have been avoided? It has been shown that the availability of NIPS and the high expectations of public in its reliability enhanced this type of social pressure. Pressure to test adequately also applying NIPS, and in case of a positive test pressure to terminate, is subjectively growing for "a large minority of pregnant women" [37].

14.3.5 What About Access to the Test?

Based on, at least questionable idea, that NIPS is able to replace all traditional prenatal care scenarios, there can also be observed 'ethical concerns' about the justice and injustice in access to the test. While no such discourse was going on about financing of the first-trimester screening program [2], suddenly there are big concerns that poorer families could have no access to NIPS and only richer ones could do the test; and thus, there would be urgent needs to finance NIPS testing by state [38–40]. The real ethical issue is not about financing the NIPS by the society, but how such a problematic test could be hyped alike.

14.4 Conclusions

At the end of these discussions on challenges and ethical issues surrounding NIPS, it has to be stated that there are yet more technical, biological, and medical challenges, than ethical issues. As we wrote earlier in [2]: "The theoretical possibilities of NIPS are promising; however, it is essential to consider the problems and limitations.

Overall, NIPS results are not well understood, and the reliability of the data is poorly communicated to both the MDs ordering the test and the pregnant women undergoing it. Hessel and Henn [41] recently noted, 'The fact is that many pregnant women perceive NIPT as a standard measure and undergo it without question.' This attitude can lead to a rude awakening. Therefore, appropriate training and continuous education for physicians providing NIPT counseling are urgently needed. As highlighted in other sources, this can only be achieved through unbiased, ongoing education offered by medical associations or independent researchers, rather than NIPT providers."

References

1. Darouich AA, Liehr T, Weise A, Schlembach D, Schleußner E, Kiehntopf M, et al. Alpha-fetoprotein and its value for predicting pregnancy outcomes—a re-evaluation. J Prenatal Med. 2015;9:18–23.
2. Liehr T, Harutyunyan T, Williams H, Weise A. Non-invasive prenatal testing in Germany. Diagnostics (Basel). 2022;12:2816.
3. Benn KN, Benn P, Campbell WA, Moaddab A, Shamshira AA. Genetic sonogram: components and role in the era of prenatal screening. Fetal Maternal Med Rev. 2014;25:214–31.
4. Liehr T, Lauten A, Schneider U, Schleussner E, Weise A. Noninvasive prenatal testing (NIPT)—when is it advantageous to apply? Biomedicine Hub. 2017;2:458–32.
5. Lo YM, Corbetta N, Chamberlain PF, Rai V, Sargent IL, Redman CW, et al. Presence of fetal DNA in maternal plasma and serum. Lancet. 1997;350:485–7.
6. Tabor A, Alfirevic Z. Update on procedure-related risks for prenatal diagnosis techniques. Fetal Diagn Ther. 2010;27:1–7.
7. Liehr T. What about noninvasive prenatal testing (NIPT) from 'cell-free fetal DNA' (cffDNA)? http://atlasofscience.org/what-about-noninvasive-prenatal-testing-nipt-from-cell-free-fetal-dna-cffdna/ (accessed 14 January 2023).
8. Liehr T. False-positives and false-negatives in non-invasive prenatal testing (NIPT): what can we learn from a meta-analyses on>750,000 tests? Mol Cytogenet. 2022;15:36.
9. Zaninović L, Bašković M, Ježek D, Katušić BA. Validity and utility of non-invasive prenatal testing for copy number variations and microdeletions: a systematic review. J Clin Med. 2022;11:3350.
10. Liehr T. Non-invasive prenatal testing, what patients do not learn, may be due to lack of specialist genetic training by gynecologists and obstetricians? Front Genet. 2021;12:682980.
11. Joynson C. BLOG Our concerns about non-invasive prenatal testing (NIPT) in the private healthcare sector. Published 02/08/2019. https://www.nuffieldbioethics.org/blog/nipt-private (accessed 14 January 2023).
12. Liehr T. What you should definitely know about the limits and possibilities of noninvasive prenatal testing (NIPT). Pädiatrische Praxis. 2022;99:7–14.
13. MedlinePlus. Bethesda (MD): National Library of Medicine (US); What is noninvasive prenatal testing (NIPT) and what disorders can it screen for? https://medlineplus.gov/genetics/understanding/testing/nipt/ (accessed 14 January 2023).
14. Hudecova I, Sahota D, Heung MM, Jin Y, Lee WS, Leung TY, et al. Maternal plasma fetal DNA fractions in pregnancies with low and high risks for fetal chromosomal aneuploidies. PLoS One. 2014;9:e88484.
15. Safari S, Baratloo A, Elfil M, Negida A. Evidence based emergency medicine part 2: positive and negative predictive values of diagnostic tests. Arch Emerg (Tehran). 2015;3:87–8.
16. Kozlowski P, Burkhardt T, Gembruch U, Gonser M, Kähler C, Kagan KO, et al. Recommendations for the implementation of first-trimester screening, detailed ultrasound, cell-free DNA screening and diagnostic procedures. Ultraschall Med. 2019;40:176–93.

17. Chen Y, Lu L, Zhang Y, Wang F, Ni Y, Wang Q, et al. Clinical application of expanded noninvasive prenatal testing for fetal chromosome abnormalities in a cohort of 39,580 pregnancies. Am J Med Genet A. 2022;188:1426–34.
18. American Society for Reproductive Medicine. Multiple pregnancy and birth: Twins, triplets, and high-order multiples. A guide for patients. 2012. https://www.reproductivefacts.org/globalassets/rf/news-and-publications/bookletsfact-sheets/english-fact-sheets-and-info-booklets/booklet_multiple_pregnancy_and_birth_twins_triplets_and_high-order_multiples.pdf.
19. Zamani Z, Parekh U. Vanishing twin syndrome. In: StatPearls. Treasure Island (FL): StatPearls Publishing; 2022. Available from: https://www.ncbi.nlm.nih.gov/books/NBK563220/. [Updated 2022 Jul 25].
20. Batsry L, Yinon Y. The vanishing twin: diagnosis and implications. Best Pract Res Clin Obstet Gynaecol. 2022;84:66–75.
21. Heesterbeek CJ, Aukema SM, Galjaard RH, Boon EMJ, Srebniak MI, Bouman K, et al. Noninvasive prenatal test results indicative of maternal malignancies: a nationwide genetic and clinical follow-up study. J Clin Oncol. 2022;40:2426–35.
22. Lone SN, Nisar S, Masoodi T, Singh M, Rizwan A, Hashem S, et al. Liquid biopsy: a step closer to transform diagnosis, prognosis and future of cancer treatments. Mol Cancer. 2022;21:79.
23. Lüthi T, Baby OK. Mutter krank (translated: Baby o. k., mother sick). Neue Züricher Zeitung, 23.08.2015. https://www.nzz.ch/vorgeburtliche-bluttests-baby-o-k-mutter-krank-ld.1565 (accessed 14 January 2023).
24. Wan J, Li R, Li F, Yu Q, Wang D, Sun X, et al. Contribution of maternal mosaicism to false-positive chromosome X loss associated with noninvasive prenatal testing. J Maternal-Fetal Neonatal Med. 2022;35:9647–53.
25. Pellestor F. Chromothripsis: how does such a catastrophic event impact human reproduction? Hum Reprod. 2014;29:388–93.
26. Kleinfinger P, Brechard M, Luscan A, Trost D, Boughalem A, et al. Case report: how whole-genome sequencing-based cell-free DNA prenatal testing can help identify a marker mhromosome. Front Genet. 2022;13:926290.
27. Yang J, Qi Y, Guo F, Hou Y, Peng H, Wang D, et al. A case of placental trisomy 18 mosaicism causing a false negative NIPT result. Mol Cytogenet. 2017;10:40.
28. Kantor V, Mo L, DiNonno W, Howard K, Palsuledesai CC, et al. Positive predictive value of a single nucleotide polymorphism (SNP)-based NIPT for aneuploidy in twins: experience from clinical practice. Prenatal Diagn. 2022;42:1587–93.
29. Grati FR, Bestetti I, De Siero D, Malvestiti F, Villa N, Sala E, et al. Positive predictive values and outcomes for uninformative cell-free DNA tests: an Italian multicentric cytogenetic and cytogenomic audit of diagnostic testing (ICARO study). Prenatal Diagn. 2022;42:1575–86.
30. D'Ambrosio V, Squarcella A, Vena F, Di Mascio D, Corno S, Pajno C, et al. Update in non-invasive prenatal testing. Minerva Ginecol. 2019;71:44–53.
31. Yang L, Tan WC. Prenatal screening in the era of non-invasive prenatal testing: a nationwide cross-sectional survey of obstetrician knowledge, attitudes and clinical practice. BMC Pregn Childbirth. 2020;2:579.
32. Thomas J, Harraway J, Kirchhoffer D. Non-invasive prenatal testing: clinical utility and ethical concerns about recent advances. Med J Aust. 2021;214:168–170.e1.
33. Foth H. Avoiding 'selection'?—references to history in current German policy debates about non-invasive prenatal testing. Bioethics. 2021;35:518–27.
34. Deans Z, Clarke AJ, Newson AJ. For your interest? The ethical acceptability of using non-invasive prenatal testing to test 'purely for information'. Bioethics. 2015;29:19–25.
35. Stanić Ž, Hrgović Z, Fureš R, Vusić I, Lagančić M. Ethical aspects of prenatal DNA testing. Z Geburtshilfe Neonatol. 2022;226:300–3.
36. Shang W, Wan Y, Chen J, Du Y, Huang J. Introducing the non-invasive prenatal testing for detection of down syndrome in China: a cost-effectiveness analysis. BMJ Open. 2021;11:e046582.

37. Perrot A, Horn R. Preserving women's reproductive autonomy while promoting the rights of people with disabilities?: the case of Heidi Crowter and Maire lea-Wilson in the light of NIPT debates in England, France and Germany. J Med Ethics. 2023;49(7):471–3.
38. Benoy ME, Iruretagoyena JI, Birkeland LE, Petty EM. The impact of insurance on equitable access to non-invasive prenatal screening (NIPT): private insurance may not pay. J Commun Genet. 2021;12:185–97.
39. Löwy I. Non-invasive prenatal testing: a diagnostic innovation shaped by commercial interests and the regulation conundrum. Soc Sci Med. 2022;304:113064.
40. Vinante V, Keller B, Huhn EA, Huang D, Lapaire O, Manegold-Brauer G. Impact of nationwide health insurance coverage for non-invasive prenatal testing. Int J Gynecol Obstet. 2018;141:189–93.
41. Hessel L, Henn W. Nichtinvasive Pränataltests—Fragwürdige Parameterauswahl (translation: non-invasive prenatal testing—questionable parameter selection). Deutsches Ärzteblatt. 2022;24:1076–7.

Implementation of Noninvasive Prenatal Screening (NIPS) in Clinical Practice: Comparison of Developed Versus Developing Countries

15

Shifali Gupta, Sujata Bhardwaj, and Inusha Panigrahi

15.1 Introduction

The 1980s and 1990s saw the development of noninvasive serum screening techniques, such as combining ultrasonography and maternal serum analyte measurements [1]. In order to perform karyotyping on fetal samples, intrusive procedures like amniocentesis or chorionic villi sampling (CVS) were sometimes necessary due to uncertain results. There is a chance of miscarriage with both of those procedures [2, 3]. The development of a different screening technique using cell-free fetal DNA (cff-DNA) for noninvasive prenatal screening (NIPS) was made possible by the discovery of cells and fetal DNA in maternal blood in 1997 [4]. It was shown in 2008, more than 10 years later, that trisomy could be detected in mother's blood using next-generation sequencing. With the successful application of massively parallel sequencing (MPS) for NIPS, a new era emerged in 2008 [5]. NIPS became accessible for clinical use after 2011 [6]. The clinical use of NIPS as a sensitive screening method for common fetal aneuploidies has been established by several studies [7, 8]. To mitigate the little risk of procedure-related miscarriage, NIPS has been presented as an alternative to conventional invasive testing techniques (CVS/amniocentesis) [9, 10]. However, before making a final choice that cannot be changed, screen-positive NIPS results should always be confirmed by invasive diagnostics [11]. The scope of this chapter is to give an overview of NIPS, its indications, strengths and limitations, application and current scenario, and barriers to use in developing countries.

S. Gupta · I. Panigrahi (✉)
Genetic Metabolic Unit, Department of Pediatrics, PGIMER, Chandigarh, India
e-mail: panigrahi.inusha@pgimer.edu.in

S. Bhardwaj
Department of Obstetrics and Gynecology, PGIMER, Chandigarh, India

R. A. Rather, S. C. Saha (eds.), *Non-invasive Prenatal Screening (NIPS) in Clinical Practice*, https://doi.org/10.1007/978-981-97-6402-0_15

15.2 What Is NIPS?

The basic understanding that has led to this revolutionary technique is the presence of cff-DNA in maternal circulation (Fig. 15.1). Fetal DNA is derived in maternal circulation when placental microparticles are shed into the maternal bloodstream during pregnancy. Fetal fragments are significantly smaller (<200 base pairs) than maternal DNA fragments and this difference in size allows cff-DNA to be distinguished from maternal DNA. Approximately 10–15% of the total cff-DNA in the mother's plasma after 10 weeks of gestation is placental, or generated from trophoblasts, which is adequate to test for certain fetal diseases [12].

Apart from the length of DNA fragments, specific methylation patterns, genetic polymorphisms, and quantitative differences in chromosomes at genomic level as evidenced in mother's plasma are used for the final analysis of data [13]. For instance, it has been demonstrated that the presence of a trisomy 21 fetus increases this representation of 21st chromosome fragments in plasma of mother by 5% when the fractional fetal DNA concentration is 10%.

15.3 Molecular/Sequencing Techniques as Basis for NIPS

The methodology involves the use of Illumina Inc.'s short read platforms for massively parallel DNA shotgun sequencing, which allows for the sequencing of all cfDNA isolated from a maternal sample [5] (Fig. 15.1). The acquired reads are

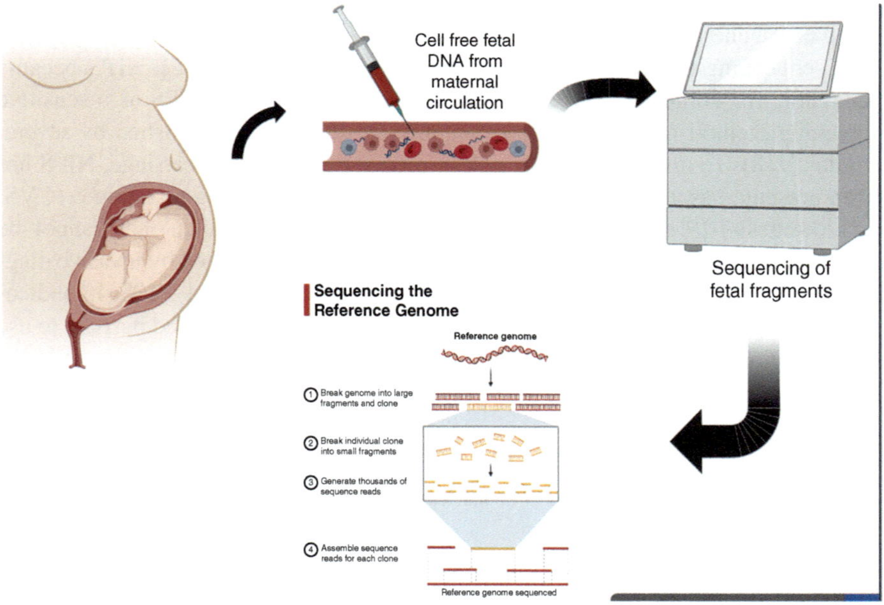

Fig. 15.1 Principles of noninvasive prenatal screening for chromosomal abnormalities

mapped and then aligned to the human reference genome. Each chromosome's (relative) fragment count is calculated. When compared to a diploid fetus, more trisomy chromosomal fragments should be present in a fetus with a trisomy. Nevertheless, chromosome 21 accounts for <1.5% of the total in normal individual, demonstrating that massive parallel sequencing (MPS) does not enrich any of the fragments. Therefore, to differentiate the fetal trisomy 21 from normal, many million DNA pieces coming from the whole human genome must be sequenced. The number of samples that may be sequenced in parallel and, consequently, the minimum read number, is a critical factor in terms of finances, effectiveness, and applicability. Because there won't be as many sequenced reads, any variation in the number of reads will have less statistical significance, which lowers reliability. However, as fewer samples may be multiplexed per run with too many reads, the expenses will rise dramatically. The number of DNA fragments that need to be sequenced and measured in order to produce a trustworthy result is still up for dispute [14]. Although selected alternatives to MPS technique have been developed, they do not appear to reduce cost or the turn-around-time for single chromosome of interest.

Targeted sequencing employing the idea of digital analysis of selected regions (DANSR) is a further strategy [15]. It entails the MPS analysis that comes after the selective amplification of particular areas on chromosomes 21 and 18. Reliability in detecting fetal trisomy 18 or 21 requires a substantially smaller number of sequenced reads than whole genome MPS techniques. The 576 nonpolymorphic loci on chromosomes 18 and 21, respectively, are the target of the DANSR assay. To detect and measure the fetal fraction, assays are created against loci on chromosomes 1 through 12 that contain single nucleotide polymorphisms (SNPs). Fetal trisomy 21 and 18 cases can be identified with high accuracy using ~1 million raw reads per sample when DANSR and fetal-fraction optimized risk of trisomy evaluation are combined. This results in a five to tenfold reduction in read count when compared to the whole genome approach, as well as a decrease in sequencing and data storage costs.

In our daily work, we encounter a few variables that impact NIPS outcomes and should always be remembered. Few examples are, the quantity of fetal DNA present, the likelihood that a chromosome anomaly is present beforehand (i.e., the woman's risk based on the results of previous screening or the age of the mother), multifetal gestation, or the presence of placental mosaicism in a nonviable second embryo [8, 16–19]. It is important to evaluate the results of such clinical settings cautiously.

15.4 Clinical Applications of NIPS

NIPS has various clinical applications which are listed as follows:

1. Foetal sex determination to evaluate risk for X-linked diseases [20–22].
2. Blood type: RhD gene from a fetus that is Rhesus D-positive found in the maternal plasma in pregnancy of mother who is Rhesus D-negative [23–27].

3. Specific microdeletions/duplications: Chr.1p36 deletion, terminal 5p loss in Cri du Chat syndrome, Di-George syndrome (deletion chr.22q11.2), Angelman/Prader–Willi syndromes (deletion chr.15q11.2-q13), Langer–Geidon syndrome (chr.8q24 deletion), Wolf–Hirschhorn syndrome (WHS, terminal deletion chr.4p) and Jacobsen syndrome (deletion terminal chr.11q) [28].
4. Chromosome disorders: trisomy affecting chromosomes 13 and 18, sex chromosomes, and Down syndrome [29, 30].
5. Single-gene disorders (autosomal) [31–34].
6. Selected genome-wide copy number variations, or CNVs.

Out of these, the most widespread usage is for autosomal and sex chromosome aneuploidies, followed by selected microdeletions/duplications. It is well validated in common autosomal trisomies.

15.5 Limitations of NIPS

Certain clinical scenarios or factors may compromise the reliability of NIPS and results may not be accurate [12]. The limitations of the use of NIPS for antenatal testing are:

1. Null calls [35].
2. Gestation less than 10 weeks [18, 36].
3. Obesity [37].
4. Bone marrow transplant recipients.
5. Recent blood transfusion in the last 1 month.
6. Screening test only.
7. It can unmask hidden malignancies, hence pretest counseling is highly important [38, 39].
8. Post-test: needs confirmatory testing [40, 41].
9. Expensive test [42, 43].
10. Not clinically recommended to check for fetal single-gene disorders [44, 45].
11. Not recommended for late pregnancy complications [46].
12. No information on applicability in anatomical abnormalities and neural tube defects.

15.6 Current American College of Medical Genetics (ACMG) Guidelines

There remains ambiguity and difficult decision-making for using NIPS in routine clinical practice. To bring uniformity American College of Medical Genetics has put forward some recommendations, which can help healthcare providers and policymakers implement this latest technology for larger population strata based on their requirements [47].

The ACMG advises informing and presenting NIPS as the most sensitive screening alternative to all expectant mothers for conventionally screened aneuploidies, such as Down syndrome (T21), Patau (T13), and Edwards (T18) [47]. All expectant mothers should be provided NIPS rather than conventional tests for sex chromosomal aneuploidy (SCA) and common trisomy (in singleton and twin pregnancies).

Additionally, ACMG advises providing diagnostic testing in the event that NIPS is followed by a positive screening test result. When a fetus is identified as having a chromosomal or genomic variation, accurate, balanced, and current information should be provided to the couple, taking into account the educational background of the expectant parents, in order to enlighten and educate them about the condition of concern.

ACMG also emphasizes that to help patients and providers make decisions and interpret results, laboratories should provide easily visible and unambiguously stated DR (detection rate), SPEC (clinical specificity), PPV (positive predictive value), and NPV (negative predictive values) for conditions being screened, in pretest marketing materials, and when reporting laboratory results. If laboratories are unable to report DR, SPEC, and PPV for these disorders, they should not be allowed to offer screening for the three common trisomies.

The recommendations are summarized in Table 15.1. A number of NIPS implementation-related issues require more research, such as the screening process for copy number variants (CNVs), which includes microdeletion and microduplication syndromes. Furthermore, as new advancements are made, the education and training provided by organizations that offer NIPS have to be updated and standardized. The development and validation of educational resources that are enlightening, devoid of commercial agendas, and that offer reasonable advantages and restrictions on screening is necessary.

Additionally, ACMG stresses the importance of pre- and post-test counseling, which is essential whenever any genetic testing is provided. In the case of NIPS, this

Table 15.1 Summary of the ACMG recommendations/guidelines: Noninvasive prenatal screening (NIPS) for chromosome abnormalities in fetus in a normal-risk population [47]

Guideline	Recommendation
Singleton pregnancies, ACMG advises using NIPS rather than standard screening techniques to check for common fetal trisomies, namely, involving chr. 21, 18, and 13	Strong recommendation
Twin pregnancies, ACMG advises using NIPS rather than conventional techniques for screening of trisomies	Strong recommendation,
In order to screen for fetal SCA, ACMG advises offering NIPS to patients who are having a singleton gestation*	(strong recommendation)
Suspected 22q11.2 deletion syndrome (Di-George syndrome) b in pregnancy, should be offered NIPS	(recommendation based on some evidence)
Currently, there is not enough data to support NIPS's recommendation for routine screening for other copy number variants (CNVs) apart from deletion 22q11.2	Not recommended
Currently, not enough data to determine whether or not NIPS should be used to identify RATS (rare autosomal trisomies)	Not recommended

*Local/national PNDT guidelines should be followed when dealing with sex chromosomal abnormalities (SCA)

counseling must cover features such as NIPS's optional and screening features, circumstances for which screening is possible and ineffective, types of test results that can be obtained, such as incidental findings and no-call results, details regarding predictive values that are both good and negative. It is advised that any unusual results be confirmed. In mosaic or complete states, previously unidentified paternal chromosomal abnormalities, such as 22q11.2DS, may be discovered. NIPS is not intended to identify structural problems in fetuses. Prior to making any decision regarding pregnancy termination, prenatal diagnostic confirmation should be offered.

Access to educational resources produced through partnerships between advocacy groups and healthcare professional associations should be granted to patients. NIPS may lead to the identification of cancers and maternal genetic abnormalities. According to estimates, between 25 and 27 out of every 100,000 pregnant women will get a cancer during their pregnancy, most often breast cancer. Using NIPS, one can find the genetic abnormalities present in cancerous cells. If NIPS report shows several aneuploidies or monosomy that is not confirmed by fetal diagnostic testing, the chance of maternal malignancy is at its highest.

15.7 Clinical Utility and Impact of NIPS on Invasive Testing

The diploid set of chromosomes present in human cells is reduced to haploid at the time of gametogenesis. Aberrations during this process may lead to an extra or missing chromosome in the haploid set which is called aneuploidy [48]. Most aneuploid fetuses get aborted spontaneously and go undetected. Some of the aneuploidies for example trisomy 13, 18, 21, sex chromosome aneuploidies are compatible with survival to some extent. However, the fetuses that have these aneuploidies have several morbidities—neurological, cardiovascular, and renal issues [49]. Hence the quality of life is hampered, and it is desirable to detect them before birth and decide for the continuation of pregnancy. According to ACOG, 1 in 150 pregnancies are affected by aneuploidies [50]. As a result, it has been considered for incorporation into universal screening for antenatal check-ups. Major impacts seen after implementation of NIPS are:

1. Reduced need for diagnostic testing in high-risk pregnancies as a result of older mothers: The single biggest reason in the past for intrusive diagnostic testing, which had an inherent risk of miscarriage, was the old age of the mother [3]. In these high-risk patients, intrusive testing has significantly decreased since the use of NIPS [51, 52]. The American College of Obstetricians and Gynecologists (ACOG) recommended screening tests for all pregnant women, and therafter a study was done to determine rates of invasive testing after that recommendation. It was found that the introduction of noninvasive prenatal tests (NIPT)/NIPS significantly reduced amniocentesis rate in pregnancies at advanced maternal age in Korea [51].

2. Utility in cases where a soft marker is found on fetal ultrasound: Patients who would normally be offered amniocentesis today but had not previously undergone screening may instead be offered NIPS. More testing is typically not advised if NIPS shows decreased risk [53–55]. The probability ratio—rather than the absolute risk—is linked to the existence of a soft marker, and none of these ratios is significant enough to offset the degree of risk reduction linked to negative NIPS results. Multiple soft markers may sometimes suggest a single gene disorder like RASopathy or ciliopathy.
3. Improved screening for the identification of aneuploidies over conventional methods: The main objective of this screening test is to detect aneuploidies with a higher degree of sensitivity and specificity than with conventional methods. With a low percentage of false positives (0.04%), NIPT has been demonstrated to be very reliable in predicting the existence of chromosomal abnormalities [8, 56]. Trisomy 21 has a 99.2% detection rate, with a little 0.09% false positive rate (FPR), according to a meta-analysis. While trisomy 13 had a lower detection rate of 91.0% and a FPR of 0.13%, for trisomy 18 the detection rate was 96.3% and FPR of 0.13%. In terms of other sex chromosomal abnormalities, the detection rate was 93% with a FPR of 0.14%, and incase of monosomy X, it was 90.3% with a FPR of 0.23%. As a result, NIPS for aneuploidies was preferable to other screening methods [57].

15.8 Current Status of NIPS in Developed and Developing Countries

Despite being accessible globally, NIPS has not been widely accepted or incorporated into national rules in other nations. NIPS is frequently used for prenatal screening in wealthy nations. Following first-trimester screening in 14 countries, NIPS has been enacted as a national policy or program for higher-risk women in Europe [58]. Every pregnant woman in Belgium and the Netherlands is eligible for NIPT [59]. The primary prenatal screening test in Australia is either NIPT or combined first-trimester screening (cFTS) [60]. NIPS is used extensively in the USA, yet it is not listed as a universal screening test in national guidelines. Because of this, between 25% and 50% of pregnant females in Italy, Spain, Netherlands, Austria, and most Australian and American states have NIPTs done. However, in Belgium, the uptake of testing exceeds 75%. For common autosomal and sex chromosome aneuploidies, NIPS is used in the majority of countries. On the other hand, NIPT is primarily available in the form of wider test (including microdeletions or full genome coverage) in certain countries like Belgium, Cyprus, Italy, Netherlands, Lithuania, and Greece. Despite the fact that NIPS is accessible in poor nations as well, a number of obstacles prevent its broad use. Important limiting constraints include availability, cost, and patients' and doctors' lack of awareness.

The following list includes a few instances of various emerging or developing country situations. In response to worries about middle-aged and older women intending to become pregnant again, the National Health Commission in China released clinical recommendations for NIPT at 12–22 weeks of gestation in the year 2016, the same year that China eliminated the one-child policy. Nevertheless, government support for universal NIPT is not available in China [61].

In Japan, NIPT and prenatal testing policies are quite clear. Furthermore, although NIPT is among the most expensive in the world, most people who require it pay for it out of pocket. Therefore, in Japan, NIPT is only advised for expectant mothers who are known to be at high risk of having anomalies in the fetus. This covers individuals who are expecting and have a mother age of 35 or older, those who have a history of having a child with known chromosomal aberration/aneuploidy, and those who have fetal chromosomal abnormalities found by maternal serum marker testing or soft markers on fetal ultrasound imaging [62, 63].

In India, NIPS is primarily used as an out-of-pocket expense and is not included in official guidelines for prenatal screening measures. The primary goal of the Preconception and Prenatal Diagnostic Techniques (PC-PNDT) Act which was initiated in 1994 is to stop parents or doctors from abusing prenatal screening methods for sex-selective abortion and fetal sex determination. This act regulates prenatal screening and diagnosis, including NIPT [64]. The T13, T18, T21, and microdeletions can all be screened for with NIPT; however, parents can only be informed about the sex of the fetus if prenatal testing reveals genetic abnormalities.

T13, T18, and T21 detection rates were higher in the NIPT screening group in a Belgian study comparing pregnant women who had primary NIPT, to combined first-trimester screening test, or triple test in second-trimester. Additionally, NIPT was shown to successfully cut the number of invasive tests required to identify these trisomies to half [65].

Prenatal diagnostic test accessibility differs across developed and developing nations, as well as between rural and urbanized regions. The lack of information, inadequate health care professionals' sensitization, infrastructural deficiencies, and, of course, the astronomical cost in comparison to other conventional screening methods are the primary causes of these discrepancies. Routine prenatal screening tests are inaccessible in rural locations and low-resource countries. Urbanized areas also have low levels of awareness of NIPS and the availability and timing of these specialized testing [66, 67].

15.9 Barriers to Implementation in Developing Countries

The main obstacle is that in many developing nations, there is less awareness and emphasis on screening for aneuploidies in pregnancy. People find these screening tests unacceptable, and one of the main factors is women in developing nations have children at a younger age. Compared to the 8.3% of women in the USA at the time of delivery, 2–5% of Indian women were over 35 [68]. The bulk of health care is supplied by commercial organizations or public-private partnerships, which bear the infrastructure cost and are often funded by grants from various national and municipal sources or by self-funding. There are only few centres or laboratories with the advanced technology needed for NIPS. Occasionally, some centers are unable to deliver the samples to labs equipped to carry out particular testing. The transport adds to exorbitant cost and hence becomes an additional limiting factor for widespread testing.

Genetic literacy, or knowledge of the tests, is another crucial component in underdeveloped nations, genetic counseling is only available to qualified professionals, most of whom work at tertiary care facilities in metropolitan areas. As a result, only a small portion of the population receives pre- and post-test support for interpreting results, which increases the risk of misinterpreting deviant results and, ultimately, inadequate clinical management [69].

Due to unequal insurance coverage, low-income families in developing nations have restricted access to NIPS in terms of financial constraints. Prenatal sonography is occasionally performed on the pregnant woman, who frequently arrives in the second trimester of her pregnancy. NIPS is more expensive than any other screening regimen in many developing nations, which puts a financial strain on low-income households. Obtaining NIPS would still be expensive, even if costs were lowered, and it might raise the disease burden among less fortunate groups. Programs for subsidies are required to deal with this problem. Legal limits on abortion and reproductive technology are among the legal factors to be taken into account, as are the absence of regulatory control and certification procedures for laboratories doing NIPT in underdeveloped nations. Abortion laws and the reasons behind them differ greatly throughout nations. Restrictive rules in certain countries raise maternal death rates because they make abortions risky.

Furthermore, in order to discourage sex-selective abortion, certain nations with a cultural predilection for male offspring have passed legislation making it illegal to disclose fetal sex information to expectant mothers; however, many developing nations do not regularly enforce these laws. Therefore, emerging nations must have more effective regulatory structures and effective oversight. There are practical and ethical issues with the growth of NIPS, especially in environments with limited resources.

These days, chromosomal microarray (CMA) is available, and NIPS is being compared to determine which is better. However, CMA necessitates invasive testing, which is limited to the use of amniotic fluid or chorionic villus samples. 1.4% of cases in a recent Israeli research had aberrant findings found by CMA [70]. More than 30% of these may have been selected by NIPS. Advanced maternal age and pregnancies with abnormal serum screening had greater detection rates of CMA (1.8% and 2.8%, respectively). Even though less than 1% of fetuses with normal ultrasound results can have chromosomal abnormalities, not all pregnancies can undergo invasive testing. In the presence of abnormal soft markers on ultrasonography or advanced maternal age, the yield of CMA is expected to be higher.

NIPS is not effective in detecting other microdeletion or microduplication syndromes, despite its great sensitivity in detecting common aneuploidies and the common microdeletion condition Di-George syndrome [71]. The detection of deletion duplications may be further improved with additional approach and/or analytic refinement. Nonetheless, an African study found that 8.94% of infants with congenital abnormalities had chromosomal problems; of these, Down syndrome accounted for a sizable share [72]. The latter can be identified prenatally on NIPS with a high sensitivity and specificity.

15.10 Strategies for Overcoming Barriers to NIPS Implementation in Developing Countries

All of these drawbacks may be mitigated and broad use of these straightforward, noninvasive, and efficient screening methods aided by systematic methodologies and strategies. It should be mentioned that if early care is given in response to the timely discovery of fetuses with aneuploidy or other genetic abnormalities, overall costs would be significantly lower. Affordable diagnostics for identifying single-gene diseases and chromosomal aneuploidies are also needed, as well as education and training for patients and medical personnel. NIPS ought to be implemented in public health sector organizations in low-income and developing nations, but doing so will necessitate funding for sequencing capacity and public health professional training [7]. To guarantee better prenatal care and health outcomes globally, all stakeholders should be involved in the global adoption of NIPS and confirmatory testing in the public health system, with guidelines tailored to each nation.

To increase awareness and NIPS uptake, several models with cultural and regional sensitivity should be created [7, 73]. More literate women had 15% more prenatal visits, according to a Pakistani study [7]. In nations where the majority of people have low or middle-class wages, the stigma attached to genetic diseases, religious concerns, and widely held cultural practices and beliefs can all have an impact on how acceptable genetic testing is. Therefore, when providing genetic testing or NIPS and counseling in various developing nations, all of these aspects should be taken into consideration.

15.11 Future Perspectives for the Implementation of NIPS in Developing Countries

As more and more data is coming and guidelines are becoming clear, there is an immense need to sensitize healthcare workers in a streamlined way so that this highly useful screening test can be implemented even in developing countries. As already discussed most important factor are lack of awareness of the availability and interpretation of NIPS in many regions across the world especially developing countries, more so in rural areas. As the fertility rate decreases, and the tendency for less number of children becomes widely accepted, campaigns to highlight the usefulness of NIPS to the masses would help overcome several barriers to implementing NIPS in developing countries [74].

Another major factor is the cost and availability of this test in several parts of the world. Inclusion of testing in Government set up for economically backward families or inclusion of the testing in the gamut of tests by health insurance companies would help overcome this major barrier. Negotiation of rates when used on a mass scale or availability on public–private partnerships in government healthcare institutions or hospitals can ease the cost factor.

Most of the common chromosomal disorders adding to the public health burden can be picked up early in pregnancy using NIPS. Once confirmatory tests establish a diagnosis, appropriate counseling for a couple can help make informed decisions. This is likely to be cost-effective if long-term health care expenditure is considered owing to the large amount of health budget going into the management of children with genetic disorders.

References

1. Graves JC, Miller KE, Sellers AD. Maternal serum triple analyte screening in pregnancy. Am Fam Physician. 2002;65(5):915–21.
2. Alfirevic Z, Navaratnam K, Mujezinovic F. Amniocentesis and chorionic villus sampling for prenatal diagnosis. Cochrane Database Syst Rev. 2017;9(9):CD003252.
3. Akolekar R, Beta J, Picciarelli G, Ogilvie C, D'Antonio F. Procedure-related risk of miscarriage following amniocentesis and chorionic villus sampling: a systematic review and meta-analysis. Ultrasound Obstet Gynecol. 2015;45(1):16–26.
4. Lo YM, Corbetta N, Chamberlain PF, Rai V, Sargent IL, Redman CW, et al. Presence of fetal DNA in maternal plasma and serum. Lancet Lond Engl. 1997;350(9076):485–7.
5. Fan HC, Blumenfeld YJ, Chitkara U, Hudgins L, Quake SR. Noninvasive diagnosis of fetal aneuploidy by shotgun sequencing DNA from maternal blood. Proc Natl Acad Sci U S A. 2008;105(42):16266–71.
6. Agarwal A, Sayres LC, Cho MK, Cook-Deegan R, Chandrasekharan S. Commercial landscape of noninvasive prenatal testing in the United States. Prenat Diagn. 2013;33(6):521–31.
7. Zhong A, Darren B, Loiseau B, He LQB, Chang T, Hill J, Dimaras H. Ethical, social, and cultural issues related to clinical genetic testing and counseling in low- and middle-income countries: a systematic review. Genet Med. 2021;23(12):2270–80.
8. Gil MM, Accurti V, Santacruz B, Plana MN, Nicolaides KH. Analysis of cell-free DNA in maternal blood in screening for aneuploidies: updated meta-analysis. Ultrasound Obstet Gynecol Off J Int Soc Ultrasound Obstet Gynecol. 2017;50(3):302–14.
9. Tabor A, Alfirevic Z. Update on procedure-related risks for prenatal diagnosis techniques. Fetal Diagn Ther. 2010;27(1):1–7.
10. Wulff CB, Gerds TA, Rode L, Ekelund CK, Petersen OB, Tabor A, et al. Risk of fetal loss associated with invasive testing following combined first-trimester screening for down syndrome: a national cohort of 147,987 singleton pregnancies. Ultrasound Obstet Gynecol Off J Int Soc Ultrasound Obstet Gynecol. 2016;47(1):38–44.
11. Anon. Practice Bulletin No. 163: screening for fetal aneuploidy. Obstet Gynecol. 2016;127(5):e123–37.
12. Suzumori N, Sekizawa A, Takeda E, Samura O, Sasaki A, Akaishi R, et al. Classification of factors involved in nonreportable results of noninvasive prenatal testing (NIPT) and prediction of success rate of second NIPT. Prenat Diagn. 2019;39(2):100–6.
13. Coppedè F, Bhaduri U, Stoccoro A, Nicolì V, Di Venere E, Merla G. DNA methylation in the fields of prenatal diagnosis and early detection of cancers. Int J Mol Sci. 2023;24(14):11715.
14. Hu P, Liang D, Chen Y, Lin Y, Qiao F, Li H, et al. An enrichment method to increase cell-free fetal DNA fraction and significantly reduce false negatives and test failures for non-invasive prenatal screening: a feasibility study. J Transl Med. 2019;17(1):124.

15. Sparks AB, Wang ET, Struble CA, Barrett W, Stokowski R, McBride C, et al. Selective analysis of cell-free DNA in maternal blood for evaluation of fetal trisomy. Prenat Diagn. 2012;32(1):3–9.
16. Luo Y, Hu H, Zhang R, Pan Y, Ma Y, Long Y, Xu J, Xu L, Hu B, Yao H, Chang Q. Factors affecting the failure of non-invasive prenatal testing and the feasibility analysis of retesting. Zhonghua Yi Xue Yi Chuan Xue Za Zhi. 2020;37(6):603–8.
17. Samura O, Okamoto A. Causes of aberrant non-invasive prenatal testing for aneuploidy: a systematic review. Taiwan J Obstet Gynecol. 2020;59(1):16–20.
18. Deng C, Liu S. Factors affecting the fetal fraction in noninvasive prenatal screening: a review. Front Pediatr. 2022;10:812781.
19. Benn P, Rebarber A. Non-invasive prenatal testing in the management of twin pregnancies. Prenat Diagn. 2021;41(10):1233–40.
20. Ormstad SS, Stoinska-Schneider A, Solberg B, Fure B, Juvet LK. Non-invasive prenatal testing (NIPT) for fetal sex determination. Health technology assessment. Oslo, Norway: Knowledge Centre for the Health Services at the Norwegian Institute of Public Health (NIPH); 2016. Report from the Norwegian Institute of Public Health No. 2016-03
21. Lu X, Wang C, Sun Y, Tang J, Tong K, Zhu J. Noninvasive prenatal testing for assessing foetal sex chromosome aneuploidy: a retrospective study of 45,773 cases. Mol Cytogenet. 2021;14(1):1.
22. Lewis C, Hill M, Skirton H, Chitty LS. Non-invasive prenatal diagnosis for fetal sex determination: benefits and disadvantages from the service users' perspective. Eur J Hum Genet. 2012;20(11):1127–33.
23. Runkel B, Bein G, Sieben W, Sow D, Polus S, Fleer D. Targeted antenatal anti-D prophylaxis for RhD-negative pregnant women: a systematic review. BMC Pregnancy Childbirth. 2020;20(1):83.
24. Yang H, Llewellyn A, Walker R, Harden M, Saramago P, Griffin S, et al. High-throughput, non-invasive prenatal testing for fetal rhesus D status in RhD-negative women: a systematic review and meta-analysis. BMC Med. 2019;17(1):37.
25. Eryilmaz M, Müller D, Rink G, Klüter H, Bugert P. Introduction of noninvasive prenatal testing for blood group and platelet antigens from cell-free plasma DNA using digital PCR. Transfus Med Hemother. 2019;47(4):292–301.
26. Alford B, Landry BP, Hou S, Bower X, Bueno AM, Chen D, et al. Validation of a non-invasive prenatal test for fetal RhD, C, c, E, K and Fya antigens. Sci Rep. 2023;13(1):12786.
27. Lo YMD. Recent developments in fetal nucleic acids in maternal plasma: implications to non-invasive prenatal fetal blood group genotyping. Transfus Clin Biol J Soc Francaise Transfus Sang. 2006;13(1–2):50–2.
28. Christiaens L, Chitty LS, Langlois S. Current controversies in prenatal diagnosis: expanded NIPT that includes conditions other than trisomies 13, 18, and 21 should be offered. Prenat Diagn. 2021;41(10):1316–23.
29. Van Opstal D, van Maarle MC, Lichtenbelt K, Weiss MM, Schuring-Blom H, Bhola SL, et al. Origin and clinical relevance of chromosomal aberrations other than the common trisomies detected by genome-wide NIPS: results of the TRIDENT study. Genet Med. 2018;20(5):480–5.
30. Pertile MD, Halks-Miller M, Flowers N, Barbacioru C, Kinnings SL, Vavrek D, et al. Rare autosomal trisomies, revealed by maternal plasma DNA sequencing, suggest increased risk of feto-placental disease. Sci Transl Med. 2017;9(405):eaan1240.
31. Deans Z, Hill M, Chitty LS, Lewis C. Non-invasive prenatal testing for single gene disorders: exploring the ethics. Eur J Hum Genet. 2013;21(7):713–8.
32. Adams S, Llorins H, Maher O, Dean M, Dobson LJ, Gbur S, et al. Single gene non-invasive prenatal screening for autosomal dominant conditions in a high-risk cohort. Prenat Diagn. 2023;43(9):1110–9.
33. Pacault M, Verebi C, Champion M, Orhant L, Perrier A, Girodon E, et al. Non-invasive prenatal diagnosis of single gene disorders with enhanced relative haplotype dosage analysis for diagnostic implementation. PLoS One. 2023;18(4):e0280976.

34. Mohan P, Parmar S, Saucier J, Jelsema R, Eaves K, Thomsen A, et al. 912: skeletal dysplasias screening by NIPT for single-gene disorders: clinical value of narrowing the differential diagnosis. Am J Obstet Gynecol. 2019;220(1):S589.
35. Jay AM, Mason E, Lebovic D, Chuba P. More than an incidentaloma: the nonreportable NIPT. Case Rep Obstet Gynecol. 2022;2022:2496057.
36. Qiao L, Zhang Q, Liang Y, Gao A, Ding Y, Zhao N, et al. Sequencing of short cfDNA fragments in NIPT improves fetal fraction with higher maternal BMI and early gestational age. Am J Transl Res. 2019;11(7):4450–9.
37. Juul LA, Hartwig TS, Ambye L, Sørensen S, Jørgensen FS. Noninvasive prenatal testing and maternal obesity: a review. Acta Obstet Gynecol Scand. 2020;99(6):744–50.
38. Bianchi DW, Chudova D, Sehnert AJ, Bhatt S, Murray K, Prosen TL, et al. Noninvasive prenatal testing and incidental detection of occult maternal malignancies. JAMA. 2015;314(2):162–9.
39. Romero R, Mahoney MJ. Noninvasive prenatal testing and detection of maternal cancer. JAMA. 2015;314(2):131–3.
40. Kater-Kuipers A, de Beaufort ID, Galjaard RJH, Bunnik EM. Rethinking counselling in prenatal screening: an ethical analysis of informed consent in the context of non-invasive prenatal testing (NIPT). Bioethics. 2020;34(7):671–8.
41. Bedei I, Wolter A, Weber A, Signore F, Axt-Fliedner R. Chances and challenges of new genetic screening technologies (NIPT) in prenatal medicine from a clinical perspective: a narrative review. Genes. 2021;12(4):501.
42. García-Pérez L, Linertová R, Álvarez-de-la-Rosa M, Bayón JC, Imaz-Iglesia I, Ferrer-Rodríguez J, et al. Cost-effectiveness of cell-free DNA in maternal blood testing for prenatal detection of trisomy 21, 18 and 13: a systematic review. Eur J Health Econ. 2018;19(7):979–91.
43. Zhang W, Mohammadi T, Sou J, Anis AH. Cost-effectiveness of prenatal screening and diagnostic strategies for down syndrome: a microsimulation modeling analysis. PLoS One. 2019;14(12):e0225281.
44. Luo Y, Jia B, Yan K, Liu S, Song X, Chen M, et al. Pilot study of a novel multi-functional noninvasive prenatal test on fetus aneuploidy, copy number variation, and single-gene disorder screening. Mol Genet Genomic Med. 2019;7(4):e00597.
45. Shaw J, Scotchman E, Chandler N, Chitty LS. Preimplantation genetic testing: non-invasive prenatal testing for aneuploidy, copy-number variants and single-gene disorders. Reprod Camb Engl. 2020;160(5):A1–11.
46. Guo FF, Yang JX, Huang YL, Qi YM, Hou YP, Peng HS, et al. Association between fetal fraction at the second trimester and subsequent spontaneous preterm birth. Prenat Diagn. 2019;39(13):1191–7.
47. Dungan JS, Klugman S, Darilek S, Malinowski J, Akkari YMN, Monaghan KG, et al. Noninvasive prenatal screening (NIPS) for fetal chromosome abnormalities in a general-risk population: an evidence-based clinical guideline of the American College of Medical Genetics and Genomics (ACMG). Genet Med. 2023;25(2):100336.
48. Kliegman RM, St Geme JW. Nelson textbook of pediatrics. 21st ed. Elsevier Health Sciences; 2019. p. 3063–80.
49. Audibert F, De Bie I, Johnson JA, Okun N, Wilson RD, Armour C, et al. No. 348-joint SOGC-CCMG guideline: update on prenatal screening for fetal aneuploidy, fetal anomalies, and adverse pregnancy outcomes. J Obstet Gynaecol Can. 2017;39(9):805–17.
50. Early pregnancy loss. [cited 2023 Nov 23]. Available from: https://www.acog.org/clinical/clinical-guidance/practice-bulletin/articles/2018/11/early-pregnancy-loss.
51. Kim SM, Kim HH, Han YJ, Choi JS, Ryu HM, Yang S, et al. Change in rates of prenatal tests for chromosomal abnormality over a 12-year period in women of advanced maternal age. Obstet Gynecol Sci. 2018;61(4):453–60.
52. Verma IC. Noninvasive prenatal testing: the Indian perspective. J Fetal Med. 2014;1(3):113–8.
53. Liu Y, Jing X, Xing L, Liu S, Liu J, Cheng J, Deng C, Bai T, Xia T, Wei X, Luo Y, Zhou Q, Zhu Q, Liu H. Noninvasive prenatal screening based on second-trimester ultrasonographic soft markers in low-risk pregnant women. Front Genet. 2021;12:793894.

54. Liu C, Zhou Y, Liu P, Geng Y, Zhang H, Dun Y, et al. Application of ultrasound combined with noninvasive prenatal testing in prenatal testing. Transl Pediatr. 2022;11(1):85–98.
55. Media BKK. Publications & Guidelines | SMFM.org—The Society for Maternal-Fetal Medicine. [cited 2023 Nov 25]. Available from: https://www.smfm.org/publications/394-smfm-consult-series-57-evaluation-and-management-of-isolated-soft-ultrasound-markers-for-aneuploidy-in-the-second-trimester.
56. Bianchi DW, Rava RP, Sehnert AJ. DNA sequencing versus standard prenatal aneuploidy screening. N Engl J Med. 2014;371(6):578.
57. Gil MM, Quezada MS, Revello R, Akolekar R, Nicolaides KH. Analysis of cell-free DNA in maternal blood in screening for fetal aneuploidies: updated meta-analysis. Ultrasound Obstet Gynecol Off J Int Soc Ultrasound Obstet Gynecol. 2015;45(3):249–66.
58. Gadsbøll K, Petersen OB, Gatinois V, Strange H, Jacobsson B, Wapner R, et al. Current use of noninvasive prenatal testing in Europe, Australia and the USA: a graphical presentation. Acta Obstet Gynecol Scand. 2020;99(6):722–30.
59. Lannoo L, van der Meij KRM, Bekker MN, De Catte L, Deckers S, Devreindt K, et al. A cross-country comparison of pregnant women's decision-making and perspectives when opting for non-invasive prenatal testing in The Netherlands and Belgium. Prenat Diagn. 2023;43(3):294–303.
60. Bowman-Smart H, Savulescu J, Mand C, Gyngell C, Pertile MD, Lewis S, et al. 'Small cost to pay for peace of mind': Women's experiences with non-invasive prenatal testing. Aust N Z J Obstet Gynaecol. 2019;59(5):649–55.
61. Wu L, Wu Y, Zou S, Sun C, Chen J, Li X, et al. Eliciting women's preference for prenatal testing in China: a discrete choice experiment. BMC Pregnancy Childbirth. 2020;20(1):604.
62. Takahashi M, Linh LK, Sayed AM, Imoto A, Sato M, Dila KAS, et al. Non-invasive prenatal testing (NIPT) implementation in Japan: a comparison with the United Kingdom, Germany, Italy, Sweden, and Taiwan. Int J Environ Res Public Health. 2022;19(24):16404.
63. Kimura M. Experiences related to prenatal testing among Japanese mothers of children with disabilities. Women's Reprod Health. 2018;5(3):183–203.
64. Aggarwal S, Phadke SR. Medical genetics and genomic medicine in India: current status and opportunities ahead. Mol Genet Genomic Med. 2015;3(3):160–71.
65. Lee DE, Kim H, Park J, Yun T, Park DY, Kim M, et al. Clinical validation of non-invasive prenatal testing for fetal common aneuploidies in 1,055 Korean pregnant women: a single center experience. J Korean Med Sci. 2019;34(24):e172.
66. Lu J, Shen JJ, Chen G, Moseley CB, Sun M, Gao F, et al. Regional disparities in prenatal care services in rural China. Asia Pac J Public Health. 2011;23(5):682–9.
67. Tischler R, Hudgins L, Blumenfeld YJ, Greely HT, Ormond KE. Noninvasive prenatal diagnosis: pregnant women's interest and expected uptake. Prenat Diagn. 2011;31(13):1292–9.
68. Matthews TJ, Hamilton BE. Delayed childbearing: more women are having their first child later in life. NCHS Data Brief. 2009;21:1–8.
69. Christianson A, Modell B. Medical genetics in developing countries. Annu Rev Genomics Hum Genet. 2004;5:219–65.
70. Sagi-Dain L, Cohen Vig L, Kahana S, Yacobson S, Tenne T, Agmon-Fishman I, et al. Chromosomal microarray vs. NIPS: analysis of 5541 low-risk pregnancies. Genet Med. 2019;21(11):2462–7.
71. Liang D, Cram DS, Tan H, Linpeng S, Liu Y, Sun H, et al. Clinical utility of noninvasive prenatal screening for expanded chromosome disease syndromes. Genet Med. 2019;21(9):1998–2006.
72. Gebremeskel Aragie T, Seyoum GG. Proportion of chromosomal disorders and their patterns among births with congenital anomalies in Africa: a systematic review and meta-analyses. ScientificWorldJournal. 2022;2022:6477596.
73. Aslam M, Sadiq M, Mehmood T. Assessment of maternal health services utilization in Pakistan: the role of socio-demographic characteristics. Asian Biomed. 2020;14(1):3–7.
74. Allyse M, Minear MA, Berson E, Sridhar S, Rote M, Hung A, Chandrasekharan S. Non-invasive prenatal testing: a review of international implementation and challenges. Int J Womens Health. 2015;7:113–26.

Sequencing: A Promising Path in the Detection of Fetal Health

16

Indu Sharma, Neha Negi, and Subhas Chandra Saha

16.1 Introduction

Congenital anomalies cause anatomical or functional problems, affecting 2–4% of pregnancies [1]. According to estimates from the World Health Organisation (WHO), congenital anomalies contributed to 270,000 worldwide deaths in the initial 28 days of birth in 2010 [2]. According to global research on birth defects published by March of Dimes (MOD), nearly eight million infants annually (or 6% of all pregnancies) have serious birth defects, with 94% occurring in countries with lower or economically disadvantaged nations. According to a report from a joint meeting of the WHO and MOD, states that 3.3 million under-five fatalities and 7% of all newborn deaths are caused by birth abnormalities [3].

Based on statistics, there are over 1.7 million birth problems per year in India, where the prevalence of birth malformations is 6–7%. The most frequent birth defects are neural tube defects (4.11–11.4 per 1000 live births), Congenital hearing loss (5.6–10 per 1000 live), and congenital cardiac disease (8–10 per 1000 live births). Although approximately 50% of the time, certain recognized risk factors can be linked to the cause of malformation, it is challenging to pinpoint the etiology of congenital anomalies. Congenital anomalies can be caused by a variety of factors, including specific gene deficiencies, chromosomal variations, multivariate lineage, extrinsic contaminants (substances that may trigger congenital anomalies), and a lack of nutrients [4, 5].

Over the past few years, the arsenal of technology at our disposal for performing prenatal genetic diagnosis on fetuses has grown quickly. The prognosis of the fetus can be determined with the help of genetic information, which can also be used to guide

I. Sharma (✉) · N Negi
Department of Zoology, Panjab University, Chandigarh, India
e-mail: indu2702@pu.ac.in

S. C. Saha
Deaprtment of Obstetrics and Gynaecology, PGIMER, Chandigarh, India

R. A. Rather, S. C. Saha (eds.), *Non-invasive Prenatal Screening (NIPS) in Clinical Practice*, https://doi.org/10.1007/978-981-97-6402-0_16

prenatal care decisions, such as those relating to prenatal care, delivery preparation, and infant care, which may lower fatalities and morbidity. Targeted testing (QF-PCR, gene panels) and genomic techniques with various resolution levels are part of the prenatal genetic toolbox (karyotype, chromosomal microarray, whole genome, and whole exome sequencing). Chromosomal research using chromosomal microarray analysis (CMA), fluorescent in situ hybridization (FISH), and karyotyping are available options for focused gene-specific molecular testing for suspected illnesses [6, 7].

Traditionally prenatal diagnosis of monogenic diseases depends on fetal DNA collected using invasive procedures that appear to be harmful to the fetus including chorionic villous sampling (CVS) and amniocentesis [8]. Recently noninvasive prenatal screening (NIPS) has become an appealing and trustworthy way to find foetal abnormalities. Pregnant women's peripheral blood has been found to contain fetal DNA fragments, which account for 20–30% of all cell-free fetal DNA (cffDNA), leading to the development of NIPS [9]. Broader use of NIPS in clinical practice has been made possible by identifying cffDNA in conjunction with advancing the NGS technique. Additionally, extracted fetal-specific data from a maternal background can be used by researchers to assess copy numbers and nucleotide sequences [10–12].

16.2 Advancements in Sequencing for Fetal Health Detection

Before the early 2000s, prenatal genetic testing and screening aimed to identify women who were more likely to give birth to infants with trisomies 13, 18, and 21—common aneuploidies. As screening methods, a variety of permutations based on family history, mother's age, concentrations of specific mother's blood analytes, plus outcomes from gestational ultrasounds were employed. [13], but during the past decade, CMA has become more widely used in prenatal diagnoses [14].

To assess and clinically prioritize prenatal defects, genetic investigations are crucial. Several approaches are being used to systematically evaluate fetal material collected using invasive methods such as CMA [15]. Even though karyotyping is still the predominant cytogenetic approach for prenatal diagnosis, CMA are favored when a fetus has one or more substantial structural abnormalities that have been detected by ultrasound since it has a greater resolution. Karyotyping can help with up to 30% of diagnoses, and CMA can help with an additional 4–6% [16]. The gradual acceptance of NGS over the past 10 years has resulted in a shift in the way that genetic diagnoses are made, whether through exome sequencing (ES), a targeted gene panel strategy, or an extensive genomic approach, depending on whether an identification control can be found.

16.3 NIPT-Based Cell-Free Fetal DNA (cffDNA) Screening

Modern genomic medicine has succeeded in sequencing cffDNA fragments that circulate throughout the bloodstream of pregnant women. This technique is used to detect fetal chromosomal abnormalities. Prenatal cffDNA screening has transformed prenatal care on a global scale in <10 years, starting with tiny, proof-of-concept research.

The sequencing of cffDNA was initially investigated in pregnant women who were at a high risk of giving birth to a child with trisomy 13, 18, or 21. According to a recent meta-analysis, DNA from the plasma of 4–6 million pregnant women have been examined to check for prenatal aneuploidy and the sensitivity and specificity for detecting trisomy 21 in vulnerable women were 99.7%, accordingly, for trisomy 18, 93% and 99.7%, and 95% and 99.9% for trisomy 13. The positive predictive value for trisomy 21 is 91%, compared to T18 (84%) and T13 (87%) [17–19]. NIPT is currently routinely used to screen for the trisomies because of its great precision and susceptibility (Fig. 16.1).

16.3.1 Genome-Wide Sequencing

Fetal morphological deformities can be caused by any sort of molecular shifts, notably aneuploidy, uniparental disomy, CNVs, as well as intragenic alterations. Congenital abnormalities found during pregnancy are a growing area of interest for genome-wide sequencing techniques because of their reduced cost, the need for less fetal DNA, the potential for relatively quick turnaround, and comprehensive DNA sequencing. Previously, prenatal whole-genome sequencing (WGS) was used, but nowadays whole-exome sequencing (WES) and targeted gene panels have received greater interest [7, 20].

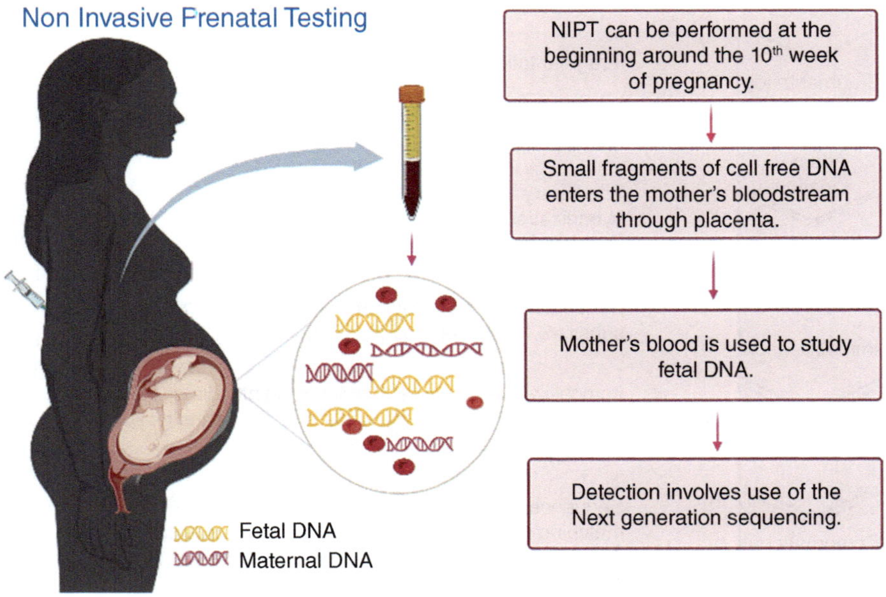

Fig. 16.1 Noninvasive prenatal testing (NIPT)

16.3.2 Next-Generation Sequencing (NGS)

NGS a, high-throughput sequencing, is used to describe several modern sequencing technologies. Because of NGS's versatility, prenatal diagnostics for Mendelian and chromosomal illnesses have been created and swiftly incorporated into the medical setting. This method is regarded as a game-changing innovation that has fundamentally altered prenatal diagnosis and detection of genomic and other inherited disorders. According to studies targeted, NGS is helpful for detecting individuals with spontaneous, congenital heart defects, which are the most common form of birth defect and are believed to account for 30–50% of premature death in infants [21]. This technique is also used for detecting copy number variants (CNVs). Recent research performed a large invasive CNV-seq cohort, and the results showed that CNV-seq had a high level of accuracy and reliability for detecting clinically relevant CNVs associated with fetal abnormalities in prenatal samples [22] (Fig. 16.2).

16.3.3 Whole-Exome Sequencing (WES)

Exome sequencing does not need to know the putative gene responsible for an observed trait. As a result, it appears to be in a very advantageous position for pregnancy. Its use in adult and pediatric genomics is universally recognized as a powerful tool, and it has increasingly migrated to the field of prenatal diagnosis as well. Using WGS and exome sequencing (ES), DNA is thoroughly examined down to a single base pair without any prior selection, DNA's coding (exons) and noncoding

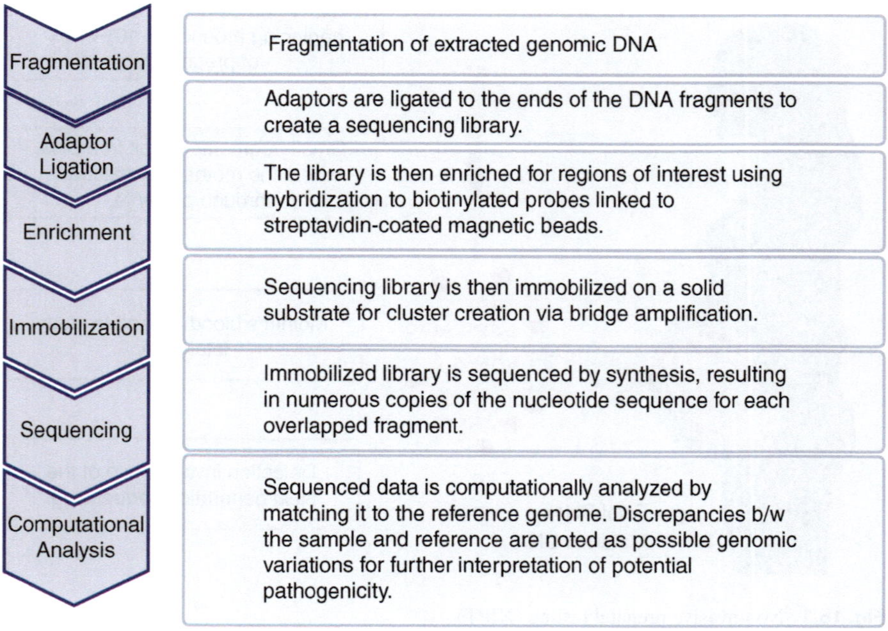

Fig. 16.2 Basic steps of NGS

(introns) regions are both sequenced during WGS. Fragments of DNA from the "protein-encoding" exons, which make up 1–2% of the genome and include over 85% of all disease-causing pathogenic mutations, are collected for ES prior to sequencing. Exome sequencing has revolutionized postnatal genetic diagnosis, but its value for prenatal screening is yet nascent. ES is successfully used in the clinical environment for prenatal anomalies and received widespread approval [23, 24].

ES is a modern substitute for identifying fetuses with structural abnormalities. It may be limited to the study of known disease-causing genes' coding regions (clinical exome) or may also contain genes with questionable clinical significance. WES has been shown to be a useful tool for prenatal diagnosis because of its capacity to detect single-nucleotide variations (SNV), minute insertions/deletions (InDel), and CNVs, which involve multiple exons. Given this, exonic CNVs with SNV/InDel participation in a trans-phase may result in genetic disorders. When karyotype analysis and CMA produce negative results for fetal structural abnormalities, WES yields a diagnostic yield of 8.5–10% (Monaghan et al. 2020). ES may be considered a new diagnostic tool in fetal medicine when an assessment is futile in a developing baby exhibiting several serious defects using conventional prenatal methods. With information from prenatal ES, an array of abnormalities in uncommon diseases or even well-known familial conditions without recorded fetal characteristics may be better understood [22]. According to studies, the overall detection output for prenatal testing using microarray (CMA), ES, and NGS technology (gene panels) was 30.4%; NGS provided 42.2% of diagnoses via gene panels and 16.6% through ES and CMA provides 43.2% of diagnoses [1] (Fig. 16.3).

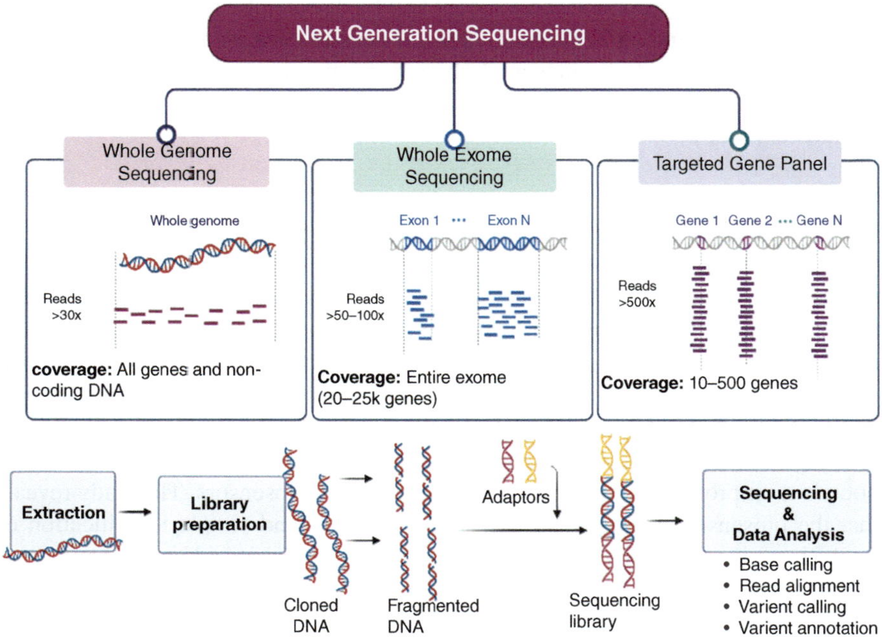

Fig. 16.3 NGS types and process or various steps

16.4 Sequencing Compared to Other Prenatal Screening Methods

16.4.1 Ultrasound

With improvements in ultrasound technology, structural defects have reportedly been shown to be detectable by ultrasonography as early as 11 weeks or at the end of pregnancy. A previous study found that screening using ultrasonography during the third trimester helped identify 33.35% of fetuses with structural defects [25]. With continuously improving resolution, prenatal ultrasonography is currently acknowledged in obstetrics as a screening method that can identify fetal structural abnormalities in about 5% of deliveries. Ultrasound can detect prenatal structural abnormalities in about 3% of pregnancies, which may range from a single tiny error to major multisystem deformities which can be lethal [26].

16.4.2 Karyotyping and Chromosomal Microarray Analysis (CMA)

The two main techniques for detecting chromosomal aberrations over the past 10 years are G-banding karyotyping and CMA. G-banded karyotyping has traditionally been used in prenatal diagnosis to identify chromosomal abnormalities. Studies show that karyotyping and chromosomal microarray analysis add a 6% yield in diagnosing fetal abnormalities [27]. The fetuses with aberrant ultrasonography findings had a 32% diagnostic accuracy for the karyotype. Furthermore, aneuploidy, translocation, and genomic inversion could all be detected by karyotypes [22].

When CMA came into existence, prenatal genetic testing entered the "genomic era," which can detect chromosomal aberrations and aneuploidies. When a pregnancy exhibits abnormalities on an ultrasonogram, CMA is typically employed diagnostic [28, 29].

16.4.3 Determination of Fetal Rh Status via Biosensor Technology

Currently, a variety of healthcare settings use biosensors extensively. Glucometers and pregnancy tests are two prominent examples of highly effective biosensor technologies. In immunological biosensors, various transducing methods are applied depending on the signal production following the creation of antigen-antibody complexes (for instance, a visual or biochemical stimulus). The transducer can be paired with high-affinity agents like immunoglobulins enzymatic agents and artificial compounds to improve the specificity associated with the biosensors. The study reveals that the biosensor technique for the quick, accurate, and simple identification of fetal Rh-positive antigens in pregnant women with Rh-negative blood is an alternative to managing Rh incompatibility. Early in pregnancy, the established biosensor screening for Rh antigen can detect fetal Rh antigens in the mother's blood (8th week of pregnancy). With the aid of a Surface plasmon resonance (SPR)-based biosensor, researchers set up a study to identify fetal Rh genotyping from cfDNA [30, 31]. These biosensor technologies have greater benefits than NIPD of Rh using real-time qPCR since it uses fewer samples, is less expensive, and can diagnose the Rh status of a fetus quickly [31].

16.4.4 Noninvasive Prenatal Testing (NIPT)

The analysis of cffDNA using NGS technology opens the possibility of adopting genomics-based noninvasive prenatal testing (gNIPT) potentially a more reliable diagnostic method to detect fetal chromosomal problems in maternal blood. A billion of these strands of DNA may now be sequenced simultaneously, enabling the development of a nonintrusive more efficient psychological tool that may be able to detect fetal aneuploidies with greater precision and sooner than existing screening methods [32].

International medical organizations such as the American College of Obstetrics and Gynaecology and the American College of Medical Genetics and Genomics encourage making genetic testing during pregnancy accessible to all pregnant women, regardless of the stage of pregnancy and mother age, and this includes examining with NIPT and nonintrusive diagnostics [33].

16.5 Applications of Sequencing in Prenatal Screening

16.5.1 Detection of Chromosomal Abnormalities

In the past 50 years, a lot of changes have been made to chromosomal abnormality screening, particularly for trisomy 21. Throughout the 1970s and 1980s, the standard was based on the monitoring of the mother's age; however, biochemical testing started to be used in the 1990s. Combination first-trimester screenings between 6 and 11 weeks of conception were the most popular and effective method of screening until 2010. A disruptive diagnostic (such as amniocentesis) is offered to pregnant women who are found to be at high risk of fetal aneuploidy although there is a probability of miscarriage. The present-day preferred method for screening is cell-free fetal DNA analysis.

Chromosomal abnormalities are one of the most severe birth disorders. Because of no effective treatment, they frequently harm the fetus significantly. Common fetal aneuploidies include Down's syndrome (trisomy 21), Edward's syndrome (trisomy 18), Patau syndrome (trisomy 13), Turner's syndrome (45, X), Klinefelter Syndrome (47, XXY), triple-X syndrome (47, XXX), and 47, XYY syndrome [34] care in many countries, even though the biochemical and ultrasound tests now in use have large rates of false negatives and false positives. CffDNA screening is currently the most reliable nonintrusive approach to estimating the probability of trisomy. Although cffDNA technology is convincing when compared to the other scanning methods (like first-trimester screenings) for trisomy, it must be emphasized that the approach has shortcomings, such as the diagnostic failure rate of roughly 3% [32, 35].

Studies have demonstrated that NIPT can be used to detect sex chromosome aneuploidy (SCAs) by examining cffDNA from the maternal circulation via extremely sequential WGS. The NIPT's overall accuracy in diagnosing SCAs was 54.5% [36]. Due to the lack of fetal-specific genetic markers and the possibility for higher test accuracy compared to current fetal aneuploidy screening methods, these advances in NGS have made it feasible for the establishment of gNIPT [32]. The accuracy of the gNIPT as a fetal screening tool is evaluated for the clinical diagnosis to determine the risk of prenatal anomalies in fetuses so that one can avoid invasive procedures [37].

16.5.2 Noninvasive Prenatal Testing (NIPT) for Trisomies

Recent changes to screening and antenatal care have been brought about by the development in NIPT. The therapeutic application of NIPT in prevalent trisomies and sex chromosomal aneuploidies has been the subject of numerous investigations during the past 5 years. Trisomy 21 has a very high detection rate (DR) (above 95% in all studies), while the accuracy of the DR for trisomy 18 and trisomy 13 has been questioned, particularly if one looks at subgroups rather than high-risk populations [38]. Additionally, CMA and amniocentesis were used to analyze fetuses with suspected CNVs via NIPT by NGS. These studies suggest that birth defects may be less common and can be accurately recognized with NIPT by NGS [39].

16.5.3 Noninvasive Diagnosis of Fetal Single-Gene Disorders

Using cffDNA analysis, individuals who have been identified to have a risk due to their past or present family history can also conduct a noninvasive test for fetal single-gene disorders. This screening is based on the direct detection of mutant DNA sequences that are either de novo or inherited from the father. This opens a new tab in maternal plasma cffDNA sequencing [10, 11]. Mother's plasma cffDNA sequencing is a significant advancement in genetic testing, leading to more accurate screening and fewer intrusive procedures. NIPS for fetal congenital anomalies by analyzing cffDNA in maternal plasma is now made available to millions of women worldwide each year.

Despite the benefits that cffDNA brings in terms of detecting typical trisomies, it ought to be recognized that it is not a complete solution for many other chromosomal abnormalities and any congenital birth defects.

16.5.4 Maternal-Fetal Genotyping for Rh Incompatibility

Rhesus (Rh) incompatibility continues to pose serious obstetric problems. When maternal and fetal Rh types are not compatible, this is referred to as rhesus (Rh) incompatibility. The frequency of Rh-negative status varies with ethnicity. The prevalence is highest in Caucasian people (about 15%), lowest in black African communities (between 3% and 5%), around 1% in Native American populations in North America, and extremely uncommon in East Asian cultures. Almost 60% of the estimated 68,000 Rh-negative pregnancies in Canada each year are thought to result in a Rh-positive fetus [40].

Serological screening (blood test) is frequently used in pregnancy care to ascertain the mother's Rh blood type, but it has traditionally been challenging to ascertain the fetus's Rh blood type prior to birth. Rh incompatibility develops whenever an expectant mother's blood type is R-negative while the embryo is Rh-positive then in that case the blood type incompatibility may cause the immune system of the mother to produce antibodies that might attack the embryo's cells in a subsequent pregnancy. Rh-negative pregnant women are often given an injection of Rh immunoglobulin (RhIG) prophylaxis, which inhibits the production of antibodies, in order to eliminate this risk to subsequent pregnancies [41]. Pregnancy-related Rh blood group incompatibility can have a major negative impact on the fetus's health. In order to manage Rh incompatibility and identify fetuses at the peril of hemolytic illness (HDFN) which leads to death and disability in fetuses [42]. Almost 1 in 21,000 live newborns have the hemolytic illness. It affects about 4000 fetuses worldwide every year, 15% of which die before delivery [40]. Around 10% of at-risk fetuses experience severe fetal anemia, which needs an intrauterine blood transfusion and is detected by invasive prenatal diagnostics (such as cordocentesis and amniocentesis). Thus, prenatal diagnosis of the fetal Rh status might be helpful.

Most approaches used to ascertain a fetus's Rh status up until this point were based on delivery-based serological procedures. Fetal Rh blood type genotyping is a noninvasive technique that accurately identifies Rh incompatibility and provides care recommendations for Rh-negative pregnancies. Noninvasive fetal Rh genotyping is more affordable and more efficient for managing alloimmunized pregnancies than standard care. And, semi-nested PCR has accuracy and specificity of 81% and 100% in detecting fetal Rh-positive pregnant women, respectively [43]. Recently the noninvasive approach relying on the qPCR measurement of prenatal Rh status of cffDNA has become a powerful option in the early stages of pregnancy. Approx. 40% of Rh-negative pregnant women, NIPD of cffDNA in maternal plasma could prevent the needless prescription of anti-D immunoglobulin [44]. For more targeted perinatal monitoring and enhanced Rh prophylactic effectiveness, it might be suggested routinely to all Rh-negative expectant women to have the NIPD for fetal Rh genotype determination [45]. However, there are two limitations on fetal blood types in NIPD. First, sequencing errors (unexpected base calls), which can be caused by PCR amplification or sequencing artifacts, reduce sensitivity. When homopolymer sections are sequenced, sequencing errors are more likely to occur [44].

16.6 Sequencing and Recurrence

Genetic disorders can be caused by large or minor chromosomal anomalies (duplications or deletions), or at the gene level where even a single base alteration can result in serious illness [46]. Depending on the circumstances, like in the case of the apparent de novo mutation (DNM), the couples are currently given a generic, population-average recurrence risk of 1–2% [47]. According to the studies, a family with a child who has complete trisomy 18 has an estimated recurrence risk of just 1%. And in some studies, recurrent trisomies have been linked to parental mosaic trisomy. It is extremely uncommon for individuals without chromosomal mosaicism to have successively autosomal trisomy pregnancies. For an immediate prenatal diagnosis of recurrent trisomy 18, the first trimester NIPT is beneficial, and the QF-PCR serves the purpose of a quick confirmation of recurrent trisomy 18 of maternal origin, which is crucial for genetics. In order to rule out the potential of parental mosaicism for trisomies [48], a cytogenetic examination of the parental karyotypes should be part of the prenatal diagnosis of recurrent trisomies counselling [18, 19].

While NIPT has proven effective in identifying these chromosomal anomalies, its focus is on diseases brought on by aneuploidies not specifically for the recurring genetic disorders caused by mutations in a specific gene. These disorders often have a complex genetic basis and may require targeted testing or genetic sequencing techniques to identify specific mutations. Studies indicate that ES may allow the diagnosis of pathogenic gene alterations and, as a result, strengthen the prognostic profile, allowing one to forecast the likelihood of recurrence in following pregnancies. It might also make it possible to distinguish between hereditary and de novo mutations and eliminate anomalies. Since ES focuses on coding regions and has a diagnostic yield of around 25% yet it still contains roughly 85% of the mutations that are known to cause genetic diseases [49].

It appears likely that the future of prenatal genetic diagnostics will be next-generation sequencing. WGS/ES or comparable technologies could be included in NIPT or other prenatal screening tests to give a more in-depth assessment of a fetus's genetic health. The likelihood of recurring disorder detection through prenatal testing will depend on further advances in genetic sequencing technologies, prenatal testing ethics and developments in our understanding of hereditary illnesses.

16.7 Challenges in the Implementation of Sequencing for Fetal Health Detection

16.7.1 Limitations of Sequencing for Fetal Health Detection

Between 2% and 5% of pregnancies are found to have fetal abnormalities, which account for 20% of prenatal fatalities [50]. Depending on the types of anomalies found, whether they are single or many, or any potential genetic causes, different pregnancy outcomes and prognoses can be expected. Currently utilized cytogenetic tests for prenatal diagnostics have limited resolution. Over 30% of dysmorphic fetuses have trisomy, which is detected by qPCR, and according to karyotyping, 5% have harmful imbalanced chromosomal rearrangements [51]. CMA detects extra harmful copy number alterations in 3–6.5% of cases for structurally defective fetuses with normal karyotypes. Near 40% of dysmorphic fetuses can have an underlying genetic cause determined using such tests in combination, although most cases still go undetected [52].

Technologies that analyze the entire genome at the molecular levels include genome sequencing and exome sequencing. The evaluation of a genome's functional and nonfunctional sections is necessary for genome sequencing, yet it can be challenging to sequence and fully examine some areas of the genome. A mere 1–2% of the genome, or more than 20,000 genes, contain the protein-coding areas where ES is restricted. GS demands additional data analytics even though its scope may make it more insightful (Monaghan et al. 2020). However, whole-exome sequencing and next-generation sequencing (NGS) allow for the quick study of many gene sequences and make it easier to find new genes that cause numerous hereditary illnesses.

Another challenging gene to sequence is one that is found in a family of highly homologous genes or has a pseudogene [50]. Furthermore, the short turnaround time needed for prenatal screening makes it impossible to always broaden the coverage of candidate genes through alternate sequencing techniques. The inadequate coverage of pertinent genes presents a problem for WES and may make it difficult to evaluate exome sequences. Even while exome sequencing's depth of coverage is probably going to get better, some parts of the genome are almost impossible to record, for instance, due to tandem repeats and high GC content [53, 54]. Considering this, traditional testing approaches like microarrays should be considered to be taken into account in addition to prenatal WES, particularly in the event of clinical indication for established prenatal manifestations of genetic disorders.

Since the early 1980s, conclusive diagnostic testing has included traditional cytogenetic procedures including FISH and karyotyping. The drawbacks of each of these approaches, whether used alone or together, highlight the need for a new technology that can eliminate the requirement for several technologies. Although NGS methods are effective at identifying alterations in sequence [33], their capacity to discover all kinds of SVs is constrained by their inability to precisely resolve repeat sections of the genome.

16.8 Prospects of Sequencing in Prenatal Screening

Due to NGS and other genomic technologies, prenatal and postnatal examinations and diagnoses for infant genetic medical conditions are changing at a rate that was previously unheard of. Early methods of detection and screening for prenatal genetic abnormalities were centered on a few problems that might be quickly diagnosed using diagnostic techniques from the pre-genomic era. In the genomic era, prenatal detection and diagnosis were first made possible by CMA, and then NGS, which can do a more in-depth genomic analysis. Thus, by transforming neonatal genomic testing for diagnostic purposes from a monitoring approach for a small number of illnesses to a more precise, targeted analysis for a wide range of inherited anomalies NGS is growing into an essential cornerstone of prenatal medicine with high accuracy. According to the studies, DNA sequencing into fetal screening is possible within a manageable time frame and can enable accurate delineation of chromosomal rearrangements like a translocation, inversion, or insertion that cannot be detected by CMA in contrast to karyotyping at the nucleotide level as well as data regarding the addition or deletion of genetic material, that might substantially enhance the prediction of phenotypic ramifications [14].

Rapid improvements in genome-wide NIPS have been observed over the past 10 years. Evidence of NIPS implementations indicates that such a technique will probably be employed more frequently in the upcoming years for clinical diagnosis and translational research. NIPS is also relevant in circumstances when there is little adherence to invasive procedures and provides accurate data prior to the decision to undertake amniocentesis [55] additionally, NIPS can be carried out almost at any point in the pregnancy.

The identification of cffDNA has created a novel and interesting potential for the NIPS-based determination of aneuploidy, CNVs, single-gene diseases, and the fetal blood group. However, this is still in the far future if cffDNA-based studies or trophoblast cell studies continue to advance, we might ultimately be capable to accomplish identical procedures on tests that were obtained noninvasively. However, it would additionally eliminate the extremely low danger of amniocentesis and CVS [56].

Regardless of compliance, a test without risk should frequently be provided to individuals initially. However, there is a need for research into the precise advantages and accuracy, which must be supported by clinical studies and eventually by using large cohorts. Thus, according to ongoing research and technological improvements, this will undoubtedly theoretically become feasible, but the main difficulty will be managing the concomitant counseling requirements and the moral, sociological, and healthcare elements of these technological advances [57].

16.9 Conclusion

Genetic diagnosis offers crucial information, for managing and making decisions throughout pregnancy, the initial approaches to diagnostic testing of prenatal genomic abnormalities could be diagnosed using the pre-genomic era tools. However, in the genomic era, CMA is the main test (transforming the traditional karyotyping) if a structural anomaly discovered by early imaging tests is the reason for thefetal diagnostic test. In accordance to present recommendations, any woman opting to have invasive prenatal screening testing should be offered the CMA [58]. However whole-genome sequencing detects imbalanced chromosomal rearrangements with molecular-level accuracy unmatched by presently used prenatal screening techniques, permitting an assessment of the regulatory genome that may be essential for clinical interpretation. As a result of the extraordinary advancements in genetics over the decades, prenatal testing methodologies that result in diagnoses and aid in decision-making have changed dramatically.

References

1. Almubarak A, Zhang D, Kosak M, Rathwell S, Doonanco J, Eaton AJ, Kannu P, et al. Prenatal genetic testing in the era of next generation sequencing: a one-center Canadian experience. Genes. 2022;13(11):2019. https://doi.org/10.3390/genes13112019.
2. Congenital disorders, World Health Organisation, February 27, 2023. https://www.who.int/news-room/fact-sheets/detail/birth-defects.
3. Maternal mortality, World Health Organisation, August 30, 2021. https://www.who.int/europe/news-room/fact-sheets/item/maternal-mortality.
4. Congenital Anomalies, National Health Portal, January 17, 2016. https://www.nhp.gov.in/disease/gynaecology-and-obstetrics/congenital-anomalies-birth-defects.
5. Newborn and Child Health, UNICEF India, April 6, 2023. https://www.unicef.org/india/what-we-do/newborn-and-child-health.
6. Haeri M, Gelowani V, Beaudet AL. Chromosomal microarray analysis, or comparative genomic hybridization: A high throughput approach. MethodsX. 2015;3:8–18. https://doi.org/10.1016/j.mex.2015.11.005.
7. Levy B, Wapner R. Prenatal diagnosis by chromosomal microarray analysis. Fertil Steril. 2018;109(2):201–12. https://doi.org/10.1016/j.fertnstert.2018.01.005.
8. Giovannopoulou E, Tsakiridis I, Mamopoulos A, Kalogiannidis I, Papoulidis I, Athanasiadis A, Dagklis T. Invasive prenatal diagnostic testing for aneuploidies in singleton pregnancies: a comparative review of major guidelines. Medicina (Kaunas). 2022;58(10):1472. https://doi.org/10.3390/medicina58101472.
9. Wu W, Zhou X, Jiang Z, Zhang D, Yu F, Zhang L, et al. Noninvasive fetal genotyping of single nucleotide variants and linkage analysis for prenatal diagnosis of monogenic disorders. Hum Genomics. 2022;16(1):28. https://doi.org/10.1186/s40246-022-00400-4.
10. Scotchman E, Chandler NJ, Mellis R, Chitty LS. Noninvasive prenatal diagnosis of single-gene diseases: the next frontier. Clin Chem. 2020a;66(1):53–60. https://doi.org/10.1373/clinchem.2019.304238.
11. Scotchman E, Shaw J, Paternoster B, Chandler N, Chitty LS. Non-invasive prenatal diagnosis and screening for monogenic disorders. Eur J Obstet Gynecol Reprod Biol. 2020b;253:320–7. https://doi.org/10.1016/j.ejogrb.2020.08.001.
12. Young E, Bowns B, Gerrish A, Parks M, Court S, Clokie S, et al. Clinical service delivery of noninvasive prenatal diagnosis by relative haplotype dosage for single-gene disorders. J Mol Diagn. 2020;22(9):1151–61. https://doi.org/10.1016/j.jmoldx.2020.06.001.
13. Caceres V, Murray T, Myers C, Parbhoo K. Prenatal genetic testing and screening: a focused review. Semin Pediatr Neurol. 2022;42:100976. https://doi.org/10.1016/j.spen.2022.100976.
14. Lord J, McMullan DJ, Eberhardt RY, Rinck G, Hamilton SJ, Quinlan-Jones E, et al. Prenatal exome sequencing analysis in fetal structural anomalies detected by ultrasonography (PAGE): a cohort study. Lancet (London, England). 2019;393(10173):747–57. https://doi.org/10.1016/S0140-6736(18)31940-8.
15. Rodriguez-Revenga L, Madrigal I, Borrell A, Martinez JM, Sabria J, Martin L, et al. Chromosome microarray analysis should be offered to all invasive prenatal diagnostic testing following a normal rapid aneuploidy test result. Clin Genet. 2020;98(4):379–83. https://doi.org/10.1111/cge.13810.
16. Sun Y, Zhang W, Wang Z, Guo L, Shi S. Chromosomal microarray analysis vs. karyotyping for fetal ventriculomegaly: a meta-analysis. Chin Med J (Engl). 2021;135(3):268–75. https://doi.org/10.1097/CM9.0000000000001683.
17. Bianchi DW, Chiu RWK. Sequencing of circulating cell-free DNA during pregnancy. N Engl J Med. 2018;379(5):464–73. https://doi.org/10.1056/NEJMra1705345.
18. Chen CP, Wu FT, Wong CH, Chen SW, Chern SR, Pan YT, et al. Prenatal diagnosis and molecular genetic analysis of recurrent trisomy 18 of maternal origin in two consecutive pregnancies. Taiwan J Obstet Gynecol. 2023a;62(3):444–7.

19. Chen Y, Ning W, Shi Y, Chen Y, Zhang W, Li L, et al. Maternal prenatal screening programs that predict trisomy 21, trisomy 18, and neural tube defects in offspring. PLoS One. 2023b;18(2):e0281201. https://doi.org/10.1371/journal.pone.0281201.
20. Monaghan, K. G., Leach, N. T., Pekarek, D., Prasad, P., Rose, N. C., & ACMG Professional Practice and Guidelines Committee. The use of fetal exome sequencing in prenatal diagnosis: a points to consider document of the American College of Medical Genetics and Genomics (ACMG). Genetics in medicine: official journal of the American College of Medical Genetics. 2020;22(4):675–80. https://doi.org/10.1038/s41436-019-0731-7.
21. Hu P, Qiao F, Wang Y, Meng L, Ji X, Luo C, et al. Clinical application of targeted next-generation sequencing in fetuses with congenital heart defect. Ultrasound Obstet Gynecol. 2018;52(2):205–11. https://doi.org/10.1002/uog.19042.
22. Chen X, Jiang Y, Chen R, Qi Q, Zhang X, Zhao S, et al. Clinical efficiency of simultaneous CNV-seq and whole-exome sequencing for testing fetal structural anomalies. J Transl Med. 2022;20(1):10. https://doi.org/10.1186/s12967-021-03202-9.
23. Petrovski S, Aggarwal V, Giordano JL, Stosic M, Wou K, Bier L, et al. Whole-exome sequencing in the evaluation of fetal structural anomalies: a prospective cohort study. Lancet (London, England). 2019;393(10173):758–67. https://doi.org/10.1016/S0140-6736(18)32042-7.
24. Pratt M, Garritty C, Thuku M, Esmaeilisaraji L, Hamel C, Hartley T, et al. Application of exome sequencing for prenatal diagnosis: a rapid scoping review. Genet Med. 2020;22(12):1925–34. https://doi.org/10.1038/s41436-020-0918-y.
25. Kilby MD. The role of next-generation sequencing in the investigation of ultrasound-identified fetal structural anomalies. BJOG. 2021;128(2):420–9. https://doi.org/10.1111/1471-0528.16533.
26. Edwards L, Hui L. First and second trimester screening for fetal structural anomalies. Semin Fetal Neonatal Med. 2018;23(2):102–11. https://doi.org/10.1016/j.siny.2017.11.005.
27. Sparks TN, Lianoglou BR, Adami RR, Pluym ID, Holliman K, Duffy J, et al. Exome sequencing for prenatal diagnosis in nonimmune hydrops fetalis. N Engl J Med. 2020;383(18):1746–56. https://doi.org/10.1056/NEJMoa2023643.
28. Dharmadhikari AV, Ghosh R, Yuan B, Liu P, Dai H, Al Masri S, Scull J, et al. Copy number variant and runs of homozygosity detection by microarrays enabled more precise molecular diagnoses in 11,020 clinical exome cases. Genome Med. 2019;11(1):30. https://doi.org/10.1186/s13073-019-0639-5.
29. Wu X, Li Y, Lin N, Xie X, Su L, Cai M, et al. Chromosomal microarray analysis for pregnancies with abnormal maternal serum screening who undergo invasive prenatal testing. J Cell Mol Med. 2021;25(13):6271–9. Advance online publication. https://doi.org/10.1111/jcmm.16589.
30. Haleem A, Javaid M, Singh RP, Suman R, Rab S. Biosensors applications in medical field: a brief review. Sensors Int. 2021;2:100100.
31. Yenilmez ED, Kökbaş U, Tuli A. A new alternative approach for RhD Incompatibility; determination Fetal RhD Status via Biosensor technology. In: Blood groups. IntechOpen; 2019.
32. Badeau M, Lindsay C, Blais J, Nshimyumukiza L, Takwoingi Y, Langlois S, Légaré F, et al. Genomics-based non-invasive prenatal testing for detection of fetal chromosomal aneuploidy in pregnant women. Cochrane Database Syst Rev. 2017;11(11):CD011767. https://doi.org/10.1002/14651858.CD011767.pub2.
33. Sahajpal NS, Barseghyan H, Kolhe R, Hastie A, Chaubey A. Optical genome mapping as a next-generation cytogenomic tool for detection of structural and copy number variations for prenatal genomic analyses. Genes. 2021;12(3):398. https://doi.org/10.3390/genes12030398.
34. Queremel Milani DA, Tadi P. Genetics, chromosome abnormalities. In: StatPearls. StatPearls Publishing; 2022.
35. Kagan KO, Sonek J, Kozlowski P. Antenatal screening for chromosomal abnormalities. Arch Gynecol Obstet. 2022;305(4):825–35. https://doi.org/10.1007/s00404-022-06477-5.
36. Zhang B, Lu BY, Yu B, Zheng FX, Zhou Q, Chen YP, et al. Noninvasive prenatal screening for fetal common sex chromosome aneuploidies from maternal blood. J Int Med Res. 2017;45(2):621–30. https://doi.org/10.1177/0300060517695008.

37. Vanstone M, Cernat A, Majid U, Trivedi F, De Freitas C. Perspectives of pregnant people and clinicians on noninvasive prenatal testing: a systematic review and qualitative meta-synthesis. Ontario Health Technol Assess Ser. 2019;19(5):1–38.
38. Suciu ID, Toader OD, Galeva S, Pop L. Non-invasive prenatal testing beyond trisomies. J Med Life. 2019;12(3):221–4. https://doi.org/10.25122/jml-2019-0053.
39. Chen S, Zhang L, Gao J, Li S, Chang C, Chen Y, et al. Expanding the scope of non-invasive prenatal testing to detect fetal chromosomal copy number variations. Front Mol Biosci. 2021;8:649169. https://doi.org/10.3389/fmolb.2021.649169.
40. Costumbrado J, Mansour T, Ghassemzadeh S. Rh incompatibility. In: StatPearls. StatPearls Publishing; 2022.
41. Kent J, Farrell AM, Soothill P. Routine administration of Anti-D: the ethical case for offering pregnant women fetal RHD genotyping and a review of policy and practice. BMC Pregnancy Childbirth. 2014;14:87. https://doi.org/10.1186/1471-2393-14-87.
42. Pegoraro V, Urbinati D, Visser GHA, Di Renzo GC, Zipursky A, Stotler BA, Spitalnik SL. Hemolytic disease of the fetus and newborn due to Rh(D) incompatibility: a preventable disease that still produces significant morbidity and mortality in children. PLoS One. 2020;15(7):e0235807. https://doi.org/10.1371/journal.pone.0235807.
43. Al-Yatama MK, Mustafa AS, Al-Kandari FM, Khaja N, Zohra K, Monem RA, Abraham S. Polymerase-chain-reaction-based detection of fetal rhesus D and Y-chromosome-specific DNA in the whole blood of pregnant women during different trimesters of pregnancy. Med Princ Pract. 2007;16(5):327–32. https://doi.org/10.1159/000104803.
44. Wienzek-Lischka S, Bachmann S, Froehner V, Bein G. Potential of next-generation sequencing in noninvasive fetal molecular blood group genotyping. Transfusion Med Hemother. 2020;47(1):14–22. https://doi.org/10.1159/000505161.
45. Minon JM, Gérard C, Dricot JF, Neve C, Senterre JM, Schaaps JP, et al. Nouvelles stratégies dans la prise en charge de l'allo-immunisation foeto-maternelle anti-RHD (Rhésus) [New approaches to prenatal diagnosis of rhesus incompatibility]. Rev Med Liege. 2006;61(11):756–62.
46. Johnson K, Eason J. Prenatal diagnosis of single-gene disorders. Obstet Gynaecol Reprod Med. 2023;20(5):155–60.
47. Kay AC, Wells J, Hallowell N, Goriely A. Providing recurrence risk counselling for parents after diagnosis of a serious genetic condition caused by an apparently de novo mutation in their child: a qualitative investigation of the PREGCARE strategy with UK clinical genetics practitioners. J Med Genet. 2023;60(9):925–31.
48. Grati FR. Chromosomal Mosaicism in human feto-placental development: implications for prenatal diagnosis. J Clin Med. 2014;3(3):809–37. https://doi.org/10.3390/jcm3030809.
49. Mackie FL, Carss KJ, Hillman SC, Hurles ME, Kilby MD. Exome sequencing in fetuses with structural malformations. J Clin Med. 2014;3(3):747–62. https://doi.org/10.3390/jcm3030747.
50. Best S, Wou K, Vora N, Van der Veyver IB, Wapner R, Chitty LS. Promises, pitfalls and practicalities of prenatal whole exome sequencing. Prenat Diagn. 2018;38(1):10–9. https://doi.org/10.1002/pd.5102.
51. Cavadino A, Morris JK. Revised estimates of the risk of fetal loss following a prenatal diagnosis of trisomy 13 or trisomy 18. Am J Med Genet A. 2017;173(4):953–8. https://doi.org/10.1002/ajmg.a.38123.
52. Chang Q, Yang Y, Peng Y, Liu S, Li L, Deng X, Yang M, et al. Prenatal detection of chromosomal abnormalities and copy number variants in fetuses with ventriculomegaly. Eur J Paediatr Neurol. 2020;25:106–12. https://doi.org/10.1016/j.ejpn.2020.01.016.
53. Exome Sequencing vs. Whole-Genome Sequencing, Biocompare, October 30, 2018. https://www.biocompare.com/Editorial-Articles/354629-Exome-Sequencing-vs-Whole-Genome-Sequencing/.
54. Sims D, Sudbery I, Ilott NE, Heger A, Ponting CP. Sequencing depth and coverage: key considerations in genomic analyses. Nat Rev Genet. 2014;15(2):121–32. https://doi.org/10.1038/nrg3642.

55. Rabinowitz T, Shomron N. Genome-wide noninvasive prenatal diagnosis of monogenic disorders: current and future trends. Comput Struct Biotechnol J. 2020;18:2463–70. https://doi.org/10.1016/j.csbj.2020.09.003.
56. Alfirevic Z, Navaratnam K, Mujezinovic F. Amniocentesis and chorionic villus sampling for prenatal diagnosis. Cochrane Database Syst Rev. 2017;9(9):CD003252. https://doi.org/10.1002/14651858.CD003252.pub2.
57. Sabbagh R, Van den Veyver IB. The current and future impact of genome-wide sequencing on fetal precision medicine. Hum Genet. 2020;139(9):1121–30. https://doi.org/10.1007/s00439-019-02088-4.
58. Ordulu Z, Kammin T, Brand H, Pillalamarri V, Redin CE, Collins RL, et al. Structural chromosomal rearrangements require nucleotide-level resolution: lessons from next-generation sequencing in prenatal diagnosis. Am J Hum Genet. 2016;99(5):1015–33. https://doi.org/10.1016/j.ajhg.2016.08.022.

GPSR Compliance

The European Union's (EU) General Product Safety Regulation (GPSR) is a set of rules that requires consumer products to be safe and our obligations to ensure this.

If you have any concerns about our products, you can contact us on ProductSafety@springernature.com

In case Publisher is established outside the EU, the EU authorized representative is:

Springer Nature Customer Service Center GmbH
Europaplatz 3
69115 Heidelberg, Germany

Batch number: 10726060

Printed by Printforce, the Netherlands